Energy Storage

Robert A. Huggins

Energy Storage

Fundamentals, Materials and Applications

Second Edition

 Springer

Robert A. Huggins
Department of Materials Science
and Engineering
Stanford University
Stanford, CA, USA

ISBN 978-3-319-33108-9   ISBN 978-3-319-21239-5 (eBook)
DOI 10.1007/978-3-319-21239-5

Springer Cham Heidelberg New York Dordrecht London
© Springer International Publishing Switzerland 2010, 2016
Softcover re-print of the Hardcover 2nd edition 2016

Springer International Publishing AG Switzerland is part of Springer Science+Business Media
(www.springer.com)

Preface

Introduction

Energy is necessary for a number of reasons; the most basic and obvious involve the preparation of food and the provision of heat in order to make life comfortable, or at least, bearable. Subsequently, a wide range of technological uses of energy have emerged and been developed, so that the availability of energy has become a central issue in society.

The easiest way to acquire useful energy is to simply find it in nature as wood or as a hydrocarbon fossil fuel. But it is advantageous to simply convert what is available in nature into more useful forms. The processing and conversion of raw materials, especially petrochemicals, has become a very large industry.

Wood

Wood has been used to provide heat for millennia. It can be acquired as needed by foraging or cutting. When it is abundant, there is relatively little need for it to be stored. However, many societies have found it desirable to collect more wood than is immediately needed during warm periods during the year, and to store it up for use in the winter, when the needs are greater, or its collection is not so convenient. One can still see this in some locations, such as the more remote communities in the Alps. One might think of this as the oldest and simplest example of energy storage.

It was discovered long ago that it is possible to heat wood under oxygen-poor conditions so that some of its volatile constituents are driven off, leaving a highly porous carbon-rich product called charcoal. Charcoal has a higher heating value per unit weight than the wood from which it was produced, approximately $30,400 \text{ kJkg}^{-1}$, instead of $14,700 \text{ kJkg}^{-1}$. Thus it is more efficient to store and to use to produce heat. This is an example of the conversion of a simple fuel into one with a higher energy value before storage.

In the nineteenth century 84 % of the energy worldwide, roughly 10^{18} megajoules, or about 1 yottajoule (YJ) was supplied by wood, charcoal, and crop residues.

Coals

Natural deposits of carbon were also discovered long ago; it was found that they can likewise be readily burned to produce heat. These solid, carbon-rich materials are often described as various types of coal, with different energy contents. The lowest energy content form is called peat, followed by lignite (brown coal), sub-bituminous coal, and hard coal (or anthracite). Their approximate specific energy contents are shown in Fig. 1.

Fig. 1 Energy content and carbon, and oxygen contents of the different types of coal. Based on US DOE data

The harder forms have sufficient energy contents that it is economical to not only store them, but also transport them to other locations. Coals constitute the largest fossil fuel resource in the world, about 5.3 YJ, and are now the most important energy source in a number of places. Where it is available, coal is the least expensive fuel, less than oil or natural gas.

As in the case of wood, coals can be heated in an air-poor environment to produce a more valuable product, called coke. Coke can then be burned to produce more heat per unit weight and volume than the coal from which it was made, while also producing very little smoke. Because of its porosity, relatively high hardness, and higher energy content, coke is used in blast furnaces to reduce iron oxide to molten iron, an early step in the production of steel.

There is an increasing concern about the toxic contaminants that are originally contained as minor constituents in coals. Because of the concern about such particles getting into the atmosphere, they are often trapped in equipment that is designed to reduce air contamination from coal-burning power plants. They also appear in the coal ash, the non-combustible components in the coal, that are generally stored in open surface ponds or landfills. About 130 million tons of coal ash are produced per year in the United States.

Other Possible Solid Petrochemical Energy Sources

In addition to coal and oil, other related natural carbon sources are known, and some are evidently very abundant in selected locations. These include large tar sand and oil-containing shale deposits in Canada. The pursuit of the possibility of their extraction and use has elicited strong, and sometimes politically charged, emotions.

One of the issues has been the desire to transport the products across the United States to the Gulf Coast, where they would be refined, and then shipped overseas, rather than being made available for consumption domestically.

These are comparatively dilute sources of carbon, and their extraction and conversion into useful fuels is relatively expensive. In addition, they involve the use and consumption of other materials, such as natural gas and water, in the conversion process.

Although very large amounts of this family of solid hydrocarbons are known in a number of places, it must be recognized that they are also depletable resources. The time required for their regeneration in nature is immense, so they can also only be counted upon once.

Crude Oil

Petroleum, or crude oil, is also a fossil fuel, similar to coal; but it has the advantage that it is liquid. This makes it much more versatile for a number of applications. The major use in the United States at the present time is for transportation. It is readily transported using vehicles (ships, trucks, and rail), as well as pipelines, and it can be easily stored in tanks. For a number of years it has been less expensive to extract crude oil, rather than coal, from the earth, although this disparity has varied with both location and time. As readily extracted natural crude oil supplies are consumed, it is generally necessary to dig ever deeper, with greater associated costs.

The specific energy of typical crude oil is about 42 $KJkg^{-1}$. This is higher than any other fossil fuel. It has become a world, commodity, and is shipped all over the world. Refineries convert it to a variety of products, such as heavy fuel oil, diesel fuel, gasoline, kerosene, etc. Subsequently, significant amounts of these liquid materials are also converted into a variety of solid plastics.

Mankind's use of petrochemical fuels based on crude oil as energy sources is actually only quite recent. Table 1 lists a number of the major crude oil discoveries and their dates. These have all occurred in the last century and a half. Thus the "oil age" has been just a recent episode in the history of modern civilization.

Table 1 The discovery of oil—locations and dates

Early discoveries		
	Near Bakku, on the Caspian Sea	About 1849
	Bend, North of Bucharest, Rumania	1857
	Oil Springs, Ontario, Canada	1858
	Drake well near Titusville, PA, USA	1859
First major oil fields discovered		
	Spindletop, near Beaumont, TX, USA	1901
	Others in Oklahoma and California, USA	Shortly thereafter
Discoveries in the Middle East		
	Bahrain	1920
	Kirkuk in Iraq	1927
	Gachsaran in Iran	1935
	Dammam in Saudi Arabia	1938
	Abqaiq in Saudi Arabia	1940
	Ghawar in Saudi Arabia	1948–1949
Alaska, USA		
	Prudhoe Bay	1968
North Sea areas		
	Forties, UK	1970
	Ekofisk, Norway	1971
	Brent, UK	1971
South America		
	Venezuela	1988

The Problem of the Depletion of Fossil Fuels

As mentioned earlier, fossil fuels are not infinitely available, and the sources that are found become depleted as their contents are removed. It was long thought that this would not become a problem, for new oil fields would be found to replace those that became depleted. Contrary to this, M. King Hubbert published predictions of future oil production in both the United States [1] and the world [2] that indicated that annual oil production would follow bell-shaped curves, reaching a peak, followed by decline in each case.

His predictions were not taken very seriously for some time, all seemed to be going well, with oil production growing every year. But US production peaked in 1970 and decreased as existing fields depleted faster than the rate of discovery.

This is depicted in Fig. 2, which compares actual production to the 1956 prediction by Hubbert that US production would peak about 1970. It is seen that the correspondence is remarkable. Obviously, the United States would be more and more dependent upon imported oil, not just because of its increasing energy demand, but also the decline in its domestic production.

Fig. 2 Relationship between actual annual oil production and the prediction made by Hubbert in 1956. Adapted from [4]

The same question naturally arose concerning the world supply of oil, and many people took the view that there is no problem in this case, as new sources will always be found. Some, however, including M.R. Simmons [3] and K.S. Deffeyes [4]. Believe that the same peak phenomenon is bound to occur, and that the world production peak had already been reached. Hubbert [2] predicted the world production to peak in 2000 and the Deffeyes estimate [4] was 2005. A number of the major known oil fields are now well beyond their production peaks.

The expectation that the world production of readily accessible crude oil would reach a peak in the future, if it had not already done so, provided a great incentive for the development of alternative sources of energy. This is compounded by the fact that the world's population is rapidly expanding. The world's production of oil on a per person basis from the traditional sources, which are primarily in the Middle East, have already reached a peak and is declining.

Thus the long-term crude oil supply situation had been considered until recently to be very bleak. Unlike wood, carbon and petrochemical fossil fuels must be considered to be non-renewable resources; they take millions of years to form. Conservation measures and the search for alternatives to crude oil has become an imperative.

As will be discussed briefly later in this Preface, this situation has changed immensely in the last few years, and these dire predictions have been forgotten.

Natural Gas

Natural gas, which is used for both the generation of electrical power and the production of petrochemicals, consists primarily of methane, CH_4. It is found associated with liquid fossil fuels and in coal beds. It is also present under the ocean as methane clathrates, as will be mentioned in the next section.

At high pressure underground, natural gas dissolves in oil, and it is often emitted when oil is pumped to the surface, where the pressure is lower. This associated gas is sometimes simply burned at the well sites as "gas flares," for its value is considered to be considerably less than the associated oil. Anyone flying over the Middle East at night can readily identify the oil well areas from the light emitted from these flares. As the price of natural gas has risen, there has been an increasing tendency to capture, rather than simply burning, it.

Oil is typically recovered at depths between 7500 and 15,000 ft below the surface. The deeper the pressure, the higher the temperature, and the larger molecules in the oil become cracked into smaller units. Below about 15,000 ft the oil is cracked down to molecules that have only one carbon atom, i.e., methane, CH_4.

Thus gas wells are often drilled to depths considerably deeper than those aimed at the extraction of oil.

Early gas wells were actually drilled before the first oil wells. One early successful one was drilled in New York State in 1821. That was 35 years before the famous Drake oil well was established in Northern Pennsylvania.

Drilling gas wells is generally much faster and less expensive than drilling for oil, and there have traditionally been many more gas wells than oil wells in the United States. Information about natural gas production and available reserves is much more difficult to obtain than that for oil. Thus there are no predictions available for natural gas comparable to those using the Hubbert method for the predictions about the oil supply.

In addition to methane, natural gas can contain significant quantities of heavier hydrocarbons, such as ethane, propane, butane, and pentane. Besides these related materials, CO_2, nitrogen, hydrogen sulfide, and helium are also often present. A significant effort is often undertaken to separate some of these more valuable components.

Methane-rich gases can also be produced by the anaerobic decay of non-fossil organic matter, such as manure and the waste in landfills. These are typically called *bio-gas*.

Methane released into the atmosphere is a potent greenhouse gas and is generally considered to be a pollutant. It gradually becomes oxidized to form CO_2 and water, which are also greenhouse gases. The kinetics of this process are such that methane that enters the atmosphere can be considered to have a half-life of about 7 years.

Natural gas can be used as a fuel for electricity generation in steam turbines and higher temperature gas turbines. It burns cleaner than either petroleum or coal and produces less CO_2 per unit energy obtained. For an equivalent amount of heat, natural gas produces 30 % less CO_2 than burning petroleum, and 45 % less than burning coal. It is therefore generally considered to be the cleanest of the fossil fuels. It can be used for heating and cooking in homes, and also as a vehicle fuel instead of gasoline or diesel fuel. Natural gas consumption surpassed that of crude oil in the United States in 1984.

The standard unit of volume of natural gas is 1000 cubic feet (28.3 m^3) at room temperature and pressure. This gives about one million BTUs, which is approximately 1 GJ of energy. The energy content of 6000 cubic feet (169.8 m^3) of gas is equivalent to one barrel of crude oil.

Some hydrogen produced from natural gas is also reacted with nitrogen in the high pressure Haber process to produce ammonia, an important feedstock for the production of fertilizers. Some 3–5 % of the world natural gas production is used for this purpose.

Natural gas can be readily transported in pipelines. It also can be liquefied, and stored and transported in refrigerated tanks, even across oceans in ships. Russia is currently the largest producer of natural gas in the world, and a large source of oil. Some of this gas, as well as oil, is exported to Western Europe through several large pipelines. In 2007 the European Union imported 100.7 million tons of oil-equivalent natural gas, 38.7 % of its total gas import, and 185 million tons of crude oil, 32.6 % of its total oil imports, from Russia. Two countries, Germany and Italy, accounted for about half of these totals.

About 80 % of these travel through pipelines that cross Ukrainian soil. This leads to the potential for serious problems for countries further to the West. There have been significant disputes between Russia and the Ukraine in recent years that have led to the cut-off of supplies to both the Ukraine and countries in Western Europe.

As in the case of oil, the supply of natural gas in the United States has changed greatly in the last few years, due to the development of fracking technology. This will be briefly discussed later in this Preface.

Hydrogen

Hydrogen is not actually an energy source, as it is obtained by using energy. Instead, it is generally described as being an energy carrier. Some 9–10 million tons of hydrogen are now produced in the United States per year. It is used for many purposes, as a feedstock and intermediate in chemical production, petroleum refining, and metals treating. Its use to convert heavy petroleum into lighter fractions is called *hydrocracking*. The current applications in transportation and as the fuel in fuel cells represent a very minor fraction of the total.

There are several methods by which hydrogen is currently obtained. The least expensive of these, which accounts for some 95 % of the total, is the production of hydrogen by the treatment of natural gas at elevated temperatures. The composition of the natural gas obtained from the earth varies from place to place. It can contain significant amounts of sulfur in some locations. In some cases it is obtained as a by-product of the extraction of petroleum. In the Middle East the natural gas that is extracted along with the petroleum is often thrown away, for its value may not be sufficient to justify the cost of containing and transporting it. As a result, it is burned, or "flared." The lights from this process can be clearly seen when flying over this part of the world at night.

Hydrogen can also be obtained in other ways. One of the most prominent is by the electrolysis of water, which currently accounts for about 4 % of the total. These alternatives will be discussed in Chap. 8.

Methane Clathrates

Another potentially immense source of energy that might be used in the future is related to the presence of large amounts of *methane clathrate* deep in the oceans. This material is also sometimes known as *methane hydrate* or as *methane ice*. It can be described as a crystalline solid H_2O-ice structure within which methane is trapped.

The widespread existence of such materials was first recognized during the 1960s and 1970s, and they are now recognized to be common constituents of the shallow (<2000 m) marine geosphere, where they occur both as deep sedimentary deposits and as outcrops on the ocean floor. They are typically found in the ocean at depths greater than 300 m, and where the bottom water temperature is around 2 °C. They have been found to be widespread along the continental shelves and are also apparently present in deep lakes, such as Lake Baikal in Siberia.

In addition, methane clathrates have been found trapped in continental rocks of sandstone and siltstone at depths of less than 800 m in cold areas, such as Alaska, Siberia, and Northern Canada.

They are generated as the result of bacterial degradation of organic matter in low oxygen aqueous environments. It is believed that large volumes of methane may occur as bubbles of free gas below the zone of clathrates. They are stable up to about 0 °C, above which they decompose to form liquid water and gaseous methane, CH_4. At higher pressures they can remain stable to higher temperatures.

A typical deposit composition has 1 mol of methane for 5.75 mol of water. As a result, the melting of 1 L of solid clathrate produces 168 L of methane gas at 1 atm pressure.

There are two general methods that have been employed to extract the methane from these clathrates. One is by heating, and the other is by reducing the pressure. The latter evidently requires significantly less energy.

Estimates of the amount of clathrate hydrates on the earth have varied widely and have decreased somewhat with time. Recent values run from 1 to 5×10^{15} cubic meters. These values are equivalent to 500–2500 gigatons of carbon, values smaller than the estimate of 5000 gigatons of carbon for all other fossil fuel reserves, but substantially greater than current estimates of natural gas (also mostly methane) reserves.

Although methane clathrates are potentially a very important source of fuel, there has evidently been only one commercial development to date, near Norilsk in Russia. Research and development projects are current underway in both Japan and China.

Since these deposits are very large in some cases, they certainly represent potentially important energy resources. The incentives to develop economical methods to recover the trapped fuel within them are immense.

Chemically Derived Fuels

In addition to the depletable energy sources found in nature, a significant amount of attention is now being given to the production of liquid fuels from renewable sources besides wood.

It has been known for hundreds of years in Asia and Europe that an oil can be produced that can be used to burn in lamps, as a machine lubricant, and also for cooking, by simply pressing rapeseed. More recently, it was found that a modification of this natural oil can be used as a diesel fuel. This plant, called "Raps" in German, has beautiful bright yellow blossoms in the spring. They provide a striking vista in large areas of the landscape in Northern Germany, where the author lived for a number of years. This material is called "Canola" in the United States and Canada.

There are a number of other biological materials that can provide useful fuels. There are two general types of such *agrofuels*. One is to use oilseed crops, such as rapeseed, soy beans, palm oil, and some seeds and nuts that contain high amounts of vegetable oil. Jatropha nuts are up to 40 % oil.

The other type includes crops that are high in sugar (e.g., sugar cane, sugar beet, and sweet sorghum) or starch (e.g., corn or switch grass (*panicum virgatum*)). These go through a yeast fermentation process to produce ethyl alcohol that can be used as a liquid fuel. Such biofuels currently provide 45 % of Brazil's fuel, where all autos run on ethanol.

When the prices of petrochemical fuels are high, the possibility of producing an economically competitive alternative fuel using corn becomes attractive. This has resulted in the diversion of a substantial fraction of the food corn production in that direction. In turn, the costs of related foods to the consumer have escalated, and become a political problem.

A further fuel source that is beginning to become economically attractive on a modest scale involves the use of animal fats. Whale oil, obtained from the fat (blubber) of whales, was the first animal oil to become commercially viable for use as a fuel, being used as a lamp oil, and a candle wax. Transesterification (catalytically reacting fats with short-chain aliphatic alcohols) can be used to produce useful liquid fuels from a number of different types of animal waste. One of the attractive features of this process is that it represents a convenient method for disposing of otherwise unusable animal waste from such places as meat processors and high volume chicken restaurants.

But ramping up the availability of these alternatives to the point that they make major contributions to the overall energy supply picture takes an appreciable period of time.

Recently Discovered Large Natural Gas and Oil Sources

Although the subject of this book is energy storage, not energy sources, a very important recent development in the latter area deserves mention here. It will surely have a very significant influence upon the energy supply, distribution, and storage in the United States.

Advances in drilling techniques recently resulted in the recognition that it is possible to obtain large amounts of natural gas and/or oil from underground shale layers. The existence of shale layers, which typically lie more than a mile below the surface, and that are known to contain gas, and also oil in some locations, has been known for some time. But it was long considered to be too difficult and expensive to extract significant quantities of useful fuels from them using normal drilling methods.

Fig. 3 Schematic representation of the difference between normal vertical drilling and horizontal drilling

The change in this picture resulted from the development of a new drilling and extraction strategy, called "fracking," which involves drilling horizontally into a shale layer, rather than through it [5,6]. This is illustrated schematically in Fig. 3.

A mixture of water and sand is forced into the shale formation at very high pressure. This causes many small fractures, or cracks, in the shale. These fractures would normally be expected to close as soon as the water pressure is decreased, but the presence of the fine sand, which goes into the cracks, causes them to remain open. As a result, significant quantities of the trapped gas and/or oil become accessible.

Regular oil and gas production using this method started in 2005. The oil and gas industry is now making very large investments to get access to this potentially large domestic source of natural gas and oil, and the rate of production is increasing rapidly.

This involves the injection of large quantities of water into the shale—up to 4–6 million gallons of fresh water per well. Various estimates indicate 15–35 % of the water and chemicals that are injected return to the surface as "flowback." This water has picked up some chemicals, generally sodium, magnesium, and calcium chlorides, which results in a significant increase in its salinity. Ground water tends to get saltier at greater depths. As a result, this water is not suitable for surface discharge or agricultural use. The least costly method for its disposal is permanent injection into a nearby deep well, but such wells must be carefully cased and cemented. There have been numerous examples of small earthquakes related to this water injection.

There are also some additional tricks that make this process much more efficient. They are typically proprietary, but often include materials such as a cross-linking "thickening agent," typically a sugar, to increase the viscosity, a biocide such as glutaraldehyde, a polyacrylamide "friction reducer," a phosphonate detergent scale inhibitor, and a surfactant.

Well construction can be quite fast, with an average time to production being less than a month. A method called "pad drilling," in which a number—15 or more—of wells are drilled in different directions from a single pad, has become common.

Production from shale gas wells typically falls off rapidly with time—often dropping 60–80 % in the first year, followed by a slower asymptotic decline. Conventional oil wells typically decline more slowly, some 25–40 % in the first year. Further drilling and re-fracking at greater horizontal distances is often done as production declines, for this is much less expensive that drilling new wells.

As was discussed earlier, the combustion of natural gas produces only about half as much carbon dioxide as the combustion of oil. Thus it is often considered to be a "bridge" or "transition" fossil fuel during the transition from the dependence upon imported oil to increased use of more environmentally friendly local energy sources.

The recent development of these large new oil and gas sources in the United States has also caused the construction of a number of previously planned new coal-fired power plants to be postponed or cancelled, for they cannot compete economically. Gas plants cost less to build and are cheaper to operate. In addition, it is much easier to get permits for new construction.

The introduction of this technology has caused economic windfalls and sudden labor shortages in some previously quiet locations. The extreme current example of this is in North Dakota, but similar economic rejuvenation is occurring in other areas, including some in the Midwest, where the metals-based manufacturing industry has been in serious decline for a number of years. North Dakota is now the second-biggest oil producing state, after Texas, in the United States.

There is not enough pipeline capacity to handle the rate of oil production, so a large amount of the oil is carried by the local railroads, which have had a number of serious spills and fires. This competition for the available transportation mechanisms has also resulted in difficulties getting the normal local farming products to market.

The development of fracking technology has resulted in a major change in the international energy picture. Instead of being dependent upon imported oil from the Middle East, the United States is now beginning to become a major oil exporter. This is expected to continue for a long time. Recent data on the production of gas and oil in the United States are shown in Fig. 4.

Fig. 4 Recent growth of gas and oil production in the United States

Other Alternative Energy Sources

A number of other energy sources are being pursued, and several of these are gradually taking over some of the energy demand burden. Among the prominent alternatives are solar and wind energy technologies. Others involve the use of geothermal sources, collected rainfall, and making use of both tidal variations and currents in the world's oceans. A further approach to the extraction of energy from natural phenomena involves various schemes to harness the power of ocean waves [7].

There is continued interest in the use of nuclear fission to produce heat, and its use to supply electricity. In this latter case, in addition to cost, there is the persistent, and extremely serious, problem of what to do with the inevitable long-lived radioactive waste. Until the waste problem is solved, this approach will continue to be extremely dangerous.

There is also great concern about the safety of large nuclear energy facilities. There was an accident at the Three Mile Island plant in the United States in March 1979 that involved a partial nuclear meltdown of one of the two nuclear reactors at that site. But, fortunately, this did not result in a major disaster, as it was relatively well contained, although there was the release of a significant amount of radiation. The cleanup cost was estimated to have been about one billion dollars.

A much larger disaster occurred at Chernobyl in the Ukraine, which was then part of the Soviet Union, in April 1986. A fire and an explosion caused large quantities of radioactive particles to go high into the atmosphere. This radioactive cloud moved to the North, across Scandinavia, and then down into Northern Europe. Radioactivity from this nuclear disaster was first detected in Sweden, 2 days before it was publicly reported in the USSR.

High levels of radioactivity soon were found in large areas of Western Europe, especially in Sweden and Norway, as well as in the USSR. The Sami (also known as Lapps, or Laplanders), who live in Northern Sweden and Norway, and whose livelihood has traditionally been based upon reindeer—herding, were especially affected by the radiation. In addition to the health risks to people, lichen, the food eaten by the reindeer, became seriously contaminated, decimating their herds. There was also widespread contamination of other food sources. Radioactive fallout got into freshwater lakes and inland forests, contaminating sheep, cows (and their milk), fish, wild game, and berries, especially in Sweden. As a result, many people were moved large distances, and forced to take up other activities.

The most dangerous pollutant was cesium 137, which has a half-life of 30 years, and it is expected that the radioactivity in the most contaminated areas will not drop to "safe" levels for at least 20 years. Eventual full disposal of the radioactivity will take up to 50 years.

In an attempt to prevent further escape of radioactive materials from the Chernobyl atomic power site some 5 million kg of lead and stone were dropped onto the reactor as rapidly as possible. A massive steel and concrete structure, called a sarcophagus, which was constructed to cover the reactor and to limit radioactive contamination of the environment material is gradually decaying, and in danger of collapsing. In 1996 it was decided to replace it by a much larger two-part containment structure by 2006. This is currently still under construction nearby, where the radiation level is significantly lower than in the reactor area. The two parts will be moved over the top of the original structure when they are completed; this is now scheduled to happen in 2015. This new containment structure, one of the largest engineering projects ever undertaken, will be 360 ft tall and 843 ft wide, and it is estimated that it will cost some 1.5 billion Euros, or about 2 billion US dollars.

After the Chernobyl explosion a large "exclusion zone," which could no longer be inhabited, was defined. It had an initial radius of about 20 miles around the nuclear power plant. This non-habitable area has subsequently been expanded and now covers about 1000 mile^2. Over 120,000 people were permanently evacuated from Chernobyl and nearby Pripyat, as well as many smaller communities in that area.

Another major nuclear disaster happened in the Fukushima Daiichi coastal area of Japan in March 2011, where a large tsunami wave, caused by a nearby magnitude 9 earthquake, overwhelmed the plant's seawall, and resulted in the meltdown of three of the power plant's six nuclear reactors. Fortunately, the other three were not in service at that time.

During the first day after this incident 134,000 people were evacuated from the nearby area, and this number was later increased by an additional 354,000 from a 12 mile radius exclusion zone.

Relatively little radioactive material went into the atmosphere in this case, compared to what happened in the Chernobyl disaster. However, the power plant was flooded by the tsunami wave, and a large amount of water swept into the site. This water became radioactive and is seriously hindering the cleanup process. More water is continuing to flow through the site and into the nearby ocean, and there is much concern about radioactivity in both the water and the fish in it. The ocean currents in this area are relatively strong, and it is expected that slightly radioactive water will be found as far away as the Western United States.

In order to stop the continued flow of this radioactive material into the ocean, there have been plans to construct a 2400 ft long steel and concrete wall between the site and the ocean, reaching 100 ft below ground. An alternative, the establishment of an underground ice wall, has also been discussed. Cleanup is expected to take decades. And as in the Chernobyl case, the major concern has been contamination by radioactive caesium 137, because of its long half-life.

These disasters have caused a number of nations to become wary of nuclear energy. As a result of the serious meltdown of the reactors in the Fukushima nuclear plant, the Japanese government decided to shut down its remaining nuclear power generators, which had been providing about 30 % of its total electrical power. This has resulted in increased dependence upon coal and natural gas energy sources, and a corresponding increase in the emission of greenhouse gases. Other technologies are also being promoted, including large groups of offshore wind turbines, and increased use of solar panels.

Germany decided to shut down eight of its nuclear reactors and to close the remaining nine by 2022. It is planned to have 30 % of its electric grid converted to solar (mostly in the South) and Wind (mostly in the North) sources in about 15 years. Residents receive a guaranteed high price for the energy that they generate, and over 1.4 million German homes currently generate electricity from either solar or wind sources. The local prices of solar and wind energies have already dropped 80 and 55 % from what they were in the year 2000.

Italy held a national referendum in which 94 % voted against the government's plan to build new nuclear power plants. The same happened in Switzerland, and later, in Belgium.

In France, the strongly pro-nuclear government was defeated in a national election, with 70 % of the public opposing nuclear energy in some polls. It was replaced by a government that promised to radically reduce France's reliance on nuclear power.

Plans for the introduction of nuclear power were abandoned in Malaysia, the Philippines, Kuwait, and Bahrain, or radically changed, as in Taiwan. China suspended its nuclear development program, but restarted it on a reduced basis in late 2012, with the government approving a "small number" of projects in each of the following 5 years. The initial plan had been to increase the nuclear contribution from 2 to 4 % of electricity by 2020, but renewable energy already supplied 17 % of

China's electricity and, post-Fukushima, it seemed likely that most of the 15 % of non-fossil energy that China aims to use by 2020 will be from renewables.

On the other hand, new nuclear energy projects are proceeding in some countries. The United Kingdom is planning a significant expansion, as are Russia, India, and South Korea.

A significant aspect of most of these alternative energy sources is that they tend to be time-dependent. The availability of water for hydroelectric and other uses varies with the time of the year, as well as the weather. The sun rises and sets, winds come and go, and tidal flows are periodic. This leads to the problem of either matching the time dependence of the energy source with the time dependence of energy needs, or the development and use of effective energy storage methods.

On the macroscopic scale, one of the ways that this is currently accomplished is to use the large-scale electrical transmission grid as a buffer. This grid is generally supplied by the use of fossil fuels such as coal or crude oil, although nuclear or hydroelectric sources are also employed in some locations. Energy from intermittent sources can be fed into this large system when its capacity is available, reducing the requirements from the normal grid-supplying sources. This can be especially useful if energy from such alternative sources is available at times when the grid's customers' requirements are high, and the cost of additional generating capacity is quite high.

The need for energy from the electrical grid generally depends upon the time of day and the day of the week. It also can vary greatly with the weather and time of year. The large-scale energy storage problem of matching supply with need, sometimes called "load leveling" and "peak shaving," will be dealt with later in this text. There are many other non-steady-state uses.

An entirely different problem arises when the energy is required for vehicles, or smaller, and likely portable uses, such as computers and telephones. In such cases, energy transfer occurs in one direction, and the storage device is likely charged, either directly, or indirectly, from the electrical grid.

Thus it can be seen that there are many different types of needs for energy storage, requiring a variety of solutions. These topics will be addressed in this book.

The very serious problem of the depletion of the traditional energy sources and the need to find and develop alternatives has attracted the attention of, and led to action by, governmental bodies worldwide. An important example in the United States was the Energy Independence and Security Act (EISA) that was passed by the Congress and became law in December 2007 [8]. The intent of this law was to increase energy efficiency and the availability of renewable energy. It included provisions to improve the efficiency of vehicles, appliances, and lighting, and to increase the production of biofuels. It also called for accelerated research and development on other energy sources and on related energy storage methods.

Further action in these directions was mandated in the American Recovery and Reinvestment Act of 2009 [9]. In addition to large increases in funding for work on energy efficiency and renewable energy, as well as research and development in energy-related areas, a considerable amount of money was allocated for the development of manufacturing facilities in the United States related to electrochemical energy storage.

It is clear that energy storage will be an increasingly important component of the overall energy supply and use picture in the future. This will be particularly critical as the alternative technologies, such as solar and wind sources, where energy production is intermittent, become more widespread.

It is the goal of this book to provide an understanding, rather than a mere description, of the various mechanisms by which energy can be stored for later or alternate use.

Stanford, CA, USA   Robert A. Huggins

Acknowledgements

The author gladly acknowledges with gratitude the important contributions made to the development of the understanding in this area by his many students and associates in the Solid State Ionics Laboratory of the Department of Materials Science at Stanford University over many years, as well as those in the Center for Solar Energy and Hydrogen Research (ZSW) in Ulm, Germany, and in the Faculty of Engineering of the Christian Albrechts University in Kiel, Germany.

References

1. Hubbert MK (1956) Nuclear Energy and the Fossil Fuels. American Petroleum Institute Drilling and Production Practice Proceedings. p 5
2. Hubbert MK (1969) Energy Resources. In: Freeman WH (ed) Resources and Man. National Research Council, Committee on Resources and Man. p 196
3. Simmons MR (2005) Twilight in the Desert. John Wiley
4. Deffeyes KS (2005) Beyond Oil. Hill and Wang
5. Rao V (2012) Shale Gas. RTI Press
6. Zuckerman G (2013) The frackers. Penguin
7. Cruz J (ed) (2008) Ocean Wave Energy: Current Status and Future Perspectives. Springer
8. Energy Independence and Security Act of 2007, which became law in the United States in December, 200.
9. American Recovery and Reinvestment Act of 2009, which became law in the United States in February, 200.

Contents

Chapter 1
Introduction

1.1 Introduction

Concern about the inevitable depletion of the fossil fuels that are now the major sources of energy has been greatly reduced, due to the recent development of fracking technology, as described in the Preface, and which will be further discussed in Chap. 23. However, there is another matter that is very important in considering the effective use of the energy that is available. This involves the relationships between the several types of energy, and also the various different uses of energy.

Worldwide energy consumption is between 500 and 600 EJ ($5-6 \times 10^{20}$ J). In terms of consumption rate this is 15–18 TW ($1.5-1.8 \times 10^{13}$ W). The USA consumes about 25 % of the total, although its share of the world's population is only about 5 %.

A recent estimate of the major US sources of energy is shown in Table 1.1. The distribution of energy use, by major category, is indicated in Table 1.2.

These different types of applications have different requirements for access to energy, and different characteristics of its use. One of the important problems with the effective use of available energy supplies is that the schedule of energy use is often not synchronous with its acquisition, even from natural sources. Thus buffer, or storage, systems are necessary.

This requirement for storage mechanisms is highly dependent upon the type of use. Those that consume fossil fuels, or their derivatives, for combustion purposes, such as for space heating or internal combustion-powered automobiles, require one type of storage and distribution system. Another, quite different, category involves the various applications that acquire their energy from the large-scale electric power transmission and distribution (T&D) grid. In that case, there are two types of storage systems to consider. One involves the electric power grid system itself, and the time dependence of its energy supplies and demands, and the other has to do

© Springer International Publishing Switzerland 2016
R.A. Huggins, *Energy Storage*, DOI 10.1007/978-3-319-21239-5_1

Table 1.1 Major sources of energy used in the USA in 2007 [1]

Source	Percent of total
Petroleum	39.4
Natural gas	23.3
Coal	22.5
Renewable energies	6.6
Nuclear electric	8.2
Total	*100*

Table 1.2 Major uses of energy [1]

Type of use	Percent of total energy use
Transportation	28.6
Industrial	21.1
Residential, commercial buildings	10.3
Electric power	40.0
Total	*100*

with storage mechanisms applicable to the various systems and devices that acquire their energy from the grid.

There are mechanisms whereby one type of source or storage technology is converted another. One example is the use of pumped-hydro storage to both take electricity and return electricity to the grid, as will be described later. Pumped-hydro technology actually operates by a mechanical storage mechanism, as it is based upon the gravitational difference between two different water storage reservoirs. Another is the use of flywheels to reversibly convert mechanical energy into electrical energy.

It will be seen that there are many different types of energy storage methods and technologies. Some of the largest of these are owned and/or operated by energy suppliers. Smaller ones are primarily related to energy users.

1.2 Storage in the Fuel Distribution System

In the simplest case, natural fuels such as wood, coal, or crude oil can be stored locally or within the transportation system, in piles, tanks, ships, and pipelines. For example, crude oil and some of the lighter petrochemical products are stored in tanks in oil depots (sometimes called oil terminal tank farms) in the vicinity of the refineries. These are often near marine tanker harbors or pipelines. This type of transient buffer storage generally has a relatively small capacity, however, and must be supplemented by other methods that can handle much larger amounts of energy.

1.3 Periodic Storage

A number of energy sources do not provide energy at a constant rate, but instead, are intermittent. Sometimes, they have a good measure of periodicity. As an example, some of the biofuels, such as switch grass, sugar cane, corn, and oilseeds are only available during part of the year. Likewise, solar, wind, and ocean motion energy sources have roughly daily cycles.

The time dependence of the uses of energy often does not correspond to the time variance of such sources. If these were to match perfectly, there would be no need for a storage mechanism. Perfect time-matching is not likely, however. Instead, at least part of the energy is supplied into some type of storage mechanism, from which it is (later) extracted when it is needed.

What is preferable is a situation in which the combination of current energy production and stored energy match the energy and power requirements at all times. This is sometimes called "load management," but it should actually be called "resource management." Overcapacity in either energy sources or storage systems is expensive, and a great deal of attention is given to how to most effectively and inexpensively meet the needs of energy users.

1.3.1 Long-Term, or Seasonal, Storage

Although there are exceptions, such as the local storage of wood mentioned in the Preface, seasonal storage generally involves very large installations, such as reservoirs and dams that accumulate water primarily during the rainy (or snowy) season of the year. Electric power is produced in hydroelectric facilities by passing water through large turbines. The water collected in such facilities is often used for agricultural purposes, as well as for energy production. Agricultural needs also vary with time during the year.

1.3.2 Daily and Weekly Storage

A number of energy sources produce energy on a daily cycle, related to the periodic characteristics of the sun, tides, and sometimes, wind. It is thus necessary to have mechanisms whereby this energy can be available when needed, and temporarily stored when not needed. But energy from these sources can also vary with the time of the year, and can be significantly affected by changes in the weather, on both long and short time scales. It is common for these sources to be connected to the large electrical transmission grid, providing energy to it when they are in operation. Thus the grid also acts as a buffer and storage medium.

1.4   The Problem of Load Leveling

The electrical load varies significantly with both the location and the time of day.
An example is shown in Fig. 1.1. The major components, industrial, commercial
and residential, require energy at different times during the day. For example,
residential use includes lighting, space heating and air conditioning, and often
electric water heaters and cooking appliances.

The magnitude of the electrical power demand also varies with the time of the
year. There is more power used in the winter for heating, and in the summer for air
conditioning. The variation with time during the day is also generally greater in the
summer than in the winter.

The use of energy also varies with the day of the week in many cases. Whereas it
is easy to understand that there is a daily pattern of energy use, the needs are not the
same every day of the week because many activities are different on weekends than
they are during workdays. This can be seen in Fig. 1.2, which shows a typical
pattern of weekly energy use [2].

Whereas both the time dependence and the magnitude can vary appreciably with
both location, the weather, and the time of year, these general patterns are almost
always present, and pose a serious problem for the electric utility firms that both
supply and manage the transmission and distribution electric power grid.

The electric utilities can supply this power to the grid from a number of different
sources. Often two or three different technologies are employed, depending upon
the load level. The least expensive is the use of coal or oil in large base-load
facilities. Thus, the utilities try to cover as much of the need as possible from such

Fig. 1.1 Example of the time dependence of the daily electrical power demand

Fig. 1.2 Hourly data on electric load during a full week, showing possible use of a storage system. After [2]

sources. However, these facilities are not very flexible, requiring 30–60 min to start up. In addition, utilities typically have a modest amount of *operating reserve*, additional capacity that is available to the system operator within a modest amount of time, e.g., 10 min, to meet the demand if a generator fails, there is another disruption in the supply, or there is a sudden additional large.

This operating reserve can be divided into two types. One, called *spinning reserve*, is extra generating capacity that can be made available by a relatively simple modification of the operational parameters of the major turbines that are in use.

The other is *supplemental*, or *non-spinning*, *reserve*. This label describes capacity that is not currently connected to the system, but that can be brought online after only a short delay. It may involve the use of fast-start generators, or importing power from other interconnected power systems. Generators used for either spinning reserve or supplemental reserve can typically be put into operation in 10 or so minutes.

In addition, there are additional secondary source technologies that are more flexible, but significantly more expensive, that can be used to handle any need for extra capacity. In some cases these involve the use of gas turbines, similar to the engines that are used to power airplanes, to drive generators.

1.5   Methods That Can Be Used to Reduce the Magnitude of the Variations in Energy Demand

It is self-evident that if the rapid large-scale variations in energy demand can be reduced, the less expensive base load technologies can play a greater role. An obvious solution would be to use *load shifting*, in which some energy needs are shifted from times when the overall demand is large to periods in which other needs are reduced. One way to encourage this is to use time-of-day pricing, a method that is used much more in Europe than in the USA at the present time.

Fig. 1.3 Ideal situation in which energy storage methods flatten the time dependence of the energy supply requirement

The author lived for a number of years in Germany, and spent some time in a house that used hot water for space heating. A large amount of water was stored in a large insulated vessel that was heated electrically at night, when the power costs were very much lower than during the daytime. This hot water was circulated as needed throughout the day.

Another approach is to use energy storage methods to absorb electrical energy when it is available and inexpensive, and to supply it back into the grid system when the demand is higher. This is shown schematically in Fig. 1.3 for the ideal situation, in which the load curve would be entirely flattened.

The dominant means of electricity storage for daily load shifting now is the use of pumped hydro facilities. They are also used for both spinning reserve and operating reserve applications, and typically produce hundreds of MW for up to 10 h. Another, but not so wide-spread, approach involves the use of compressed air energy storage in underground caverns. Both of these technologies are dependent upon the availability of appropriate sites.

About 2.5 % of the total electric power delivered in the USA is currently cycled through such a large-scale storage facility, most commonly pumped hydro. This is different from the practices in Europe and Japan, where about 10 % and 15 %, respectively, of the delivered power is cycled through such storage facilities. In Japan this is a reflection of both higher electricity prices, and a much greater difference between peak and off-peak energy costs.

The development of additional pumped hydro facilities is very limited, due to the scarcity of further cost-effective and environmentally acceptable sites in the USA. Countries with more mountainous terrain have an obvious advantage.

Several additional advanced medium and large-scale energy storage technologies are being developed to help with this problem, including Na/Na$_x$S and flow batteries. These will both be discussed later in this text.

1.6 Short-Term Transients

In addition to these major variations in the energy demand, there are also many short-term transients. This is illustrated in Fig. 1.4.

These transients can lead to generator rotor angle instability, leading to oscillations and unstable operating conditions. Voltage instability can also occur when the load and the associated transmission system require a large amount of reactive, rather than real, power. This can result in a sudden and drastic voltage drop. As a result, short-term (less than 5 min) power outages can also occur, which can be very costly.

A different type of technology is necessary to take care of this problem, which is currently handled by making small adjustments in the frequency. However, fast-reaction high-power storage mechanisms would be ideal for this application.

As will be discussed later, the rapid response characteristics of flywheel systems make it possible to use them to reduce the problem of short-term transients in the load upon the electrical grid. Other options that are being used in some places include Cd/Ni, hydride/Ni, and even Pb-acid battery systems. Super-capacitors are also of interest in situations in which their limited energy storage capacity is not a problem. These systems are all discussed in later chapters in the text.

Fig. 1.4 Short-time transients superimposed upon the load curve. After [2]

1.7   Portable Applications That Require Energy Storage

In addition to these traditional medium- and large-scale considerations, attention has to be given to the fact that a large number of additional uses of energy, with quite different energy storage requirements, have become very important in recent years. Very obvious examples are the many small-to-medium-size electronic devices that one now encounters everywhere. A large fraction of these are portable, and thus require batteries for their operation. Their important parameters are quite different from the stationary applications mentioned above.

An additional area of application for energy storage that has become especially visible in recent years involves vehicles, and both full-electric and hybrid internal combustion—electric vehicles are becoming increasingly numerous. In these cases, the electrical energy storage components are critical with respect to both performance and cost.

The important characteristics required by these various types of applications vary widely. As a result, a number of different technologies play major roles.

1.7.1   Storage Methods for Use with Portable
Electronic Devices

Devices such as computers, telephones, music players, camcorders, and personal digital assistants, as well as electronic watches and hearing aids are now ubiquitous. Almost all of these now are powered by one or another type of electrochemical battery. There are several reasons for this. One is that the need for energy is sporadic, rather than steady state. Thus there is no need to be continually connected to a fixed energy source. Batteries can either be recharged or replaced as needed.

The amount of energy stored per unit weight, the specific energy, is quite high for some types of batteries. This is obviously important for applications in which weight is important. But in some other cases the volume of the energy source, the energy density, is critical. A great deal of progress has been made in these directions in recent years. Then there is the issue of price, and especially, safety. The latter has become more of a problem the greater the amount of energy stored in small packages. These matters, and some of the important battery technologies, are discussed in some detail in later chapters in the text.

1.7.2   Energy Use and Storage in Vehicles

Vehicle propulsion currently mostly involves the consumption of gasoline or diesel fuel in internal combustion engines. *Hybrid autos* are rapidly being introduced, but still represent only a small fraction of the total. They combine an internal combustion engine with a relatively small high-rate battery that can acquire some of the vehicle's kinetic energy during deceleration. This combination serves to reduce the amount of fuel used, but results in no change in energy source.

A relatively small number of *all-electric* passenger and commercial vehicles are being produced, as well. While most of the attention has been focused on the future of larger vehicles, some 60 million electric scooters and bicycles are already being used. They get their energy from the electric grid, typically by charging overnight. Another variant that is getting increased attention is the category called *plug-in hybrids*. These are different from the normal hybrids in that the on-board battery can be recharged (again, typically overnight) from the electric grid. The charging of the battery component allows these vehicles to travel a limited range on electricity alone. Thus if they do not drive very far, they get their energy from the grid. If they travel farther, the energy input is shared between the electrical grid and a petrochemical fuel.

As the number of plug-in hybrid and all-electric vehicles increases, the overall transportation energy demand will gradually move more toward electricity, and away from liquid fuels. But in addition, this part of the electric load will be mostly at night, where the other demands are reduced, and thus this will contribute to load leveling.

The energy storage devices used in both hybrid autos and plug-in hybrids are batteries. The desirable characteristics are different in the two cases, however. Hybrids have relatively small batteries, for the amount of energy that must be stored is limited. Instead, the rate of energy absorption during braking can be quite high. Thus these batteries must be able to operate at high rates, or high power. Metal hydride/nickel batteries are now primarily used for this purpose. However, almost all of the auto manufacturers who are involved in the development of these types of vehicles indicate that they expect to use lithium-ion batteries in the future, for they generally store considerably more energy per unit weight than the batteries currently used in vehicles, and also be able to operate at high power.

The all-electric and plug-in hybrid applications are different, for in order for the vehicles to have an appreciable range, a large amount of energy must be stored. This requires large, and thus heavy, batteries. And they must be optimized in terms of energy storage, rather than high power. As a result, it is reasonable to expect that this type of vehicles will be designed to employ two different types of batteries, one optimized for energy, and the other for power.

Another obvious factor is cost, especially for the potentially more desirable lithium-ion batteries. This is an area receiving a large amount of research and development attention at the present time, and is discussed in some detail later in this text.

1.8   Hydrogen Propulsion of Vehicles

There has recently been a good deal of interest in the use of hydrogen as a fuel for vehicular propulsion. There are actually two different versions of this topic. One is the direct combustion of hydrogen in internal combustion engines. This can easily be done, as it requires only slight modifications of current gasoline or diesel engines and the injection of gaseous hydrogen at relatively high pressures. The German auto firms BMW and Daimler Benz have demonstrated such vehicles over a number of years. In the BMW case, the hydrogen is stored as a liquid at a low temperature in an insulated tank. Early vehicles from Daimler Benz stored the hydrogen in metal-hydrogen compounds known as metal hydrides some time ago. These are essentially the same materials as the hydrides used in the negative electrodes of the common hydride/nickel batteries. Upon heating, the hydrogen is released so that it can be used in the engine.

The other hydrogen-powered vehicle approach, which was heavily promoted by the United States government for a number of years, is to use hydrogen-consuming fuel cells for propulsion. Fuel cell–propelled vehicles are currently being developed in Japan by Honda and Toyota, Hyundai in South Korea, as well as Audi, BMW, and Mercedes Benz in Germany. Other companies indicate that they are moving in this direction, but none in the USA at the present time.

But again, the hydrogen must be carried, and stored, in fuel cell-powered vehicles. Because of the metal component, the current metal hydrides are quite heavy, storing only a few percent hydrogen by weight. The United States Department of Energy has been supporting a substantial research program aimed at finding other materials that can store at least 6 % hydrogen by weight. This target is based upon the assumption that consumers will require a hydrogen-powered vehicle that has approximately the same driving range as the current internal combustion autos. It is obvious that the increasing interest in reduced-range vehicles that is so evident in connection with those that are electrically powered does not yet seem to have migrated to the hydrogen fuel cell community.

It is interesting that a firm in Germany is producing military submarines that are powered by (quiet) hydrogen-consuming fuel cells, and metal hydrides are used to store the necessary hydrogen. The weight of the hydride materials is not a problem in the case of submarines, and some of the materials that are currently known are evidently satisfactory for this purpose.

1.9 Temperature Regulation in Buildings

As indicated in Table 1.2, a significant amount of energy is used in residential and commercial buildings. In addition to lighting, much of this is for the purpose of the regulation of the temperature in living and work spaces. In the winter, this involves the supply of heat, and in the summer it often requires temperature reduction by the use of air conditioning equipment.

In addition to the use of improvement thermal insulation, the magnitude of the related energy requirements can be reduced by the use of thermal storage techniques. There are several types of such systems that can be used for this purpose, as will be discussed in Chap. 4.

1.10   Improved Lighting Technologies

It was also shown in Table 1.2 that electric power now constitutes about 40 % of the total energy use in the USA. Of this, some 22 % is consumed by lighting. Thus over 8 % of the total energy use in the USA is related to lighting. On a global basis, about 2700 TWh, or 19 % of total electricity consumption is used for this purpose [3].

Most lighting today involves the use of incandescent bulbs in which tungsten wire is electrically heated to about 3500 K. The spectrum of the emitted radiation is very broad, covering both the range to which the human eye is sensitive, and also well into the infrared range, where the result is heat. Only about 5 % of the electrical energy consumed in these bulbs is converted to visible light, the other 95 % is heat. Thus this is a very inefficient process.

Lighting technology is changing fast, and it is reasonable to expect that its contribution to the total amount of energy use will decrease significantly in the future. The first step in this transition involves greatly increased use of fluorescent lamps, which have light emission efficiencies of about 20–25 %, rather than the 5 % of traditional incandescent bulbs.

Governmental regulations are now being employed in many locations to reduce the use of energy-inefficient incandescent lighting. A law in the State of California required a minumum standard of 25 lm/W by 2013, and 60 lm/W by year 2018. Other states are surely to move in this direction also. In addition to these local initiatives, the US Federal Government effectively banned the use of some sizes of the common incandescent lights after January, 2014 by passing the Clean Energy Act of 2007.

However, it is now generally expected that fluorescent light technology will be overtaken by light sources based upon semiconductors, light-emitting diodes, or LEDs. Development efforts in this direction have been underway for a long time, with large jumps in their effectiveness [3], and some LED products are now widely used as bicycle lights, in the tail lights of automobiles, and as street lights in cities. Although still having rather high initial costs, they are beginning to be employed in household lighting.

The major breakthroughs have involved development of GaN-based LEDs and the compositions and processes for making them that produce devices with various emitted wavelength ranges [3, 4]. It is now clear that this solid-state type of technology, which is some 15 times as energy efficient as incandescent lights, and which is moving quickly onto the commercial market, will play an even greater role in the future.

The relative energy efficiencies of these three technologies can be seen from the data in Table 1.3.

Table 1.3   Energy efficiency of different lighting technologies

Light source	Lumens per watt
Tungsten incandescent bulb	17.5
Compact fluorescent tubular bulb	85–95
White light-emitting diode	170

1.11 The Structure of This Book

This book is intended to provide a basic understanding of the various mechanisms and related technologies that are currently employed for energy storage. An initial chapter introduces relevant basic concepts. This is followed by a group of chapters that describe the most important chemical, mechanical, and electromagnetic methods.

The general principles involved in the various electrochemical technologies that are becoming increasingly important are then introduced, followed by a group of chapters on the most important battery systems. These are followed by a chapter that discusses the storage methods that are most important for the major areas of application.

Much of the discussion will involve both concepts and methods of thermodynamics, mechanics, electrochemistry and other areas of materials science. However, it is not necessary to already have a significant amount of prior expertise in such areas, for the background that is necessary will be presented when they are relevant to different portions of the text. A number of general sources in which further information can be found are included as references [4–11].

References

1. Slight modification of US Energy Information Administration data for the year 2007.
2. R. Fernandes, Proc. 9th Intersociety Energy Conf. (1974), p. 413
3. Humphreys CJ (2008) MRS Bull 33:459
4. Johnstone B (2007) Brilliant!: Shuji Nakamura and the Revolution in Lighting Technology. Prometheus Books, New York
5. Jensen J, Sorenson B (1984) Fundamentals of Energy Storage. Wiley-Interscience, New York
6. Ter-Gazarian A (1994) Energy Storage for Power Systems. Peter Peregrinus Ltd., London
7. EPRI-DOE handbook of Energy Storage for Transmission and Distribution Applications (2003) Electric Power Research Institute, Palo Alto, CA
8. Energy Storage for Grid Connected Wind Generation Applications, EPRI-DOE Handbook Supplement (2004) Electric Power Research Institute, Palo Alto, CA
9. Basic Research Needs for Electrical Energy Storage (2007) Office of Science, Department of Energy, U.S. Government
10. Kalhammer FR, Kopf BM, Swan DH, Roan VP, Walsh MP (2007) Status and Prospects for Zero Emissions Vehicle Technology. State of California Air Resources Board, California
11. Huggins RA (2009) Advanced Batteries: Materials Science Aspects. Springer, New York

mechanical work done on a body and the resultant increase in its temperature was determined by James Prescott Joule in the mid nineteenth century. He found that this relation is

$$1 \; cal \; = \; 4.184 \, Joules \tag{2.1}$$

The thermochemical calorie, the unit quantity of heat, is defined as the amount of heat that must be added to one gram of water to raise its temperature 1 °C. The Joule, a measure of energy, can be expressed in either electrical or mechanical terms:

$$1 \, Joule \; = \; 1 \, watt \, sec \; = \; 1 \, volt \, coulomb \tag{2.2}$$

or

$$1 \, Joule \; = \; 1 \, Newton \, meter \; = \; 1 \, kg \, m^2 sec^{-2} \tag{2.3}$$

The amount of heat required to raise the temperature of any material 1 °C is called its *heat capacity*, or its *specific heat*. In the latter case, the amount of heat per unit weight, the dimensions are J kg^{-1} K^{-1}.

2.3   The First Law of Thermodynamics—Conservation of Energy

In a closed system energy cannot either be created or destroyed. It can only be converted from one type to another type. This is called the *first law of thermodynamics*, or the *law of the conservation of energy*. It can be expressed as

$$\Delta U \; = \; q \; + \; w \tag{2.4}$$

where U is the internal energy of a material or a system, assuming that it is not in motion, and therefore has no kinetic energy, q is heat absorbed by the system, and w is work done on the system by external forces. In the case of a simple solid, U can be thought of as the sum of the energy of all of its interatomic bonds. It does not have an absolute value, but is always compared to some reference value.

2.4   Enthalpy

Another important quantity is the *enthalpy H*, which is sometimes called the *heat content*. The name came from the Greek *enthalpein*, to warm. This is defined by the equation

$$H = U + pv \tag{2.5}$$

where p is the applied pressure and v the volume. The product pv is generally quite small for solids in the conditions that will be met in this text.

If a system (e.g., a material) undergoes a change in state, such as melting or a chemical reaction, there will be a change in enthalpy ΔH. A positive value of ΔH means that heat is absorbed, and the reaction is described as *endothermic*. On the other hand, if heat is evolved, ΔH is negative, the internal energy is reduced, and the reaction is called *exothermic*. The latter is always the case for spontaneous processes. This change in heat content when a reaction takes place is called the *latent heat* of the reaction.

H, U, and the pv product all have the dimensions of energy, kJ/mol. Their values are always compared to a reference *standard state*. For pressure, the standard state is 1 bar, or 0.1 MPa. The conventional reference temperature is 298.15 K, or 25 °C.

The standard state that is generally used for all solid substances is their chemically pure form at a pressure of 1 bar and a specified temperature. This is indicated by the use of the index "^{0}." At a temperature of 298.15 K and a pressure of one bar, the value of H^0 is zero for all pure materials.

2.5   Entropy

A further important quantity in discussions of thermodynamics is the *entropy* (from the Greek word *trope*, a deviation or change). Entropy, S, is a measure of disorder, or randomness. What this means will become evident in the examples to be discussed below.

2.5.1   Thermal Entropy

At a temperature of absolute zero (0 K) the structure of a solid material is fixed, or "frozen." When the temperature is increased by adding heat, or thermal energy, its constituent particles begin to vibrate in place, acquiring local kinetic energy, similarly to the energy in a swinging pendulum or vibrating spring. The magnitude of this vibrational, or motional, kinetic energy, or heat, is proportional to the temperature. The proportional factor is called the *thermal entropy*, S_{th}. This can be simply written as

$$q = TS_{th} \tag{2.6}$$

This can be rearranged to define the thermal entropy, the randomness of the positions of the vibrating particles at any time:

$$S_{th} = q/T \tag{2.7}$$

It can be seen that the product TS_{th} also has the dimension of energy.

One can find more sophisticated discussions of the origin and physical meaning of thermal entropy in a number of places, such as [1, 2], but such greater depth is not necessary here.

2.5.2 Configurational Entropy

A different type of randomness also has to be considered in materials systems. In addition to the vibratory motion of the fundamental particles present, there can also be some degree of disorder in the arrangement of the particles, i.e., atoms and electrons, in a material. This is sometimes called the configurational, or structural, entropy, S_{conf}. It is a measure of the uniformity or regularity of the internal structure, or the arrangement of the particles in a crystal structure, and also has the dimension of energy/T.

The magnitude of this type of entropy changes when there is a change in crystal structure of a solid, the solid melts to become a liquid, or a chemical reaction takes place in which the entropies of the reactants and the products are different.

The total entropy, indicated here simply by the symbol S, is the sum of the thermal entropy and the configurational entropy:

$$S = S_{th} + S_{conf} \tag{2.8}$$

2.6   The Energy Available to Do Work

The driving force for reactions or other changes to take place is always a reduction in energy. This is relatively easy to understand in the case of mechanical systems, where the total energy is generally divided into two types, potential energy and kinetic energy. It may seem to be a bit more complicated in chemical systems, however.

Some time ago, the driving force for a chemical reaction was called the *affinity*. For example, if A and B tend to react to form a product AB, the amount of energy released would be the *affinity* of that reaction. But since the work of J. Willard Gibbs [3, 4] it has been recognized that the driving force for any process, and also therefore the maximum amount of work that can be obtained from it, is a change in the quantity G, where

$$G = H - TS \tag{2.9}$$

The name now generally given to G is the *Gibbs free energy*, although it is sometimes called the *Gibbs energy*. In parts of Europe it is called the *free enthalpy*. It is also equivalent to the *exergy*, a term introduced in 1956 by Zoran Rant [5], that is used in some branches of engineering.

It is obvious that the energy that is available to be used to do either mechanical or electrical work is less than the heat (total energy) H present by the amount of energy tied up in the entropy, TS.

The Gibbs free energy G will appear many times in this text, for it plays a significant role in many applications. Changes or differences in G, not in H or U, constitute the driving forces for many processes and reactions.

2.7   The Temperature Dependence of G, H, and S

The temperature dependences of the Gibbs free energy G, the enthalpy H, and the total entropy S for a simple metal, pure aluminum, are shown in Fig. 2.1.

It can be seen that there is a discontinuity in both the enthalpy H and the entropy S, but not the Gibbs free energy G, at the melting point, 933.45 K. The entropy increases by 11.48 J K^{-1} mol^{-1}, for the liquid aluminum structure has a greater value of configurational entropy than solid aluminum. There is also a corresponding discontinuity in the enthalpy of 10.70 kJ mol^{-1} at that temperature since the process of melting is endothermic, requiring the input of heat at the melting point to convert aluminum from solid to liquid.

The difference between the gradual changes in both H and S with temperature and their discontinuities when melting or other phase changes occur will be discussed further in Chaps. 3 and 4, which deal with the storage of heat. In the one case one considers changes in *sensible heat*, and in the other, the important quantity is the *latent heat* of the reaction.

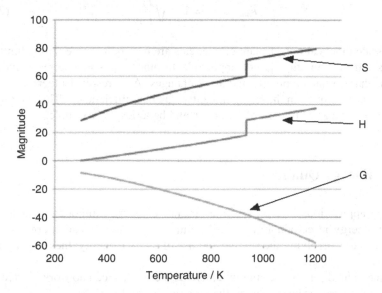

Fig. 2.1 Thermodynamic data for pure aluminum. Dimensions are J/K mol for S, and kJ/mol for H and G

2.8 Irreversible and Reversible Storage Modes

There are two general types of energy storage to be considered. In one type, energy that is already present, or stored, is available to be used once. In the other, energy can be both used and replaced, i.e., the energy is stored reversibly.

2.9 The Carnot Limitation

Considerations of the thermodynamics of energy systems often include discussions of the so-called *Carnot limitation*. It is applicable to energy conversion processes that involve heat engines, and results in the desire to have such devices operate at as high a temperature as possible, producing products at as low a temperature as possible.

The maximum efficiency that can be obtained from any process that involves the input of energy at one (higher) temperature, and its exhaust at another (lower) temperature is given by

$$\varepsilon_{max} = \frac{T_H - T_L}{T_H} = 1 - \frac{T_L}{T_H} \tag{2.10}$$

There is also a theoretical limit to the maximum power that can be obtained from such a heat engine [6]:

$$\varepsilon_{max\ power} = 1 - \sqrt{\frac{T_L}{T_H}} \tag{2.11}$$

Large-scale practical power plant efficiencies are now quite close to this limit.

This Carnot limitation is not applicable to other types of energy conversion systems that do not involve temperature changes. As a result, isothermal energy conversion systems can have significantly greater efficiencies. This is a great inherent advantage of the use of fuel cells and batteries, for example.

2.10 Energy Quality

In many applications, not only is the amount of energy important, but also its *quality*. Energy in certain forms can be much more useful than energy in other forms. In cases involving the storage and use of thermal energy, the temperature is important.

As mentioned earlier, the amount of useful energy, that can do work, G is related to the total energy or heat content, the enthalpy H, by Eq. (2.9).

The energy quality can be defined as the ratio G/H,

$$\frac{G}{H} = \frac{\Delta T}{T} = 1 - \frac{T_0}{T} \tag{2.12}$$

where T is the temperature in question, and T_0 is the ambient temperature. The energy quality is thus greater the larger the value of T.

In electrochemical systems the quality of stored energy is dependent upon the voltage. High voltage energy is more useful than low voltage energy. This will be discussed in Chap. 9.

References

1. Lewis GN, Randall M (1961) "Thermodynamics", revised by K.S. Pitzer and L. Brewer, McGraw-Hill Book Co., New York
2. L. Pauling, General Chemistry, Dover New York (1970)
3. Gibbs JW (1873) Trans Conn Acad Arts Sci 2(309):382
4. Gibbs JW (1875) Trans Conn Acad Arts Sci 3(108):343
5. Rant Z (1956) Forschung auf dem Gebiete des Ingenieurwesens 22:36–37
6. Rubin M, Andresen B, Berry R (1981) Beyond the Energy Crisis, ed. R. Fazzolare and C. Smith, Pergamon Press, New York

Chapter 3
Thermal Energy Storage

3.1 Introduction

It was mentioned in Chap. 1 that a significant portion of the total energy use is for temperature control and lighting in living and working spaces. Energy use for lighting purposes comprises between 20 and 50 % of the total energy used in homes, and varies appreciably with both location and time of the year, of course. This is expected to decrease substantially as the result of the use of fluorescent and light-emitting-diode (LED) devices in the future.

Residential and commercial heating and cooling needs are currently mostly taken care of by the use of gas or electric heating and electrically powered air conditioning. This energy requirement can also be reduced appreciably by various measures. Among these is the reduction of heat transfer to and from the environment by the use of better insulation. Another is to make use of thermal energy storage systems. Thermal energy storage is also used in large-scale power plants, although that will not be included here. Discussions of these matters can be found in several places [1–4].

Household hot water systems in the United States typically have a gas- or electrically-heated water heater and storage tank, from which water is distributed using pipes throughout the dwelling. There is thus heat loss from the distribution piping as well as through the insulation on the storage tank. A different system is used in parts of Europe, where energy is more expensive. This involves local heating of water where, and only when, it is needed. Thus there are essentially no storage losses.

In addition to the storage of heat in order to maintain something at a high temperature, the main topic of this chapter, thermal energy storage methods are also sometimes used to maintain cold temperatures for the storage of food or for other chiller applications. Information about this type of application can be found in [3].

© Springer International Publishing Switzerland 2016
R.A. Huggins, *Energy Storage*, DOI 10.1007/978-3-319-21239-5_3

There are two general types of thermal storage mechanisms. As will be seen below, one is based upon the use of the *sensible heat* in various solid and/or liquid materials. The other involves the *latent heat* of phase change reactions.

3.2 Sensible Heat

Energy can be added to a material by simply heating it to a higher temperature. The energy that is involved in changing its temperature is called "sensible heat," and its amount is simply the product of the specific heat and the temperature change.

This sensible heat can be transferred to another, cooler, material, or to the environment, by radiation, convection, or conduction. Thus this is a method for storing energy in the form of heat, and transferring it again. A simple example is the traditional procedure of using a hot rock or a hot water bottle to pre-warm a bed before going to sleep.

This type of energy storage has also been used to control the temperature in living or working spaces. In some cases the amount of storage material that is needed can be quite large, so that there is the obvious concern about its cost. This results in the use of relatively simple and inexpensive materials. Data on the thermal capacity of some examples are presented in Table 3.1.

The amount of heat q that can be transferred from a given mass of material at one temperature to another at a lower temperature is given by

$$q = \rho C_P V \Delta T \qquad (3.1)$$

where ρ is the density, C_p the specific heat at constant pressure, and V the volume of the storage material. ΔT is the temperature difference.

Each of the materials in Table 3.1 has some advantages and some disadvantages. The specific heat of water is more than twice that of most of the other materials, but it is only useful over a limited temperature range (5–95 °C). Thus water appears

Table 3.1 Thermal properties of some common materials

Material	Density $(kg\ m^{-3})$	Specific heat $(J\ kg^{-1}\ K^{-1})$	Volumetric thermal capacity $(10^6\ J\ m^{-3}\ K^{-1})$
Clay	1458	879	1.28
Brick	1800	837	1.51
Sandstone	2200	712	1.57
Wood	700	2390	1.67
Concrete	2000	880	1.76
Glass	2710	837	2.27
Aluminum	2710	896	2.43
Steel	7840	465	3.68
Magnetite	5177	752	3.69
Water	988	4182	4.17

attractive for systems involved in the control of living space heating and cooling. An example of this was mentioned in Chap. 1.

On the other hand, some of the inexpensive bulk solids can be used over a wider range of temperature, and they can be more compact, due to their higher densities. An additional factor that may be important in some cases is the thermal conductivity of the storage material, for that influences the rate at which heat can be either absorbed or extracted.

No matter how well the system is insulated, there are always some losses when using this method. Thus the insulation or thermal isolation of the storage material can be quite important, particularly if the storage period is substantial.

The rate of heat loss to the surroundings is proportional to the surface area, and also to the temperature difference, but the total amount of thermal storage is proportional to the volume of any storage container. Therefore, it is more effective to use large vessels with shapes that are not far from spherical.

3.3 Latent Heat

A different mechanism for the storage of energy involves using *phase transitions* with no change in the chemical composition. It was shown in Fig. 2.1 of Chap. 2 that there is a jump in the value of the entropy, and a corresponding change in the enthalpy, or heat content, but not of the Gibbs free energy, at the melting point of aluminum, a simple metal. This is characteristic of phase transitions in elements and compounds that melt *congruently*; that is, the solid and liquid phases have the same chemical composition. There are also phase transitions in which both phases are solids. As an example, there is a transition between the alpha and beta phases in titanium at 882 °C. The alpha, lower temperature, phase has a hexagonal crystal structure, whereas the beta phase has a body-centered cubic crystal structure. Because they have different crystal structures they have different values of configurational entropy, and there is a corresponding difference in their enthalpy values.

In these cases latent heat is absorbed or supplied at a constant temperature, rather than over a range of temperature, as it is with sensible heat. Isothermal latent heat systems are generally physically much smaller than sensible heat systems of comparable capacity.

A further simple example is water. At low temperatures it is solid, at intermediate temperatures it becomes a liquid, and at high temperatures it converts to a gas. Thus it can undergo two phase transitions, with associated changes in entropy and enthalpy. The Gibbs free energy, or chemical potential, is continuous, for the two phases are in equilibrium with each other at the transition temperature.

From

$$\Delta G = \Delta H - T\Delta S = 0 \tag{3.2}$$

the change in heat content ΔH at the transition temperature is equal to $T\Delta S$ at that temperature.

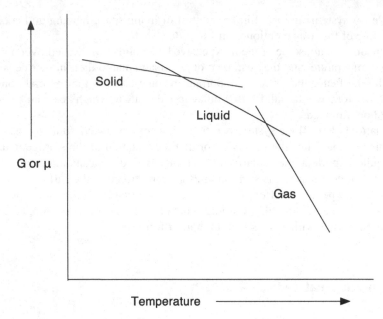

Fig. 3.1 Schematic representation of the temperature dependence of the Gibbs free energy of water

The slope of the temperature dependence of the Gibbs free energy is proportional to the negative value of the entropy, which is different in the different phases, for they have different structures. This is shown schematically in Fig. 3.1.

3.3.1 Inorganic Phase Change Materials

There are a number of other materials that also undergo congruent phase transitions. Data on some simple inorganic materials are given in Table 3.2. There are also many cases in which liquids convert isothermally into two or more phases with different compositions by means of *eutectic reactions* upon cooling, with accompanying heat evolution. Upon heating, heat is absorbed by the reverse reactions. Data on a large number of molten salt eutectic systems can be found in [5].

In addition to the phase changes involved in the conversion of solids into liquids, there are a number of cases in which there is a phase change in the solid state. These involve changes in the crystal structure, and thus changes in the entropy and the heat content at temperatures below their melting points. These materials can be used to store thermal energy at the temperatures of their solid-state phase transitions. Some examples are given in Table 3.3.

There are several features of the data in this table that deserve comment. The total entropy change, and thus the heat absorbed, between the low temperature solid state and the molten state is divided between that which occurs at the solid state

Table 3.2 Heat of fusion data for several simple inorganic materials

Phase	Melting point/°C	Heat of fusion (MJkg^{-1})
NH_4NO_3	170	0.12
$NaNO_3$	307	0.13
NaOH	318	0.15
$Ca(NO_3)_2$	561	0.12
LiCl	614	0.31
$FeCl_2$	670	0.34
$MgCl_2$	708	0.45
KCl	776	0.34
NaCl	801	0.50

Table 3.3 Data on solid-state phase transition and melting entropy for a number of materials

Material	Transition temp./°C	Melting temp./°C	Transition entropy (Jmol^{-1} K^{-1})	Melting entropy (Jmol^{-1} K^{-1})
FeS	138	1190	4.05	21.51
AgI	148	558	14.61	11.33
Ag_2S	177	837	8.86	7.01
Na_2SO_4	247	884	12.5	18.2
Ag_2SO_4	427	660	26.66	19.19
Li_2SO_4	577	860	29.2	7.9
$LiNaSO_4$	518	615	31.2	small

phase transition and that which occurs upon melting. In general, the crystal structures below and above solid state transformations are not very different. As a result, the entropy changes that are involved are not very large, and most of the total entropy changes occur at the melting point. An example of this type that is included in Table 3.3 is FeS.

But there are some materials in which the higher temperature solid phase has a crystal structure in which one of the atomic species has an unusual vibrational amplitude or inter-site mobility. When that phase is stable it has a high value of entropy. Thus the entropy change in its formation is large. Two examples of this are shown in Table 3.3, AgI and Ag_2S. Silver ions have an unusually large vibrational amplitude and high mobility in the higher temperature phases of both of these materials. AgI is called a *fast ionic conductor*, and Ag_2S is a *fast mixed conductor*. The difference is that Ag_2S is an electronic, as well as an ionic, conductor, whereas AgI is an electronic insulator.

Na_2SO_4 and Ag_2SO_4 are two additional examples of materials with this type of behavior. In these cases the sodium and silver ions are relatively mobile, and the SO_4 tetrahedral groups are essentially static parts of the crystal structure.

However, there are some other materials in which the type of structural disorder that leads to large values of entropy is quite different. Whereas atomic or ionic motion in most solids, including AgI and Ag_2S, involves species either jumping into adjacent crystallographic vacancies or between dilutely occupied interstitial positions, some materials have been found in which atomic motion occurs in coordinated groups. That is, groups of atoms move together. This type of atomic motion occurs in a group of lithium sulfates. It was found that these materials can also have unusually high values of ionic conductivity, different from that of the sodium and silver analogs [6, 7]. The structural reason for this is that there is rotation of SO_4 groups within the "static" crystal structure that assists in the longer-range transport of the lithium ions. This was first described as a *cogwheel mechanism*, but it is now generally referred to as a *paddlewheel mechanism* [8]. These materials also exhibit a high degree of mechanical plasticity, and are sometimes also called *plastic crystals*. Reviews of this topic can be found in [9, 10].

It can be seen in Table 3.3 that the entropy changes related to the solid state structural changes in the paddlewheel materials are significantly greater than the entropy changes upon melting. They are thus particularly attractive as heat storage components if the temperature is appropriate. Materials that have large values of latent heat related to solid-state phase transitions are discussed in [11].

3.3.2 Organic Phase Change Materials

In addition to these inorganic materials, it is also possible to take advantage of analogous behavior in organic materials. Data for some simple organic materials with large values of the heat of fusion are shown in Table 3.4, and the thermal properties of fatty acids and two simple aromatic materials are presented in Tables 3.5 and 3.6.

Table 3.4 Data relating to some organic phase change materials

Material	Melting temperature (°C)	Heat of fusion (kJ kg^{-1})
Paraffin wax	64	173.6
Polyglycol E 400	8	99.6
Polyglycol E 600	22	127.2
Polyglycol E 6000	66	190.0

Table 3.5 Data relating to some fatty acids used as phase change materials

Material	Melting temperature (°C)	Heat of fusion (kJ kg^{-1})
Stearic acid	69	202.5
Palmitic acid	64	185.4
Capric acid	32	152.7
Caprylic acid	16	148.5

Table 3.6 Some aromatics used as phase change materials

Material	Melting temperature (°C)	Heat of fusion (kJ kg^{-1})
Biphenyl	71	19.2
Naphthalene	80	147.7

3.4 Quasi-Latent Heat

Chemical reactions generally result in the generation or absorption of heat, similar to the thermal effects related to phase transitions in materials in which there are no changes in chemical composition. The thermal effects related to chemical reactions are often described in terms of *quasi-latent heat*. They will be discussed in Chap. 4.

3.5 Heat Pumps

Although they are not involved in the storage of heat, the topic of this chapter, it seems desirable to briefly discuss *heat pumps*. They can be used to move heat from one location to another. There are two general types of such systems. One involves the input of mechanical work to move low temperature heat from one location to another location at a higher temperature.

In a second type, which is sometimes called a chemical heat pump, the important feature is the temperature dependence of a chemical reaction. An example of this method is a system involving metal hydride materials in two locations, one that is at a high temperature in which there is heat input, and one that is cooler. Heat is moved by the movement of a chemical species, such as hydrogen, from one location to another.

References

1. Beghi G. ed. (1981) Energy Storage and Transportation. D. Reidel Pub. Co
2. Jensen J, Sorenson B. (1984) Fundamentals of Energy Storage. Wiley-Interscience
3. Dincer I, Rosen M.A., eds. (2002) Thermal Energy Storage. John Wiley
4. Ter-Gazarian A. (1994) Energy Storage for Power Systems. Peter Peregrinus Ltd.
5. Physical Properties Data Compilations Relevant to Energy Storage. 1. Molten Salts: Eutectic Data, NSRDS-NBS 61, Part 1 (1978)
6. Foerland T, Krogh-Moe J (1957) Acta Chem Scand 11:565
7. Foerland T, Krogh-Moe J (1958) Acta Cryst 11:224
8. Kvist A, Bengtzelius A. (1973) Fast Ion Transport in Solids, ed. W. van Gool. North Holland Pub. Co. p. 193
9. Lunden A, Thomas JO. (1989) High Conductivity Solid Ionic Conductors, ed. T. Takahashi. World Scientific Publishing Co. p. 45
10. Lunden A. (1992) Fast Ion Transport in Solids, ed. B Scrosati, A Magistris, CM Mari and G. Mariotto. Kluwer Academic Publishers, p. 181
11. K. Schroeder and C.-A. Sjöblom, High Temp.-High Press. 12, 327 (1980)

Chapter 4
Reversible Chemical Reactions

4.1 Introduction

In the discussion of thermal energy storage in Chap. 3, it was pointed out that energy can be stored in both the *sensible heat* that is related to changes in the temperature of materials and their heat capacities, and the *latent heat* involved in isothermal phase transitions. A common example of such an isothermal phase transition with a significant amount of stored heat is the melting and freezing of water. In such cases, there are no changes in chemical composition. The chemical species below and above the phase transition are the same. Only their physical state is different. Such reactions are said to be *congruent*.

Another type of energy storage involves reversible chemical reactions, in which there is a change in the chemical species present. In many cases, such reactions can also be reversible. An example of this would be the reaction of hydrogen and oxygen to form water, and its reverse, the decomposition of water into hydrogen and oxygen. The energy, or heat, involved in this reversible reaction is the heat of reaction, and is sometimes called *quasi-latent heat*.

As was mentioned in Chap. 3, the amounts of heat involved in latent heat reactions can be much larger than those that are involved in sensible heat. As a result they are typically used in situations in which it is desired to maintain the temperature of a system at, or near, a constant value. As will be seen in this chapter, there are a number of materials in which chemical reactions can take place that can exhibit both *enhanced sensible heat* and *quasi-latent heat*.

© Springer International Publishing Switzerland 2016
R.A. Huggins, *Energy Storage*, DOI 10.1007/978-3-319-21239-5_4

4.2   Types of Non-congruent Chemical Reactions

In addition to the *congruent* reactions mentioned above, there are a number of different types of *chemical reactions* in which one or more materials react to produce products with different compositions. Reactions in which the reactants and products have different compositions are *non-congruent*.

In some cases the result of the reaction is a change in the chemical composition of one or more of the phases present by the addition of another species to it, or deletion of a species from it. Such reactions are called by the general label *insertion reactions*.

There are also a number of important chemical reactions in which some phases grow or shrink, or new phases form and others disappear. The result is that the microstructure of the material gets significantly changed, or reconstituted. Such reactions, of which there are several types, are called *reconstitution reactions*.

In order to see how to make use of them for energy storage, it is important to understand the major types of *reaction mechanisms*, as well as the *driving forces* that tend to cause such reactions to occur.

4.2.1   *Insertion Reactions*

Changes in the chemical composition of materials can be the result of the *insertion* of guest species into normally unoccupied crystallographic sites in the crystal structure of an existing stable *host material*. The opposite can also occur, in which atoms are *deleted* from crystallographic sites within the host material.

Such reactions, in which the composition of an existing phase is changed by the incorporation of guest species, can also be thought of as a *solution* of the guest into the host material. Therefore, such processes are also sometimes called *solid solution reactions*. In the particular case of the insertion of species into materials with layer-type crystal structures, insertion reactions are sometimes called *intercalation reactions*.

Although the chemical composition of the host phase initially present can be changed substantially, this type of reaction does not result in a change in the identity, the basic crystal structure, or amounts of the phases in the microstructure.

However, in most cases the addition of interstitial species into previously unoccupied locations in the structure, or their deletion, causes a change in volume. This involves mechanical stresses, and absorbs or emits mechanical energy. The mechanical energy related to the insertion and extraction of interstitial species plays a significant role in the *hysteresis*, and thus energy loss, observed in a number of reversible battery electrode reactions.

Generally, the incorporation of such guest species occurs *topotactically*. This means that the guest species tend to be present at specific (low energy) locations inside the crystal structure of the host species.

A simple reaction of this type might be the reaction of an amount x of species A with a phase BX to produce the *solid solution* product A_xBX. This can be written as

$$xA + BX = A_xBX \tag{4.1}$$

4.2.2 Formation Reactions

A different type of reaction results in the formation of a new phase, rather than the modification of an existing one. A simple example can be the reaction of species A with species B to form a new phase AB. This is a *formation reaction*, and can be represented simply by the equation

$$A + B = AB \tag{4.2}$$

Since this modifies the microstructure, this is an example of one type of *reconstitution reaction*.

There are many examples of this type of formation reaction. There can also be subsequent additional formation reactions whereby other phases can be formed by further reaction of the products of an original reaction.

The *driving force* for the simple reaction in Eq. (4.2) is the difference in the values of the *standard Gibbs free energy of formation* of the products, only AB in this case, and the standard Gibbs free energies of formation of the reactants, A and B:

$$\Delta G_r^0 = \sum \Delta G_f^0(\text{products}) - \sum \Delta G_f^0(\text{reactants}) \tag{4.3}$$

The standard Gibbs free energy of formation of all elements is zero, so if A and B are simple elements, the value of the standard Gibbs free energy change, ΔG_r^0, that results per mol of this reaction is simply the standard Gibbs free energy of formation per mol of AB, $\Delta G_f^0(AB)$. That is

$$\Delta G_r^0 = \Delta G_f^0(AB) \tag{4.4}$$

Values of the standard Gibbs free energy of formation for many materials can be found in a number of sources, for example, [1]. These values change with the temperature, due to the change in the product of the temperature and the entropy related to the reaction, as was discussed in Chap. 2.

In addition to the change in Gibbs free energy, there is a change in enthalpy, and thus the stored heat, related to such reactions. This is the difference between the enthalpies of the products and the reactants:

$$\Delta H_r^0 = \sum \Delta H_f^0(\text{products}) - \sum \Delta H_f^0(\text{reactants}) \tag{4.5}$$

Values of the enthalpy of most materials are relatively temperature-insensitive, except when phase changes take place that involve structural changes, and therefore changes in the entropy.

4.2.3   Decomposition Reactions

A number of materials can undergo *decomposition reactions* in which one phase is converted into two related phases. This is the reverse of *formation reactions*, and can be represented simply as

$$AB = A + B \qquad (4.6)$$

There are two general types of these reactions. In one case, one phase decomposes into two other phases when it is cooled, and in the other, a single phase decomposes into two other phases when it is heated. There are specific names for these reactions. If a single liquid phase is converted into two solid phases upon cooling, it is called a *eutectic reaction*. On the other hand, if the high temperature single phase is also solid, rather than being a liquid, this is a *eutectoid reaction*.

In the case of a single solid phase decomposing into a solid phase and a liquid phase when it is heated, the mechanism is called a *peritectic reaction*. Likewise, if all phases are solids, what happens is called a *peritectoid reaction*.

These types of reactions will be discussed in more detail later.

4.2.4   Displacement Reactions

Another type of *reconstitution reaction* involves a *displacement* process that can be simply represented as

$$A + BX = AX + B \qquad (4.7)$$

in which species A displaces species B in the simple binary phase BX, to form AX instead. A new phase, consisting of elemental B, will be formed in addition. This will tend to occur if phase AX has a greater stability, i.e., has a more negative value of ΔG_f^0, than the phase BX. An example of this type is

$$Li + Cu_2O = Li_2O + Cu \qquad (4.8)$$

in which the reaction of lithium with Cu_2O results in the formation of two new phases, Li_2O and elemental copper. In this case

$$\Delta G_r^0 = \Delta G_f^0(Li_2O) - \Delta G_f^0(Cu_2O) \qquad (4.9)$$

since the standard Gibbs free energy of formation of all elements is zero.

The thermal effect of such a reaction is therefore

$$\Delta H_r^0 = \Delta H_f^0(Li_2O) - \Delta H_f^0(Cu_2O) \tag{4.10}$$

It is also possible to have a *displacement reaction* occur by the replacement of one interstitial species by another inside a stable host material. In this case, only one additional phase is formed, the material that is displaced. The term *extrusion* is sometimes used in the technical literature to describe this type of process.

In some cases, the new element or phase that is formed by such an *interstitial displacement process* is *crystalline*, whereas in other cases, it can be *amorphous*.

4.3   Phase Diagrams

Phase diagrams are *thinking tools* that are useful to help understand these various types of reactions, and associated phenomena. They are graphical representations that indicate the phases and their compositions that are present in a materials system under equilibrium conditions. They were often called *constitution diagrams* in the past. Thus it is reasonable that reactions in which there is a change in the identity or amounts of the phases present are designated as *reconstitution reactions*.

Phase diagrams are widely used in materials science and materials engineering, and are especially useful in understanding the relationships between thermal treatments, microstructures and the properties of metals, alloys, and other solids.

4.3.1   The Gibbs Phase Rule

The *Phase Rule* was proposed by J. Willard Gibbs, a physicist, chemist, and mathematician, in the 1870s. Gibbs developed much of the theoretical foundations of chemical thermodynamics and physical chemistry, and is often considered to have been one of the greatest American scientists. He introduced the concepts of chemical potential and free energy, made great contributions to statistical mechanics, and also invented vector analysis.

What is now generally called the *Gibbs phase rule* is an especially useful *thinking tool* when considering reactions within and between materials. It will appear a number of times in this text.

For present purposes it can be written as

$$F = C - P + 2 \tag{4.11}$$

in which C is the *number of components* (e.g., chemical elements), and P is the *number of phases present* in a materials system in a given experiment.

A phase is formally described as a distinct and homogeneous form of matter separated by its surface from other forms. That definition may not seem to be very useful. But consider a mixture of salt and pepper. They are different phases, for it is possible to physically separate them as black and white particles. Likewise, water and ice can be identified and separated, even though they have the same chemical composition, H_2O. Therefore, they are separate phases.

The quantity F may also be difficult to understand. It is the *number of degrees of freedom*; that means the number of *intensive thermodynamic parameters* that must be specified in order to *define the system* and *all of its associated properties*.

Intensive parameters have values that are independent of the amount of material present. For this purpose, the most useful intensive thermodynamic parameters are the temperature, the overall pressure, and either the chemical potential or the chemical composition of each of the phases present. In electrochemical systems, the electric potential can also be an important intensive parameter.

4.3.2   Binary Phase Diagrams

As mentioned earlier, *phase diagrams* are figures that graphically represent the equilibrium state of a chemical system. There are various types of phase diagrams, but in the most common case they are two-dimensional figures that indicate the temperature and compositional conditions for the stability of various phases and their compositions under equilibrium conditions.

A *binary phase diagram* is a two-dimensional plot of temperature versus the overall composition for materials (*alloys*) composed of two different components (elements). It shows the temperature-composition conditions for the stability and composition ranges of the various phases that can form in a given system, and is commonly used in materials science and engineering. It will be seen that there are temperature-composition regions in which only a single phase is stable, and regions in which two phases are stable.

A very simple binary system is shown in Fig. 4.1. In this case it is assumed that the two components (elements) A and B are completely miscible; that is, they can dissolve in each other over the complete range of composition, from pure element A to pure element B, both when they are liquids and when they are solids. Thus one can speak of a liquid solution and a solid solution in different regions of temperature-composition space. Since the elements have different melting points, the temperature above which their liquid solution is stable (called the *liquidus*) will vary with composition across the diagram. The temperature below which the material is completely solid (called the *solidus*) is also composition-dependent. At any composition there is a range of temperature between the solidus and the liquidus within which two phases are present, a liquid solution and a solid solution.

Likewise, at any temperature between the melting points of pure A and pure B, the liquid and solid phases that are in equilibrium with each other have different

Fig. 4.1 Schematic phase diagram for a binary system with complete miscibility in both the liquid and solid phases

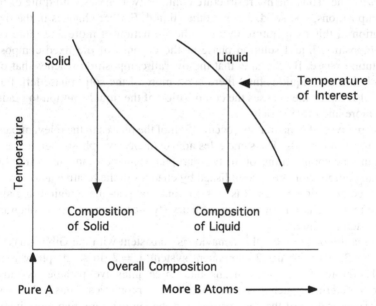

Fig. 4.2 Compositions of liquid and solid phases in equilibrium with each other at a particular temperature between the melting points of the two elements

compositions, corresponding to the compositional limits of the two-phase region in the middle part of this type of phase diagram. This is shown schematically in Fig. 4.2.

The compositions of the solid and liquid phases that are in equilibrium with each other at a particular temperature of interest can be directly read off the composition scale at the bottom of the diagram. It is also obvious that these two compositions will both be different for other temperatures, due to the slopes of the solidus and liquidus curves. At a fixed temperature, the relative amounts, but not the compositions, of the two phases depend upon the overall composition. The overall composition is, of course, not limited to the range within the two-phase mixture region of the phase diagram.

Consider an isothermal experiment, and suppose that the overall composition were to start far to the left at the temperature of interest in the case illustrated in Fig. 4.2, perhaps as far as pure A. Now if more and more atoms of element B are continually added, a solid solution of B atoms in A will be formed whose composition will gradually rise. This can continue over a relatively wide compositional range until the overall composition reaches the solidus line that signifies the one-phase/two-phase border.

Further addition of B atoms causes the overall composition to keep changing, of course. However, the composition of the solid solution can not become indefinitely B rich. Instead, when the overall composition arrives at the *solidus* line, some liquid phase begins to form. Its composition is different from that of the solid solution, being determined by the composition limit of the liquid solution at that temperature, the *liquidus* line. Thus the microstructure contains two phases with quite different local compositions, one solid, and the other liquid. Further changes in the overall composition at this temperature result in the formation of more and more of the fixed-composition liquid solution phase at the expense of the fixed-composition solid solution phase. By the time that the overall composition reaches that of the liquid solution (the liquidus line) there is no more of the solid phase left. Further addition of B atoms then causes the composition of the liquid solution to gradually become more and more B rich.

This same type of behavior also occurs if all of the phases in the relevant part of a phase diagram are solids. The same rules apply. In any two-phase region at a fixed temperature the compositions of the two end phases are constant, and variations of the overall composition are accomplished by changes in the relative amounts of the two fixed-composition phases. It is thus obvious that one-phase regions are always separated by two-phase regions in such binary (two-component) phase diagrams at a constant temperature.

It can readily be seen that this behavior is consistent with the Gibbs phase rule, $F = C - P + 2$. For a binary, 2-component system, $C = 2$. In single-phase regions, $P = 1$. Therefore, $F = 3$. This means that three intensive parameters must be specified to determine such a system, and all of its properties. Thus if the pressure is 1 atm, specification of the composition and the temperature, two more intensive variables, determines all the properties in one-phase regions of this part of the phase diagram.

On the other hand, in a two-phase region, i.e., where a mixture of a solid solution and a liquid solution are both present, $C = 2$, and $P = 2$. Therefore, $F = 2$. This means that if the pressure is fixed, for example, as one atm, only one other parameter must be specified to determine all the properties. This parameter might be the temperature, for example. As shown in Fig. 4.2, that will automatically determine the compositions of each of the two phases. Alternatively, if the composition of the solid solution were specified, that can only occur at one temperature, and with only one composition of the liquid solution.

4.3.3 The Lever Rule

The relation between the overall composition and the amounts of each of the phases present in a two-phase region of a binary phase diagram can be found by use of a simple mechanical analog, and is called the *lever rule*.

If two different masses M_1 and M_2 are hung on a bar that is supported by a fulcrum, the location of the fulcrum can be adjusted so that the bar will be in balance. This is shown in Fig. 4.3. The condition for balance is that the ratio of the lengths L_2 and L_1 is equal to the ratio of the masses M_1 and M_2. That is,

$$M_1/M_2 = L_2/L_1 \tag{4.12}$$

Analogously, the amounts of the two phases in a two-phase region can be found from the lengths L_1 and L_2 on the composition scale. The ratio of the amounts of phases 1 and 2 is related to the ratio of the deviations of their compositions L_2 and L_1 from the overall composition on the composition scale. This is illustrated in Fig. 4.4, and can be expressed as

Fig. 4.3 Mechanical lever analog

Fig. 4.4 Application of the lever rule to compositions in a two-phase region of a binary phase diagram

$$Q_1/Q_2 = L_2/L_1 \tag{4.13}$$

in which Q_1 and Q_2 represent the amounts of phases 1 and 2.

4.3.4 Three-Phase Reactions in Binary Systems

It was mentioned in Sect. 4.2.3 that a number of materials undergo decomposition reactions, i.e., peritectic, eutectic, or their all-solid analogs, peritectoid and eutectoid, reactions upon changes in the temperature. These reactions all must involve three phases, A, B, and AB, or more generally, α, β, and γ.

Again considering the Gibbs phase rule, in a binary system where $C = 2$, and $P = 3$, F must be 1. At a fixed pressure of one atmosphere, no more parameters can be varied. What this means is that three phases can only be in equilibrium under one set of conditions. The compositions of each of the phases present can only have a single value, and the temperature must also be fixed.

Thus when three phase reactions take place at a fixed pressure in binary systems all of the phases must be present and in equilibrium with each other at a specific temperature, and with particular compositions. Two will touch the isothermal composition line at the ends, and the third will do so at a single composition between them. Both above and below the unique peritectic temperature only one-phase and two-phase regions are possible.

In peritectic reactions one of the phases at the ends must be a liquid, whereas the other two phases are solids. This situation can be understood by consideration of the simple binary system containing a peritectic reaction shown in Fig. 4.5. In this case the terminal (end) phases on the A and B component sides of the phase diagram are labeled α and β.

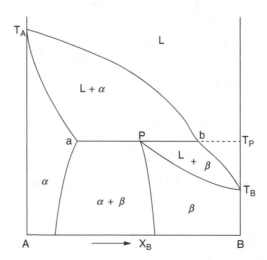

Fig. 4.5 Generalized configuration showing the isothermal arrangement of phases in a peritectic reaction

Fig. 4.6 Simple schematic representation of a peritectic reaction

Fig. 4.7 Schematic representation of a salt hydrate decomposition reaction

The melting points of components A and B are indicated by T_A and T_B, whereas the temperature of the peritectic reaction is T_P. At this temperature phase α has a composition a, phase β has a composition P, and the liquid a composition b.

The situation at the peritectic temperature can also be represented schematically as shown in Fig. 4.6.

4.3.5 Examples of Materials Systems with Peritectic Reactions

An example of a group of materials that undergo peritectic reactions is the family of salt hydrates. Upon heating, they form two product phases, a saturated aqueous solution plus a solid salt phase, at a specific transition temperature. This type of reaction could be represented as shown in Fig. 4.7. The transition temperature is the peritectic temperature.

The phase diagram for this type of system, in which the salt forms a monohydrate, is shown in Fig. 4.8.

Some examples of materials that can undergo this type of structural change at modest temperatures are included in Table 4.1. It can be seen that they can store a rather large amount of heat per unit volume.

The properties of some salt ammoniates that also undergo congruent reactions are presented in Table 4.2.

Fig. 4.8 Phase diagram of a salt-water system in which a salt monohydrate forms at low temperatures

Table 4.1 Data on several salt hydrates

Hydrate	Temperature (°C)	Heat of fusion (kJ kg^{-1})	Latent heat density (MJ m^{-3})
$CaCl_2 \cdot 6\,H_2O$	29	190.8	281
$Na_2SO_4 \cdot 10\,H_2O$	32		342
$Na_2CO_3 \cdot 10\,H_2O$	33		360
$CaBr_2 \cdot 6\,H_2O$	34	115.5	
$Na_2HPO_4 \cdot 12\,H_2O$	35		205
$Zn(NO_3)_2 \cdot 6\,H_2O$	36	146.9	
$Na_2HPO_4 \cdot 7\,H_2O$	48		302
$Na_2S_2O_3 \cdot 5\,H_2O$	48		346
$Ba(OH)_2 \cdot 8\,H_2O$	78	265.7	655
$Mg(NO_3)_2 \cdot 6\,H_2O$	89	162.8	162.8
$MgCl_2 \cdot 6\,H_2O$	117	168.6	168.6

Table 4.2 Data on several salt ammoniates

Reaction	Dissociation temp. at 1 bar (°C)	ΔH (kJ per mol)
$CaCl_2 \cdot 8\,NH_3 = CaCl_2 \cdot 4\,NH_3 + 4\,NH_3$	27	184
$BaBr_2 \cdot 4\,NH_3 = BaBr_2 \cdot 2\,NH_3 + 2\,NH_3$	40	86
$LiCl \cdot 3\,NH_3 = LiCl \cdot 2\,NH_3 + NH_3$	56	46
$BaBr_2 \cdot 2\,NH_3 = BaBr_2 \cdot NH_3 + NH_3$	66	46
$MnCl_2 \cdot 6\,NH_3 = MnCl_2 \cdot 2\,NH_3 + 4\,NH_3$	87	200
$CuSO_4 \cdot 5\,NH_3 = CuSO_4 \cdot 4\,NH_3 + NH_3$	100	60

4.3.6 Binary Systems That Contain Eutectic Reactions

The discussion thus far has been about materials systems containing peritectic reactions. There are a number of cases in which eutectic reactions are present, in which a liquid phase decomposes to form two solid phases, α and β upon cooling. This can be represented schematically as shown in Fig. 4.9.

A real example of this is the lead–tin metallurgical system, whose phase diagram is shown in Fig. 4.10. In this case, the liquid phase reacts to form the terminal lead and tin solid solutions at the eutectic temperature, 185 °C.

The discussion thus far has assumed that the overall composition of the liquid is the same as the eutectic composition. This is, of course, not necessary.

If the overall composition does not coincide with the *eutectic composition*, or the *peritectic composition* in the case of phase diagrams containing peritectic reactions,

Fig. 4.9 Simple schematic representation of a eutectic reaction

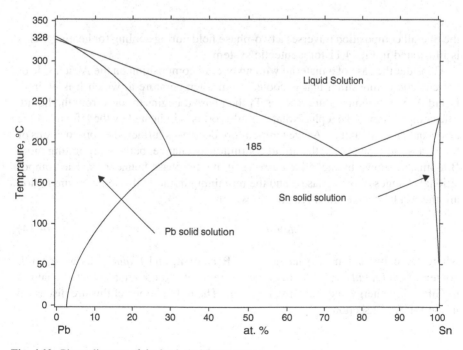

Fig. 4.10 Phase diagram of the lead–tin binary system

Chapter 2
General Concepts

2.1 Introduction

This book is about energy, and various mechanisms by which it can be stored for use at a later time, for a different purpose, or at a different place.

Energy is a key component of what is called *thermodynamics*. Thermodynamic principles are involved in considerations of the different types of energy and their relation to macroscopic variables such as temperature, pressure, volume, and chemical and electrical potentials. It is also centrally involved in the transformation of energy between different forms, such as heat and mechanical, electrical, chemical, magnetic, electrostatic, and thermal energy.

It was pointed out long ago in the very influential book by Lewis and Randall [1] that, "aside from the logical and mathematical sciences, there are three great branches of natural science which stand apart by reason of the variety of far-reaching deductions drawn from a small number of primary postulates. They are mechanics, electromagnetics, and thermodynamics." While thermodynamics (from the Greek word *therme*, for heat) is often thought by students to be quite esoteric and uninteresting, it can actually be of great practical use, as will be seen later in this text.

2.2 The Mechanical Equivalent of Heat

Thermodynamics originated from the observation that there is a relation between two different forms of energy, heat, and mechanical work.

The first step was the observation by Count Rumford (Benjamin Thompson) in 1798 that the friction of a blunt borer in a cannon caused an increase in the cannon's temperature, and that the increase in temperature was related to the amount of mechanical work done. The quantitative relationship between the amount of

© Springer International Publishing Switzerland 2016
R.A. Huggins, *Energy Storage*, DOI 10.1007/978-3-319-21239-5_2

Fig. 4.11 Illustration showing how the compositions of both the solid phase α and the remaining liquid change with temperature above the eutectic temperature

the overall composition traverses a two-phase field during cooling (or heating). This is illustrated in Fig. 4.11 for a eutectic system.

Consider the case of a material with an overall composition on the A-rich side of the eutectic composition that is cooled from a temperature at which it is entirely liquid. When its temperature reaches T_1 the phase α begins to form from the liquid. The composition of the α phase that initially forms is indicated by the left end of the arrow at that temperature. As the temperature decreases further, the compositions of both the solid α phase and the liquid continuously change, both incorporating more B atoms, as shown by the curved arrows along the phase boundaries. The compositional changes of both phase α and the remaining liquid in this temperature region are the result of a chemical reaction in which

$$Liquid_1 = \alpha + Liquid_2 \tag{4.14}$$

where the composition of α has a lower B content, and $Liquid_2$ has a higher B content, than $Liquid_1$. As with all chemical reactions, there are associated changes in Gibbs free energy and enthalpy, i.e., heat. The implications of this are discussed in the following section.

4.4 Thermal Effects Related to Liquid and Solid Reactions

An experimental method that is often used to study phase transformations of materials is commonly called *thermal analysis*. In a simple example, the temperature is raised to a high value, the material is allowed to cool, and the temperature recorded as a function of time.

Changes in the slope of the resulting temperature-time *cooling curve* give information about the type of reactions that are taking place, and over what temperature intervals they occur. If the rate of extraction of heat is maintained constant, it is also possible to get quantitative thermodynamic information in this way.

As discussed in Chap. 3, the amount of heat given off during the cooling of a given mass of material is proportional to the product of the heat capacity and the temperature change. This can be expressed as

$$\int dq = \rho V C_P \int dT \qquad (4.15)$$

where q is the heat evolved, ρ the density, V the volume, C_P the heat capacity at constant volume, and T the temperature, if no chemical changes take place.

But when a chemical change also takes place as the temperature is changed in a cooling reaction, an additional heat term must be added to this relation. If the temperature-dependent reaction causes the evolution of heat, the *effective heat capacity* will be greater, with the result that the material will cool at a slower rate. This is illustrated schematically in Fig. 4.12 for a material with an overall

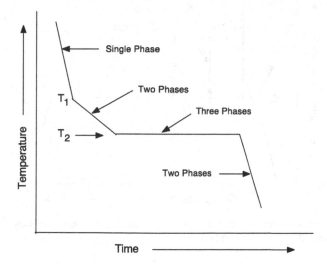

Fig. 4.12 Schematic temperature-time cooling curve showing changes in slope during a constant rate of heat extraction experiment with a material with the overall composition indicated in Fig. 4.11

composition shown in Fig. 4.11. The cooling curve has a relatively steep slope when the material is entirely liquid. When temperature T_1 is reached upon cooling, the reaction indicated in Eq. (4.14) begins to take place, causing the generation of heat. Thus the effective heat capacity is increased, and the rate of cooling, the slope of the cooling curve, decreases. When temperature T_2, the eutectic temperature, is reached, the composition of the remaining liquid is $Liquid_{eut}$, the eutectic composition, and there is no further cooling as the eutectic reaction

$$Liquid_{\text{eut}} = \alpha + \beta \tag{4.16}$$

takes place. When that reaction has come to completion, and there is no more liquid left, the slope of the cooling curve again increases.

The characteristics of the cooling curve will change as the overall composition is varied. This is shown schematically in Figs. 4.13 and 4.14 for three different values of the overall composition.

The thermal behavior of materials with this type of reaction can be used for temperature control or maintenance purposes. By changing the overall composition of the material being used, it is possible to vary the relative effects of the isothermal latent heat reaction and the enhanced sensible heat reaction that takes place over a range of temperature.

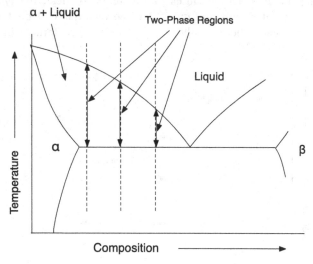

Fig. 4.13 Illustration of the temperature ranges over which the two-phase (a + liquid) structure is formed for three different overall compositions

Fig. 4.14 Changes in the cooling curve as the initial overall composition is varied

4.5 Thermal Effects Related to Reversible Gas Phase Reactions

There are a number of simple gas phase reactions of the type

$$A + B = C + D \qquad (4.17)$$

The relation between the Gibbs free energy, the enthalpy, and the temperature

$$G = H - TS \qquad (4.18)$$

was introduced in Chap. 2. Changes in G constitute the driving forces for all reactions, whereas changes in H indicate changes in heat content.

As has been mentioned earlier, the change in the Gibbs free energy resulting from a reaction of the type in Eq. (4.17) under standard conditions, ΔG_r^0, is the difference between the standard Gibbs free energies of formation of the products and of the reactants. Likewise, the change in enthalpy resulting from such a reaction, ΔH_r^0, is the difference between the standard enthalpies of formation of the products and the reactants.

Reactions such as Eq. (4.17) can generally take place over a substantial range of temperatures. The driving forces for such reactions can sometimes be reversed by changes in the temperature, depending upon the magnitudes of the entropy values of the species involved on both sides of the reaction, according to Eq. (4.18).

On the other hand, the magnitude of the heat generated or absorbed, determined by the enthalpies of the reactants and products, is relatively temperature-independent. However, it goes in the opposite direction if the reaction is reversed.

The first example of the use of this phenomenon to store and transmit heat involves the reaction of methane with steam, generally described as the *steam reforming of natural gas* [2, 3]. It is discussed further in Chap. 8, for it is commonly used to produce hydrogen from natural gas:

$$CH_4 + H_2O = CO + 3H_2 \qquad (4.19)$$

The temperature dependence of the direction of this reaction can be understood by calculation of the standard Gibbs free energy change ΔG_r^0 from information about the variation of the standard Gibbs free energies of formation of the species involved with temperature

$$\Delta G_r^0 = \Delta G_f^0(CO) + 3\Delta G_f^0(H_2) - \Delta G_f^0(CH_4) - \Delta G_f^0(H_2O) \qquad (4.20)$$

Values of the standard Gibbs free energy of formation of the species in this reaction for three different temperatures are given in Table 4.3.

From these data it is possible to obtain the standard Gibbs free energy of reaction as a function of temperature. The results are listed in Table 4.4, and are plotted in Fig. 4.15. It is seen that this reaction will go forward at temperatures above about 900 K, and in the reverse direction at lower temperatures.

The thermal effects of this reaction can be obtained from information on the standard enthalpies of the species involved. These are shown in Table 4.5. From these data, the temperature dependence of the standard enthalpy change of the reaction can be obtained. The results are listed in Table 4.6, and plotted in Fig. 4.16. As expected, there is not very much variation of the heat effect with temperature. But the important point is that the reaction tends to go forward, and is endothermic, at high temperatures, and in the reverse, exothermic, direction at low temperatures.

Table 4.3 Temperature dependence of the standard Gibbs free energies of formation of species in reaction (Eq. (4.20))

Species	ΔG_f^0 (400 K) (kJ mol^{-1})	ΔG_f^0 (800 K) (kJ mol^{-1})	ΔG_f^0 (1200 K) (kJ mol^{-1})
CO	−146.4	−182.5	−217.8
H_2O	−224.0	−203.6	−181.6
CH_4	−42.0	−2.1	+41.6
H_2	0	0	0

Table 4.4 Temperature dependence of standard Gibbs free energy of reaction (Eq. (4.20))

Temperature (K)	ΔG_r^0 (kJ mol^{-1})
400	+119.6
800	+23.2
1200	−77.8

Fig. 4.15 Temperature dependence of the standard Gibbs free energy of reaction (4.20)

Table 4.5 Temperature dependence of the standard enthalpies of species in reaction (Eq. (4.20))

Species	ΔH_f^0 (400 K) (kJ mol^{-1})	ΔH_f^0 (800 K) (kJ mol^{-1})	ΔH_f^0 (1200 K) (kJ mol^{-1})
CO	−110.1	−110.9	−113.2
H_2O	−242.8	−246.4	−249.0
CH_4	−78.0	−87.3	−91.5
H_2	0	0	0

Table 4.6 Temperature dependence of standard enthalpy of reaction (Eq. (4.20))

Temperature (K)	ΔH_r^0 (kJ mol^{-1})
400	+210.7
800	+222.8
1200	+227.3

Use can be made of the temperature-dependent reversal of this reaction, for it can be driven at high temperatures at one location, where it is endothermic, and the product gases sent through pipes to a distant location, where the reaction is caused to react exothermically in the reverse direction, giving off heat. The same thing could be done in a single location, of course, using heat when it is available, and supplying it when it is needed.

Fig. 4.16 Temperature dependence of the standard enthalpy of reaction (4.20)

Table 4.7 Reversible chemical reactions in the C-H-O system

System	Enthalpy of reaction, ΔH_r^0, kJ mol^{-1}	Temperature range (K)
$C_{10}H_{18} = C_{10}H_8 + 5\ H_2$	314	450–700
$C_7H_{14} = C_7H_8 + 3\ H_2$	213	450–700
$C_6H_{12} = C_6H_6 + 3\ H_2$	207	500–700
$CH_4 + 2\ H_2O = CO_2 + 4\ H_2$	165	500–700
$CH_4 + CO_2 = 2\ CO + 2\ H_2$	247	700–1200
$CH_4 + H_2O = CO + 3H_2$	225	700–1200

A number of other reactions in the carbon–hydrogen–oxygen system can also be used for this purpose. Some of these are listed in Table 4.7.

References

1. I. Barin, Thermochemical Data of Pure Substances, 3rd Edition, VCH 1995, Published Online 24 Apr 2008, ISBN 9783527619825
2. Schulten R, Van der Decken CB, Kugler K, Barnert H. (1974) Proc. British Nuclear Energy Soc
3. Harth R, Range J, Boltendahl U. (1981) G. Beghi, ed. Energy Storage and Transportation, D. Riedel Publishing Co. p. 358

Chapter 5
Energy Storage in Organic Fuels

5.1 Introduction

The Preface included a discussion of the several types of natural materials that can be obtained from the earth and used as fuels. The major ones are wood and the several fossil fuels, including the various types of coals, crude oil, and natural gas. The fossil fuels, which now play such a major role in the energy supply, will surely gradually become less important as they become depleted. But the recent development of fracking technology, as discussed in the Preface, will result in the access to substantial additional sources of crude oil in some locations.

There are also some organic materials that can be considered to be renewable. These include the agrofuels, which contain significant amounts of vegetable oil, and the crops that are high in sugar or starch. These are often discussed in terms of *biomass*, and *bioenergy*.

While these materials are basically fuels, and therefore energy carriers, they can also be thought of as energy storage media, for as they grow they accumulate energy that can be utilized in the future. But in addition, some of them have other characteristics, such as being nutrients, or serving as raw materials for a number of industries.

Another aspect of biomass is that this category should also include living materials, including animals. There is, of course, crossover between these categories, for animals consume agrofuels, and also contribute energy in the form of food. They can also provide mechanical energy.

5.2 Storage of Energy in Living Biomass

It has been estimated that the energy storage densities of living plants and animals are 10–30 MJ kg^{-1} of dry weight [1]. The question is how this energy can be usefully acquired. In the case of dry wood and straw, it can be converted to heat by burning. But most organic material contains a significant amount of water, and the

© Springer International Publishing Switzerland 2016
R.A. Huggins, *Energy Storage*, DOI 10.1007/978-3-319-21239-5_5

energy that has to be furnished to dry it can exceed the amount that is obtained by its burning.

There are other methods by which the energy contained in biomass can be retrieved, however. One of these is the consumption of biomass by a wide variety of living organisms as food without having to dry it.

The total amount of energy stored in living biomass on the earth is roughly 1.5×10^{22} J, and its average living residence time is about 3.5 years. There is a difference between living biomass in the oceans and on the land. Growth is generally more rapid in the oceans, mostly in the form of phytoplankton, which has a very short lifetime, of the order of a few weeks. On land the biomass growth rate is much lower than in the oceans, but the average residence time is longer.

For biomass to be considered as a renewable energy storage mechanism, the rate of growth must be at least as rapid as the rate of the extraction of the energy by harvesting. This can be influenced in various ways, such as by the use of fertilizers and/or artificial irrigation or lighting. Growth typically involves the consumption of CO_2, which is generally also considered positive for other (environmental) reasons. It has been shown that increasing the concentration of CO_2 in the air can enhance the rate of growth of many plants, in some cases up to a factor of two [2]. This is done in commercial greenhouses in some places. Pest control can also be important.

The rapid growth of plants can be very impressive. The efficiency of the use of solar radiation to produce stored energy by a number of plant types is shown in Table 5.1 [2]. The efficiency is defined as the energy content of the crop divided by the accumulated incident solar energy for a given land surface area.

These data are general averages, and actual yields can be increased by the use of fertilizers, well-designed irrigation, etc. The matter of irrigation is a critical world-wide problem, for the needs are growing more rapidly, partly because of the increase in human population, than sources are being developed.

As mentioned earlier, the situation is somewhat different for aquatic plants. Open ocean photosynthesis is dominated by phytoplankton. The limiting factor determining production is the availability of nutrients, particularly nitrogen and phosphate. Where these are readily available, for example from agricultural runoff, the growth in nearby waters can be remarkable. This is not always desirable, however.

Table 5.1 Examples of high-yield plant species

Plant	Short-term efficiency	Annual average efficiency	Annual yield ($kg\ m^{-2}years^{-1}$)
Sugarcane		0.028	11.2
Napier grass	0.024		
Sorghum	0.032	0.009	3.6
Corn	0.032		
Alfalfa	0.014	0.007	2.9
Sugar beet	0.019	0.008	3.3
Chlorella	0.017		
Eucalyptus		0.013	5.4

The author saw this happen in Clear Lake in Northern California, where there was evidently a lot of agricultural fertilizer runoff. He was there to participate in small sailboat racing. The water was so full of green algae marine growth that it was essentially opaque. It was like sailing through thick pea soup. This problem has evidently been subsequently alleviated by controlling the runoff.

The organic energy production in the North Atlantic ocean is about 0.05 Wm-2 [3]. But in coastal areas the productivity typically rises to about 0.5 Wm-2. For coral reefs and areas where currents bring nutrient-rich water, such as along the Peruvian coast, the conversion efficiency can be 2–3 % of the incident solar energy. Under ideal conditions, with an artificial nutrient supply, energy conversion efficiencies can reach 4 % [4].

Despite these attractive data, the annual marine plant harvest is only about 1 million metric tons. Most of this is consumed as food in Japan and Korea. Most algae are rich in protein, and it would seem to make sense to increase their use in food. Unfortunately, pollution interferes with this in many areas.

5.3 Storage via Animals

Domesticated animals are fed mostly by plant material that has grown by accepting energy from the sun. Animals convert and store energy acquired from the consumption of plants. This energy can be delivered in the form of food, both to humans and in some cases to other animals.

Roughly one-third of the total harvested plant material is fed to animals; yet it is estimated [1] that animal production constitutes only 14 % of the total food production. The difference between the 33 % input and the 14 % output shows that the use of plant food to raise animals is actually not a very efficient use of the energy in the plants. There are, however, other arguments why this is done. These involve concerns about the desired distribution of meat, vegetable, and milk products in the human diet, which varies with location in the world. Thus energy efficiency is not always the dominant concern.

In the past animals also have contributed mechanical energy to a number of important applications, such as transportation and the tilling of soil to assist agriculture. In the more affluent countries these applications have been mostly taken over by mechanical alternatives, such as tractors, which consume fossil fuels.

There are also some other ways in which the stored energy in animals is utilized. When travelling in the Black Forest region of Germany one notices that the older buildings were typically constructed so that the animals were kept underneath the living quarters of the farming families. This was done to utilize the heat from the animals to increase the temperature of the living quarters in the winter.

Jensen [1] provided some interesting figures on several aspects of energy consumption, storage and output related to domestic animals. He estimated that there are some 1.5×10^9 domesticated large animals in the world (cattle, horses, yaks, buffalo, donkeys, camels, etc.). About 400 million of these may be contributing mechanical work, at an average power level of 375 W per animal for some 6 h a day. If the average food intake is 600 W per animal, the efficiency of the conversion of the food energy to mechanical energy would be 16 %. In many applications, however, only a fraction of the animals' mechanical power is actually used for a useful purpose. For example, there is no useful energy output, although there is energy consumption, when a horse or cow merely walks around.

5.4   Hard Biomass

The growth of wood and other hard biomass by the absorption of solar energy, and its consumption, after at least some level of storage, by oxidation to provide heat for cooking, space heating, and other purposes, is important in many relatively less highly developed societies. But much of this heat is actually wasted. For example, in cooking, less than 10 % of the fuel's heat of combustion actually reaches the contents of a cooking pot. In addition, some of the energy that does reach the pot is given off as heat to the surroundings. As a result, only a fraction of the heat energy involved in cooking can be actually used, although this depends to a considerable extent upon the design and operation of the stove. Some of the cast iron stoves used in Europe in the past were actually quite good at converting and storing the heat of combustion, partly for cooking, and partly for space heating.

5.5   Synthetic Liquid Fuels

A number of fuels can be made by the modification of species found in nature. The most interesting of these are methanol (CH_3OH), ethanol (C_2H_5OH), methane (CH_4), ammonia (NH_3), and methylcyclohexane (C_7H_{14}).

The production of methanol involves the *steam reforming reaction*,

$$CH_4 + H_2O = CO + 3H_2 \tag{5.1}$$

followed by further reaction of CO with hydrogen

$$CO + 2H_2 = CH_3OH \tag{5.2}$$

Alternatively, it is possible to react CO_2 with hydrogen to form methanol and water

$$CO_2 + 3H_2 = CH_3OH + H_2O \tag{5.3}$$

Ethanol can be produced from CO_2 and H_2 in a similar way as methanol by upgrading the hydrogen content.

5.6   Gaseous Fuels Stored as Liquids

Some liquid fuels can readily be stored in small high-pressure tanks. This is often done in cases where a fuel is used for small scale heating or cooking, e.g., in recreational vehicles, boats, or when camping. The fuel is kept as a liquid under pressure, but as soon as the pressure is reduced by opening a valve, it emits as a gas.

The approximate pressure at ambient temperature can be calculated from the boiling point at one atmosphere pressure, T_{bp}, by use of the ideal gas equation

$$PV = nRT \tag{5.4}$$

where V is the volume, n is the number of mols, and R is the gas constant. So for a fixed volume and amount of material the pressure at ambient temperature, assumed to be 298 K, is given by

$$P_{298} = 298/T_{bp} \tag{5.5}$$

Some common examples are shown in Table 5.2.

5.7   The Energy Content of Various Materials Used as Fuels

To put some of these data in perspective, the specific energies of a number of materials that are used as fuels are shown in Table 5.3. It is interesting to compare these numbers with the specific energy of the well-known explosive TNT, which is only 3.6 MJ kg^{-1}, which, when converted to electrical units, is about 1 kW-hr kg^{-1}.

Table 5.2 Data on materials commonly used for small-scale heating

Material	Boiling temperature (K)	Pressure at 298 K (atm)
Butane	272.5	1.1
Propane	231	1.3
Methane	111	2.7

Table 5.3 Specific energy of various materials used as fuels

Fuel	Specific energy (MJ kg^{-1})
Crude oil	42
Coal	32
Dry wood	15
Hydrogen gas	120
Methanol	21
Ethanol	28
Propane	47
Butane	46
Gasoline	44
Diesel fuel	43

References

1. Jensen J, Sorensen B. (1984) Fundamentals of Energy Storage. Wiley-Interscience
2. Bassham J (1977) Science 197:630
3. Odum E (1972) Ecology. Holt-Reinhardt and Winston
4. Sorensen B (1979) Renewable Energy. Academic Press, New York

Chapter 6
Mechanical Energy Storage

6.1 Introduction

There are two basic types of energy storage that result from the application of forces upon materials systems. One of these involves changes in potential energy, and the other involves changes in the motion of mass, and thus kinetic energy. This chapter focuses upon the major types of potential energy and kinetic energy storage. It will be seen that it is possible to translate between these two types of energy, as well as to convert these energies to heat or work.

6.2 Potential Energy Storage

Potential energy always involves the imposition of forces upon materials systems, and the energy stored is the integral of the force times the distance over which it operates. Thus

$$\text{Energy} = \int (\text{force}) (\text{distance}) \qquad (6.1)$$

Consider the application of a tensile stress upon a solid rod, causing it to elongate. This is illustrated simply in Fig. 6.1.

The stress σ is the force per unit cross-sectional area, and the resultant fractional change in length $\Delta x/x_0$ is the strain ε.

In metals the strain is proportional to the force, and this can be represented as a stress/strain diagram, as shown in Fig. 6.2. The proportionality constant is the Young's modulus Y, and this linear relation is called "Hooke's Law."

© Springer International Publishing Switzerland 2016
R.A. Huggins, *Energy Storage*, DOI 10.1007/978-3-319-21239-5_6

Fig. 6.1 Simple example of the elongation of a solid rod as the result of an applied tensile force upon its ends

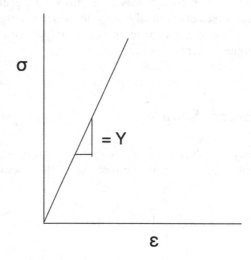

Fig. 6.2 Schematic stress/strain diagram for an elastic metal

If this mechanical deformation is elastic, the work W that is done on the spring is the area under the stress/strain curve. This is obviously proportional to the magnitude of the applied stress. That is

$$W = \frac{1}{2}\sigma\varepsilon = \frac{1}{2}Y\varepsilon^2 \qquad (6.2)$$

If this mechanical process is reversible without any losses, the work is equal to the amount of stored energy in this simple system.

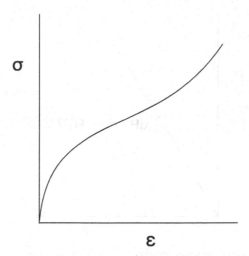

Fig. 6.3 Schematic stress/strain curve for rubber

In metals and ceramics Young's modulus is a constant up to a critical value of the stress, called the yield point. This is because the interatomic forces in such materials are linear at small displacements. At higher values of stress, however, there can be plastic (nonreversible) deformation, and then, ultimately, fracture.

In polymers and rubbers Young's modulus can vary with the value of the strain, due to the action of different physical processes in their microstructures. An example of a stress/strain curve for a common rubber is shown schematically in Fig. 6.3.

The deformation of a metallic spring in a mechanical clock, and the use of stretched rubber bands to power model airplanes are simple examples of this type of stored mechanical potential energy.

6.3 Energy Storage in Pressurized Gas

Everyone who has had to pump up a bicycle tire knows that that process requires work, and that the required force becomes greater as the pressure increases. If there is a leak, or the valve is opened, the gas stored in the tire is released. This is a simple example of the storage of energy in a gas.

It is possible to store energy by making use of the elastic properties of gases in a manner similar to that of the elastic properties of solids.

This can be readily understood by consideration of the ideal gas law, or the equation of state of an ideal gas, that can be written as

$$PV = nRT \qquad (6.3)$$

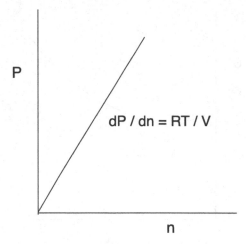

P

$dP / dn = RT / V$

n

Fig. 6.4 Elastic behavior of gas under pressure

where P is the absolute pressure of the gas, V its volume, n the number of moles, R the gas constant, and T the absolute temperature. The value of R is 8.314 J mol^{-1} K^{-1}, or 0.082 l atm K^{-1} mol^{-1}. Using this latter value, the volume of a mole of gas can be readily found to be 22.4 l at 273 K or 0 °C.

For a constant volume, such as that of a bicycle tire, the pressure is proportional to the amount of gas (air), n, that has been pumped into it. This can be simply represented as shown in Fig. 6.4, which is seen to be directly analogous to Fig. 6.2.

Gases can be compressed and stored in simple mechanical tanks, so long as the pressure is not so large as to cause mechanical damage. This is one of the ways in which the hydrogen used as the fuel in the fuel cells that are being developed for the propulsion of vehicles is contained. Such tanks have been traditionally made of high-strength metals, but carbon fiber composite materials, which can have greater strength and lighter weight, are becoming more attractive for use at high pressures.

The tanks that are used to store gaseous hydrogen to power the fuel cells in the automobiles currently under development can operate up to a pressure of 10^4 psi. One of these autos carries a total of 8 kg of hydrogen in its tanks.

Another alternative for this purpose is to store the hydrogen in solid metal hydride materials. That topic will not be discussed at this point, however. It appears in Chap. 8.

It is also possible to store gases under elevated pressure in underground cavities, if they are gas tight. This is the case for some large salt caverns, depleted oil wells, or underground aquifers.

Since this process is generally close to adiabatic if these are done rapidly, heat is given off during compression, and there is cooling during expansion. Some sort of heat transfer system must be employed to take care of this problem. This can be a serious consideration. For example, compression to a pressure of 70 atm can produce a temperature of about 1000 K. This can cause the overall efficiency of a pressure storage system to be significantly reduced.

6.4   Potential Energy Storage Using Gravity

Instead of depending upon the elastic properties of solids or gases, there are energy production and storage methods that are based upon gravitational forces

One example that is familiar to many people is a type of clock that is driven by the gravitational force on a mass, or "weight." Some of these are called "grandfather clocks," and others are "cuckoo clocks." These types of clocks evolved from the realization by Galileo in the early 1600s that the period of the swing of a pendulum is independent of its amplitude, and that this phenomenon might be used for timekeeping. This led to the invention of the pendulum clock by Christiaan Huygens in 1656, which was shortly followed by the invention of the anchor escapement mechanism by Robert Hooke in 1657.

"Cuckoo clocks" driven by weights have been produced in the Black Forest area of Germany since the middle of the 1700s. They typically include a moving bird, and a small bellows is used to make bird call sound. These have been popular tourist items for some time.

These pendulum-regulated clocks are driven by the action of gravity upon a weight, instead of a metallic spring. This mechanism is illustrated schematically in Fig. 6.5.

Fig. 6.5 Schematic illustration of the mechanism used to provide energy to pendulum-regulated clocks

In these cases the potential energy involves attractive forces between two bodies, W_{pot}, where

$$W_{pot} = -G \frac{Mm}{r} \tag{6.4}$$

G is the gravitational constant, $6.67 \times 10^{-11} \, m^3 \, kg^{-1} \, s^{-2}$, M is the mass of the earth (5.98×10^{24} kg), m is the mass being moved, and r is the distance between their centers. Thus

$$\Delta W_{pot} = -GMm \left(\frac{1}{r + \Delta r} - \frac{1}{r} \right) = mg\Delta r \tag{6.5}$$

where

$$g = GMr^{-2} = 9.81 \, ms^{-2} \tag{6.6}$$

6.5   Hydroelectric Power

Water is evaporated from the earth's surface by solar energy as part of the global climate cycle. This evaporation is partly from the land masses, but primarily from the world's oceans. The moisture rises and condenses to form clouds in the sky, which are transported by the global air circulation. This moisture can later precipitate in the form of rain or snow, sometimes at higher elevations. The water from the rain and snow that fall at high altitudes can be stored in reservoirs, from which it can be run through turbines to lower elevations, producing electricity.

This "hydroelectric power" is a major source of electrical energy in a number of countries, including Switzerland and Norway. It is also an important component of the total energy picture in parts of the USA.

One of its major advantages us that the flow through the turbines can be turned on and off in response to the current need. This is not instantaneous, however, for there is a start-up time for the turbines of the order of a few minutes.

There are some disadvantages to this method of energy acquisition and storage as well, for the large reservoirs and their related water collection areas can require a considerable amount of real estate, and this can become a political problem. In addition, the dams, themselves, can be quite expensive.

The use of the gravitation force on collected water to produce energy can also involve much smaller scale facilities. Years ago there were many small water-wheels that were driven by falling water to provide either mechanical or electrical output. These were generally located on smaller waterways, and often used modest millponds to store the water before it was fed onto the water mills.

Another form of hydroelectric energy production takes advantage of the tidal rise and fall of the ocean surface as the result of the gravitational forces of the moon and sun, coupled with the effects of the earth's rotation. Variations in the wind can also have a temporary effect.

The gravitational effect of the moon would theoretically cause the surface of the ocean to rise about 54 cm at its highest point, if it were to have a uniform depth, there were no land masses, and the earth were not rotating. The gravitational effect of the sun is somewhat smaller, theoretically producing an amplitude of about 25 cm under comparable conditions. Tides rise and fall with a cycle time near 12 h, with their magnitudes dependent upon the relative positions of the sun and the moon. Tides with the greatest amplitudes occur when the sun and moon are in line, and are called "spring tides." Those with the smallest amplitudes are "neap tides."

Tidal amplitudes vary greatly from place to place as well, depending upon variations in local ocean depth as well as the nearby underwater land mass topography. In some locations the difference between low and high tide is quite large. Examples include some areas near the English Channel, along the coast of New Zealand, and in the Bay of Fundy and Ungava Bay in Eastern Canada.

Programs have been initiated to make use of the large variations in water level is several locations by the construction of "impoundment ponds" that admit and release seawater through turbines. The power that is generated in this way is, of course, periodic, related to the timing of the tides.

6.6 Pumped-Hydro Storage

A modification of hydroelectric storage and power is called "pumped-hydro" storage. The general configuration in this case involves water storage facilities at two different elevations. They can either be natural or artificially constructed, and can have a wide range of sizes. They could include underground caverns, old mine shafts, volumes formerly occupied by oil, or newly excavated volume.

Water can be run through turbines from the upper one to the lower one, producing electricity, as in simple hydroelectric power systems. But then water can be pumped back up to the storage area at the higher elevation, effectively recharging the system. In some cases this involves the use of two-way turbines. This can make sense if the price of electricity varies significantly at different times of the day or the week. This type of energy storage can be especially useful in connection with daily peak shaving and load leveling, as well as weekly and seasonal variations in the energy demand.

This scheme is illustrated schematically in Fig. 6.6.

Whereas the efficiency of large-scale water-driven turbines can be quite high, even over 95 %, the efficiency of the dual-cycle reversible storage system typically is about 80 %. There are other losses, of course, such as water evaporation from one or both of the reservoirs, leakage around the turbine, and losses due to friction of the moving water.

There are many such pumped-hydro storage systems in the world. Some of them are listed in Table 6.1.

Fig. 6.6 Schematic illustration of a pumped hydro system

Table 6.1 Examples of large pumped hydro systems

Country	Name	Capacity (MW)
Argentina	Rio Grande-Cerro Pelado	750
Australia	Tumut Three	1500
Austria	Malta-Haupstufe	730
Bulgaria	PAVEC Chaira	864
China	Guangzhou	2400
France	Montezic	920
Germany	Goldisthal	1060
	Markersbach	1050
India	Purulia	900
Iran	Siah Bisheh	1140
Italy	Chiotas	1184
Japan	Kannagawa	2700
Russia	Zagorsk	1320
Switzerland	Lac des Dix	2099
Taiwan	Mingtan	1620
United Kingdom	Dinorwig, Wales	1728
United States	Castaic Dam	1566
	Pyramid Lake	1495
	Mount Elbert	1212
	Northfield Mountain	1080
	Ludington	1872
	Mt. Hope	2000
	Blenheim-Gilboa	1200
	Raccoon Mountain	1530
	Bath County	2710

6.7 Use of the Kinetic Energy in Moving Water

It is possible to extract power from moving water by the immersion of a water-driven propeller or turbine. This could be done in flowing rivers, where the flow is relatively constant with time. It also can be done in locations in which there are significant tidal currents. In this case, however, the current, and thus the power available, are periodic, with rather substantial periods between.

Whereas this may appear to be rather simple, some practical matters require attention. These include seawater corrosion and the growth of barnacles and other biological species on underwater surfaces. Some recent designs involve retractable propellers that can be periodically cleaned.

6.8 Kinetic Energy in Mechanical Systems

In addition to potential energy, it is also possible to store kinetic energy. This is energy that is related to the motion of mass. This will be discussed in two parts, linear motion of a mass, and rotational motion of mass. More information on this topic can be found in [1].

6.8.1 Linear Kinetic Energy

The kinetic energy E_{kin} related to a body in linear motion can be written as

$$E_{kin} = \frac{1}{2}mv^2 \qquad (6.7)$$

where m is its mass, and v its linear velocity.

A simple hammer is an example of this principle. The kinetic energy in its moving mass is utilized to drive nails, as well as for other purposes.

A somewhat more exotic example involves hybrid automobiles. In some of the current hybrid internal combustion/electric vehicles an electric motor/generator in the drive train drives the wheels. It is fed from a high-rate battery (sometimes called a "supercapacitor") that is recharged as needed by an efficient internal combustion engine.

When the vehicle is slowed down, or braked, the motor/generator operates in reverse, and some of the vehicle's kinetic energy is converted to electrical energy that is fed back into the battery. This energy can then be used subsequently for propulsion. The amount of this recovered energy is typically about 10 % of the basic propulsion energy in urban driving. This system is shown schematically in Fig. 6.7.

A variant on this model is the "plug-in hybrid," in which the net energy consumed from the battery is replaced by the use of an electrical recharger when

Fig. 6.7 Schematic representation of the interaction between the internal combustion engine and the battery in a hybrid vehicle

the vehicle is not being used. This might involve connection to the electrical system of a home overnight, for example. There is further discussion of this topic in Chap. 24.

6.8.2   Rotational Kinetic Energy

Kinetic energy is also present when a body is rotated. In this case,

$$E_{kin} = (1/2)I\omega^2 \tag{6.8}$$

where I is the moment of inertia and ω is the angular velocity.

Flywheels can be used for the purpose of storing kinetic energy, and there are ready methods whereby this mechanical energy can be converted to and from electrical energy.

The moment of inertia of a rotating body can be expressed as

$$I = \int \rho(x)r^2 dx \tag{6.9}$$

where $\rho(x)$ is the mass distribution, and r is the distance from the center of rotation.

Thus the magnitude of the kinetic energy in a rotational system is increased if the rotational velocity ω is large and I is large. The latter can be achieved by having a large mass at a large value of r.

Consideration must be given to the strength of the material from which the flywheel is constructed, for it must be able to withstand the centrifugal force. This can be represented schematically as shown in Fig. 6.8.

The centrifugal force is given by

$$\text{Force} = (\text{mass})(\text{acceleration}) = mr\omega^2 \tag{6.10}$$

Flywheels can have a variety of shapes. One optimization strategy is to use a disc design in which the stress is the same everywhere. If the material with which the flywheel is constructed is uniform, this results in a shape such as that illustrated schematically in Figs. 6.9 and 6.10.

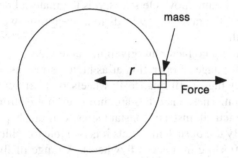

Fig. 6.8 Schematic representation of the centrifugal force operating on a mass in rotation

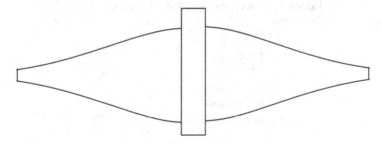

Fig. 6.9 Schematic drawing of the cross section of a disc upon a central shaft in which the centrifugal stress in the disk is the same everywhere

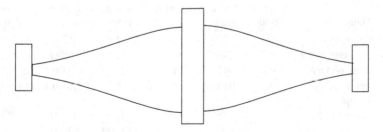

Fig. 6.10 Constant stress disk with constant-thickness outer rim

The local thickness b is given by

$$b = b_0 \exp\left[(\text{constant})(r^2)\right] \tag{6.11}$$

The radial strain causes tangential strain, and thus tangential stress. Thus if such a flywheel were to be reinforced by use of high strength carbon fibers, this is done by placement of the fibers around the outside, so that they carry the tangential force.

In this case, the specific energy, the kinetic energy per unit mass, in the flywheel is given by

$$\frac{E_{\text{kin}}}{\text{mass}} = \frac{\sigma_{\text{max}}}{\rho} K_{\text{m}} \tag{6.12}$$

where σ_{max} is the maximum allowable stress, ρ is the material density, and K_{m} the shape factor. Note that the ratio σ_{max}/ρ leads to a strength-to-weight ratio, not just the material's strength.

Some values of the shape factor are given in Table 6.2.

Because of the advantage of operation at as high a stress as possible, there is a tendency to construct high performance flywheels of fiber reinforced materials, either Kevlar or carbon. These materials are, however, not isotropic, so that simple disc shapes are not practical. Instead, constant stress central shapes with high stress rims are often used. By use of such materials it is possible to achieve energy storage values as high as 200 kJ/kg in modern flywheels. A range of flywheel types, and their characteristics are shown in Table 6.3.

Table 6.2 Values of the shape factor for several simple disc shapes

Shape	K_{m}
Brush shape	0.33
Flat disk	0.6
Constant-stress disc	1.0
Thin rim only	0.5
Thin rim on constant-stress disk	0.6–1.0

Table 6.3 Examples of flywheel characteristics

Object	K, shape factor	Mass (kg)	Diameter (m)	Angular velocity (rpm)	Energy stored	Energy stored (kWh)
Bicycle wheel	1	1	0.7	150	15 J	4×10^{-7}
Flintstone's stone wheel	0.5	245	0.5	200	1680 J	4.7×10^{-4}
Train wheel, 60 km/h	0.5	942	1	318	65,000 J	1.8×10^{-2}
Large truck wheel, 18 mph	0.5	1000	2	79	17,000 J	4.8×10^{-3}
Train braking flywheel	0.5	3000	0.5	8000	33 MJ	9.1
Electrical power backup flywheel	0.5	600	0.5	30,000	92 MJ	26

It should be pointed out that flywheels can be dangerous. If they get out of balance or begin to come apart, parts can become very-high-velocity projectiles. Thus it is safer to construct them of many small pieces. This is the reason for the circular brush shape concept. They are generally housed in very robust steel containment, and large units are placed underground for safety reasons.

Flywheels also store energy in the form of mechanical strain potential energy—like springs—due to the forces upon them. The magnitude of this potential energy is small, for example, 5 %, compared to their kinetic energy, however.

Another consideration in the use of flywheels is rate at which energy can be added or deleted. That is, their power.

The maximum power that can be applied or extracted is determined by the mechanical properties of the central shaft. The maximum torque τ_{max} that can be withstood can be expressed as

$$\tau_{max} = \tfrac{2}{3}\pi\sigma_s R_0^3 \tag{6.13}$$

where R_0 is the shaft radius and σ_s is the maximum shear strength of the shaft material.

The torque involved in a change in the rotational velocity of the flywheel

$$\tau = I\frac{d\omega}{dt} \tag{6.14}$$

for a disk of radius R and thickness T,

$$\tau = \frac{\pi}{2}\rho T R^4 \frac{d\omega}{dt} \tag{6.15}$$

The maximum possible acceleration can be obtained by setting this value of τ equal to the value of τ_{max} calculated above:

$$\frac{d\omega}{dt} = \frac{4}{3}\left(\frac{\sigma_s}{\rho RT}\right)\left(\frac{R_0}{R}\right)^3 \tag{6.16}$$

The power is the rate of change of the kinetic energy. So the maximum power is

$$P_{max} = \frac{\pi}{2}\rho\omega T R^4 \frac{d\omega_{max}}{dt} = 2\frac{\pi}{3}R_0^3\sigma_{max} \tag{6.17}$$

for the case in which the strength of material from which the shaft is made is the same as the strength of the flywheel material.

But the maximum rotational velocity of a flywheel is related to the strength of its material by

$$\omega_{max} = \frac{1}{R}\left(\frac{2\sigma_{max}}{\rho}\right)^{1/2} \tag{6.18}$$

To see the magnitudes involved, consider a flywheel with a weight of 4.54 kg, $R = 17$ cm, $T = R_0 = R/10$, $\sigma_{max} = 1.5 \times 10^6$ lb/in.^{2}, and $\rho = 3.0$.

If $\omega_{max} = 1.6 \times 10^6$ rad/s, the maximum power, P_{max} is 5×10^8 W, or 10^6 W/kg

This is a very large number. Thus flywheels are very good at handling high power, and therefore energy transients. For comparison, the power per unit weight of a typical battery might be of the order of 100 W/kg.

Aerodynamic drag can be substantial in high-velocity flywheels. As a result, they are typically operated under vacuum conditions.

6.9 Internal Structural Energy Storage

It is also possible to introduce a different type of mechanical energy into solid materials by plastically deforming them such that changes occur in their micro-structure. These can involve changes in the concentrations or distributions of dislocations or crystallographic point defects during mechanical deformation or irradiation. In some extreme cases, such as heavy forging at elevated temperatures, this can become quite evident, for it can be seen from color changes that the extensive mechanical deformation causes the internal temperature to rise. Although at least some of this energy can be recovered upon annealing, this increase of the internal energy in solids is not readily reversible, and is therefore not of interest for the types of applications that are generally considered in this text.

Reference

1. Genta G (1985) Kinetic Energy Storage. Butterworths, London

Chapter 7
Electromagnetic Energy Storage

7.1 Introduction

Several of the prior chapters in this text have shown that there is a wide range of energy storage needs with widely different time periods; some involve seasonal, weekly, and daily cycles, and others require energy intermittently, sometimes over much shorter time periods. A variety of different technologies are employed to meet these various requirements.

This chapter deals with two general mechanisms by which electrical energy can be stored. One involves capacitors, in which energy is stored by the separation of negative and positive electrical charges. The other involves the relationship between electrical and magnetic phenomena.

It will be seen that both of these mechanisms are most applicable to situations in which there is a requirement for the storage of modest amounts of energy under very transient conditions, for relatively short times and sometimes at high rates. Such applications, therefore, emphasize fast kinetics and high power, rather than focusing on the amount of energy that can be stored. A very large cycle life is also generally very important. It will be seen later that the amount of energy that can be stored by such methods is generally much less than can be stored by chemical and electrochemical methods.

The range of both current and potential future energy storage requirements and applications is very broad. In addition to the reduction in short-term transients in the large electrical power distribution grid system mentioned in Chapter 1, typical examples that are now highly visible include digital communication devices that require pulses in the millisecond range, implanted medical devices that require pulses with characteristic times of the order of seconds, and hybrid vehicle traction applications, where the high power demand can extend from seconds up to minutes, and the ability to absorb large currents upon braking is also important.

© Springer International Publishing Switzerland 2016   69
R.A. Huggins, *Energy Storage*, DOI 10.1007/978-3-319-21239-5_7

There are two general approaches to the solution of these types of requirements. One involves the use of electrical devices and systems in which energy is stored in materials and configurations that exhibit capacitor-like characteristics. The other involves the storage of energy using electromagnets. These are discussed in the following sections.

7.2 Energy Storage in Capacitors

Energy can be reversibly stored in materials within electric fields and in the vicinity of interfaces in devices called capacitors. There are two general types of such devices, and they can have a wide range of values of the important practical parameters, the amount of energy that can be stored, and the rate at which it can be absorbed and released.

7.2.1 Energy in a Parallel Plate Capacitor

In a parallel plate capacitor, such as that shown schematically in Fig. 7.1, a dielectric material sits between two metallic plates. If an electrical potential difference is applied between the plates, there will be an electric field across the material located between them that causes local displacements of the negative and positive charges within it.

The energy stored W_C in a material between capacitor plates of area A separated by a distance d is given by

$$W_c = \frac{1}{2}\varepsilon A \frac{V^2}{d} \tag{7.1}$$

where ε is the permittivity of the material between the plates, and V is the applied voltage.

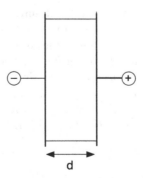

Fig. 7.1 Schematic view of a simple parallel plate capacitor

Table 7.1 Values of the relative permittivity of some materials at ambient temperature

Material	ε_r
Nylon	2.1
Teflon	2.6
Bakelite	4.9
Soft glass	6–7
Distilled water	80
High-permeability oxides	10 to 15×10^3

The capacity C of this configuration is given by

$$C = \varepsilon \frac{A}{d} \tag{7.2}$$

and

$$\varepsilon = \varepsilon_r \varepsilon_0 \tag{7.3}$$

where ε_r is the relative permittivity of the material, and ε_0 is the permittivity of a vacuum, 8.854×10^{-12} F per meter. The permittivity was sometimes called the dielectric constant in the past.

Values of the relative permittivity of several materials are shown in Table 7.1.

By substituting Eqs. (7.2) and (7.3) into Eq. (7.1), the energy stored in the capacitor can be written as

$$W_C = \frac{1}{2}CV^2 \tag{7.4}$$

Since the charge stored in a capacitor, Q, is related to its capacitance and the voltage by

$$Q = CV \tag{7.5}$$

the stored energy can also be expressed as

$$W_C = \frac{1}{2}QV \tag{7.6}$$

It is interesting that the amount of the energy stored in such a capacitor is inversely proportional to the volume of the dielectric material between the plates of the capacitor. This is counterintuitive, for we normally think that if a material carries energy, the greater the amount of the material, the greater amount of energy that it contains.

7.3 Electrochemical Charge Storage Mechanisms

The behavior of such a parallel plate configuration becomes very different if the space between the electrodes contains a material that has the properties of a liquid, or even a solid, electrolyte. In such a case the charge is stored in the interfacial region where the electronically-conducting material meets the ionically-conducting material. The physical separation between the positive and the balancing negative charges is then very small. This can be thought of as equivalent to a parallel plate capacitor with a thickness of the order of interatomic distances, and results in much greater amounts of charge storage per unit area.

Devices with this type of local structure are called electrochemical capacitors, and there are two general types. One involves the storage of charge in the electrical double-layer at or near the electrolyte/electronic material interface. Such devices are called *ultracapacitors*. The other type makes use of transient additional reversible absorption of atomic species into the crystal structure of the solid electronically-conducting electrode, and such devices are called *supercapacitors*. Both of these mechanisms can lead to much larger values of capacitance than capacitors with dielectric materials between their plates, as discussed above.

7.3.1 Electrostatic Energy Storage in the Electrical Double-Layer in the Vicinity of an Electrolyte/ Electrode Interface

As mentioned above, the interface between a chemically inert electronic conductor electrode and an adjacent electrolyte with mobile ionic charges can function as a simple capacitor with a very small distance separating two parallel plates.

The amount of charge that can be stored in such a configuration is generally of the order of 15–40 μF per cm^2 of interface [1–3]. Thus efforts are made to maximize the amount of interface in the device microstructure to produce the greatest amount of stored charge. Techniques have been devised to produce various types of carbon, as well as some other electronically conducting, but chemically inert, materials in very highly dispersed form, leading to very large interfacial areas. Typical values of specific capacitance, in Farads per gram, of a number of double-layer electrode materials are shown in Table 7.2.

Table 7.2 Characteristics of some double-layer electrode materials

Electrode material	Specific capacitance/F/g
Graphite paper	0.13
7.Carbon cloth	35
Aerogel carbon	30–40
Cellulose-based foamed carbon	70–180

The electrical potential difference that can be applied is limited by the decomposition voltage of the electrolyte, which is about 1.23 V for common aqueous electrolytes, but can be up to 4–5 V for some organic solvent electrolytes. The result is that the specific energies of simple aqueous electrolyte systems of this type are generally in the range of 1–1.5 Wh/kg, whereas those that use organic solvent electrolytes can be 7–10 Wh/kg.

A device in which this is the dominant charge storage mechanism will behave like a pure capacitor in series with its internal resistance. Its time constant is equal to the product of the capacitance and the series resistance. Thus it is important to keep the resistance as low as possible if rapid response is desired. Organic electrolytes characteristically have much lower values of ionic conductivity, and thus provide greater resistance, and longer time constants than aqueous electrolytes. The conductivity of acids, such as H_2SO_4, is somewhat greater than that of aqueous bases. Furthermore, the larger the capacitance, the greater the time constant, the slower the device, and the lower the power level.

Another important feature of devices that operate by the double-layer mechanism is that the amount of charge stored is a linear function of the voltage according to Eq. (7.5). The voltage therefore falls linearly with the amount of charge extracted. Thus voltage-dependent applications can only utilize a fraction of the total energy stored in such systems. The power supplied to resistive applications is proportional to the square of the instantaneous voltage, so this can be an important limitation.

Although there is some confusion in the literature about terminology, devices of this type that have been developed with large values of such capacitance have generally been called either EDLC (electrical double-layer capacitive) devices, or *ultracapacitors* [1–3].

Ultracapacitor devices utilizing the storage of charge in the electrochemical double layer have been developed and produced in large numbers in Japan for a considerable period of time [4]. These are primarily used for semiconductor memory backup purposes, as well as for several types of small actuators.

7.3.2 Underpotential Faradaic Two-Dimensional Adsorption on the Surface of a Solid Electrode

Due to the characteristics of the electrolyte/electrode surface structure and its related thermodynamics, it is often found that modest amounts of *Faradaic electrodeposition* can occur at potentials somewhat removed from those needed for the bulk deposition of a new phase. This results in the occupation of specific crystallographic sites on the surface of the solid electrode by species from the electrolyte. This mechanism typically results in only partial surface coverage, and thus the production of an *adsorption pseudo-capacitance* of some 200–400 mF/cm^2 of interfacial area [2]. This is significantly larger than the amount of charge stored per unit area in the electrochemical double layer. However, materials with which this mechanism can be effectively used are rare, and it is not common.

7.3.3 Faradaic Deposition that Results in the Three-Dimensional Absorption of the Electroactive Species into the Bulk Solid Electrode Material by an Insertion Reaction

As will be discussed in later chapters, many materials are now known in which atoms from the electrolyte can move into the surface of a solid electrode material, changing its local composition. These are called *solid solution electrodes*. In such cases, the *electroactive species* diffuses into and out of the interior of the crystal structure of the solid electrode as its potential is changed. Since the amount of energy stored is proportional to the amount of the electroactive species that can be absorbed by the electrode, this bulk storage mechanism can lead to much higher values of energy storage per unit volume of electrode structure than any surface-related process. Because it makes no sense to express this bulk phenomenon in terms of the capacitance per unit interfacial area, for it depends upon how far the material from the electrolyte penetrates into the solid, values of capacitance are generally given as *Farads per gram* in this case.

This mechanism was called *redox pseudo-capacitance* by the Conway group [1–3]. They also first started the use of the term *supercapacitors* to describe devices utilizing this type of charge storage. In this way, solid solution bulk storage *supercapacitors* can be distinguished from double-layer storage *ultracapacitors*, in which atoms from the electrolyte do not enter the solid electrode material.

The bulk storage *supercapacitor* mechanism is utilized in the devices that are most interesting for energy-sensitive pulse applications. Since the kinetic behavior of such devices is related to the electrolyte/electrode area, it is important that they also have very fine large surface area microstructures.

During investigations of the *dimensionally stable electrodes* that are used as positive electrodes in the *chlor-alkali process* it was noticed that RuO_2 seemed to behave as though its interface with the electrolyte had an unusually large capacitance. This swiftly led to several investigations of the capacitive behavior of such materials [5–8].

The possibility of the development of RuO_2-type materials as commercial supercapacitors began in Canada about 1975, the key players being D.R. Craig [9, 10] and B.E. Conway. This soon evolved into a proprietary development program at the Continental Group, Inc. laboratory in California, which subsequently was taken over by the Pinnacle Research Institute. The products that were initially developed and manufactured were all oriented toward the military market. This orientation is changing now, and activities are being undertaken by several firms to produce this type of product for the civilian market. Activities in this area have also been initiated more recently in Europe, including work at Daimler-Benz (Dornier) and Siemens in Germany, and at Thompson-CSF in France.

It was originally thought that charge storage in these materials is due to *redox reactions* near the interface between the electrolyte and RuO_2. Careful measurements showed that the capacitance is large, and proportional to the surface area [11].

It is now known, however, that the charge is stored by hydrogen insertion into the bulk of the RuO_2, and the capacitive behavior is not just on the surface [12]. At relatively short times the depth of hydrogen diffusion is limited, and the amount of charge stored in the bulk will not reach its ultimate saturation value.

Experiments [6, 13] showed that the chemical diffusion coefficient of hydrogen in bulk crystalline RuO_2 is about 5×10^{-14} cm^2/s. Thus the penetration into the bulk crystalline solid is rather shallow at the relatively high frequencies typically used in capacitor experiments. It was shown in later work [14] that the apparent hydrogen solubility is considerably higher in amorphous RuO_2 than in crystalline material. Experiments on hydrated RuO_2 [12, 15, 16] demonstrated that it has a substantially larger charge storage capacity than anhydrous RuO_2.

It has also been found [16] that the amount of charge that can be stored in hydrated RuO_2 is independent of the surface area, but proportional to the total mass. This is shown in Fig. 7.2. Over one hydrogen atom can be reversibly inserted into the structure per Ru atom [12]. The *coulometric titration* curve is shown in Fig. 7.3.

Crystalline RuO_2 is a very good electronic conductor, with an electronic resistivity of about 10^{-5} Ω cm. This is about a factor of 100 lower than that of bulk carbons. The hydrated material, on the other hand, has a considerably higher resistivity, and it has been found to be advantageous to add some carbon to the microstructure in order to reduce the electronic resistance in that case.

Several other materials that are *electrochromic*, i.e., change color as charge is inserted or deleted, show similar pseudo-capacitive behavior. This clearly indicates the insertion of species into the bulk crystal structure. Materials such as NiOOH and IrO_2 thus could be used in supercapacitors.

Typical values of specific capacitance of a number of insertion reaction electrode materials are included in Table 7.3.

Fig. 7.2 Apparent capacitance of RuO_2 hydrate as a function of surface area [16]

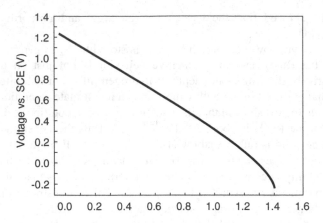

Fig. 7.3 Dependence of the potential of RuO_2 hydrate upon the amount of inserted hydrogen [12]

Table 7.3 Characteristics of some insertion reaction electrode materials

Electrode material	Specific capacitance/F/g
Polymers (e.g., polyaniline)	400–500
RuO_2	380
RuO_2 hydrate	760

The insertion of guest species into the host crystal structure in such insertion reactions generally results in some change in the volume. This can lead to morphological changes in electrodes, and a reduction in capacity upon cycling. The volume change is generally roughly proportional to the concentration of the guest species. As a result, it is often found that the magnitude of this degradation depends upon the depth of the charge-discharge cycles.

7.3.4   Faradaically-Driven Reconstitution Reactions

The electrodes in many battery systems undergo *reconstitution reactions*, in which new phases form and others are consumed. In accordance with the *Gibbs phase rule*, this often results in an open-circuit electrode potential that is independent of the state of charge. As discussed elsewhere in this text, the amount of charge storage is determined by the characteristics of the related phase diagram, and can be quite large. Some reactions of this type can also have relatively rapid kinetics. However, there is a potential difficulty in the use of this type of reaction in applications that require many repeatable cycles, for they generally involve microstructural changes that are not entirely reversible. Thus the possibility of a cycle-life limitation must be kept in mind.

A special strategy whereby this microstructural irreversibility may be avoided or reduced in certain cases has been proposed [17]. This involves the use of an all-solid electrode in which a mixed-conducting solid matrix phase with a very high chemical diffusion coefficient surrounds small particles of the reactant phases.

7.4 Comparative Magnitudes of Energy Storage

The maximum amount of energy that can be stored in any device is the integral of its voltage-charge product, and cannot exceed the product of its maximum voltage and the maximum amount of charge it can store. On this basis, we can make a simple comparison between these different types of energy storage mechanisms.

The results are shown schematically in Fig. 7.4, in which the relationship between the potential and the amount of charge delivered is plotted for three different types of systems, a double-layer electrode, an insertion reaction electrode, and a reconstitution reaction electrode. Electrodes that involve two-dimensional *Faradaic underpotential* deposition are not included, as they do not constitute a practical alternative.

In the case of a true capacitor, the amount of charge stored is a linear function of the applied voltage. Thus, as shown on the left side of Fig. 7.4, the voltage falls off linearly with the amount of charge delivered.

A single-phase solid solution insertion-reaction type of electrode characteristically has a potential-charge relation of the type shown in the middle. The thermodynamic basis for this shape, in which the potential is composition-dependent, and thus state of charge dependent, will be discussed in Chap. 13.

The characteristic behavior of a reconstitution-reaction electrode system is shown on the right side. In this case, it is assumed that the temperature and pressure are fixed, and that the number of components is equal to the number of phases, so that from a thermodynamic point of view there are no degrees of

Fig. 7.4 Comparison of the variation of the potential with the amount of charge extracted for different types of energy storage mechanisms

Fig. 7.5 The amount of energy available for materials with different types of storage mechanisms, indicated by the area under their curves

freedom. This means that all of the intensive variables, including the electrode potential, are independent of the overall composition, and thus independent of the amount of charge delivered. Thus the discharge characteristic consists of a voltage plateau.

As mentioned above, the maximum amount of energy that is available in each case is the area under the V/Q curve. This is indicated in Fig. 7.5 for the three cases of interest. It is seen that the maximum amount of energy that can be stored in an electrode that behaves as a capacitor is $1/2$ $(V_{max})(Q_{max})$. The actual amount of available energy will, of course, depend upon the power level, due to unavoidable losses, such as that due to the inevitable internal resistance of the system.

7.5   Importance of the Quality of the Stored Energy

As mentioned earlier, the *quality* of heat is a commonly used concept in engineering thermodynamics. High temperature heat is generally much more useful than low temperature heat. Thus, in considering a practical thermal system one has to consider both the amount of heat and the temperature at which it is available.

One can consider an analogous situation in the application of energy storage devices and systems. In such cases, in addition to the total amount of energy that can be stored, one should also consider its quality, the voltage at which it is available.

If this factor is taken into account, an additional difference between systems that utilize electrodes that operate by these three different types of mechanisms can be seen. This is indicated in Table 7.4, in which the amount of higher value energy available in the different cases is compared. In that case, only a simple distinction is used. Energy at a potential above $V_{max}/2$ is considered to be high value energy.

Thus there are a number of parameters that determine important properties of a transient energy storage system. These are listed in Table 7.5.

Table 7.4 Maximum amount of high-value energy available

Type of electrode	High-value energy/%
	Where $(V > V_{max}/2)$
Double-layer electrode	37.5
Insertion reaction electrode	About 80
Reconstitution reaction electrode	About 90

Table 7.5 Parameters that determine the values of maximum potential, maximum charge, and maximum energy stored

Type of electrode	V_{max} determined by
Double-layer electrode	The electrolyte stability window
Insertion reaction electrode	Thermodynamics of guest-host phase
Reconstitution reaction electrode	Thermodynamics of polyphase reactions
Type of electrode	Q_{max} determined by
Double-layer electrode	Electrode microstructure, electrolyte
Insertion reaction electrode	Mass of electrode, thermodynamics
Reconstitution reaction electrode	Mass of electrode, thermodynamics

7.6 Transient Behavior of a Capacitor

In addition to the question of the amount of energy that can be stored in a capacitor, consideration must also be given to the rate at which it can be obtained.

There is always some series resistance connected to a capacitor. This can be indicated for the case of a simple parallel plate capacitor and *ultracapacitors* in the simple equivalent circuit of Fig. 7.6. If one or both of the electrodes also undergo some insertion reaction, as is the case in *supercapacitors*, the kinetics become more complicated. The kinetic behavior in that case can be treated by the use of Laplace transform techniques, and will be discussed later in this chapter.

If such a capacitor is charged, and then shorted, which sets the external voltage V to zero, the voltage across the capacitor will be balanced by the voltage across the resistor, for

$$V = V_R + V_C \tag{7.7}$$

As the capacitor discharges, its voltage decreases, as does the voltage across, and thus the current through, the resistor.

The instantaneous current $i_{(t)}$ through the resistor decays exponentially with time according to

$$i_{(t)} = -i_0 \exp\left(-t/_{RC}\right) \tag{7.8}$$

where i_0 is the initial current at the start of discharge.

Fig. 7.6 Simple equivalent circuit of a capacitor and its associated series resistance

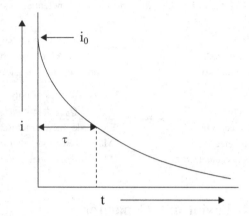

Fig. 7.7 Time-dependent current decay in a capacitor system

The product RC is called the *time constant*, τ. It is a useful parameter in understanding the rate at which energy can be obtained from capacitor-based systems.

Taking the logarithm of both sides of Eq. (7.8),

$$\ln\left(\frac{i}{i_0}\right) = -\frac{t}{RC} \tag{7.9}$$

Thus the time t has the value of the time constant RC when

$$\left(\frac{i}{i_0}\right) = \exp(-1) \tag{7.10}$$

The value of $\exp(-1)$ is approximately 0.3679.

In the case of a series arrangement of a resistor R and an inductance L the time constant is given by

$$\tau = \frac{R}{L} \tag{7.11}$$

The variation of the current through the resistor R with time in a capacitor system is shown in Fig. 7.7.

The rate at which energy can be stored—or supplied—is determined by the time constant. Therefore, it is desirable that the series resistance R is as small as possible.

But it is also obvious that if the capacitance C is large, the time constant will be large.

During operation some of the applied power goes into heating the resistor:

$$Power = i_{(t)}V_R = i_{(t)}^2 R \tag{7.12}$$

The maximum power is determined by the value of V^2/R, and can be very large for a short time, and values up to 10^9 watt/m^3 can be attained in some designs.

$$P = I_0^2 R = \left(V/R\right)^2 R = V^2/R \tag{7.13}$$

7.7 Modeling Transient Behavior of Electrochemical Systems Containing Capacitive Components Using Laplace Transforms

7.7.1 Introduction

The quantitative understanding of the application of capacitive, or other, components in actual devices requires knowledge of the relationship between component properties and system behavior, for systems typically involve more than one component. As an example, in addition to the electrode impedances, there are almost always resistive, and/or capacitive, impedances present, both relating to internal phenomena in the device and to external factors.

It is often very useful to utilize *equivalent electrical circuits* whose electrical behavior is analogous to the behavior of physical systems as *thinking tools* to obtain insight into the important parameters and their interrelationships. This allows the use of the methods that have been developed in electrical engineering for circuit analysis to evaluate the overall behavior of interdependent physical phenomena.

A useful way to do this is based upon the simple concept of a relation between a *driving force* and the *response* of a device or system to it. This relation can be written very generally as

Driving Function = (Transfer Function) × (Response Function)

In an electrochemical system, the driving function represents the current or voltage demands imposed by the application, and the response function is the output of the electrochemical system in response to these demands. The key element of this approach is the determination of the (time-dependent) transfer function of the device or system, for that determines the relationship between application demand and system output.

7.7.2 Use of Laplace Transform Techniques

The general method that has been developed for electrical device and circuit analysis involves the use of *Laplace transform* techniques. There are several basic steps in this analysis. They involve:

1. The determination of the transfer function of the individual equivalent circuit components
2. The calculation of the transfer function of the total system
3. The introduction of the driving function determined by the application
4. The calculation of the system (energy source) output

Some readers of this chapter may not be familiar with Laplace transform methods. But they can be readily understood by the use of an analogy. Consider the use of logarithms to multiply two numbers, e.g., A and B. The general procedure is to find the logarithms (transforms) of both A and B, to add them together, and then to use antilogarithms to reconvert the sum of the logarithms (transforms) into a normal number.

This method has been applied to some simple electrochemical situations, including the influence of the presence of series resistance upon the rate of charge accumulation in an insertion reaction electrode [18] and the electrical response of electrochemical capacitors [11].

The calculation of the transient electrical response of an insertion, or solid-solution, electrode involves the solution of the diffusion equation for boundary conditions that are appropriate to the particular form of applied signal. In addition, the relation between the concentration of the electroactive mobile species and their activity is necessary. This approach has been utilized to determine the kinetic properties of individual materials by employing current and/or voltage steps or pulses.

But in real electrochemical systems or devices one has to consider the presence of other components and phenomena, that is, other circuit elements. As a simple example, there is always an electrolyte, and thus a series resistance, present, and the behavior of the electrolyte/electrode interface may have to be also considered. Thus the simple solution of Fick's diffusion laws for the electrochemical behavior of the electrode alone may not be satisfactory.

Examples of the Laplace transforms of several common functions are given in Table 7.6.

7.7.3 Simple Examples

To illustrate this method, several examples will be described; the response of an insertion electrode under both a step in potential and a step in current, as well as a system consisting of a simple series arrangement of a resistance and an insertion reaction electrode that has a diffusional impedance.

Table 7.6 Examples of Laplace transforms

Function	Laplace transform
General impedance function	$Z(p) = E(p)/I(p)$
Fick's second law	$pC - c(t = 0) = D\dfrac{d^2C}{dx^2}$
Current step $d(t)$	$I(p) = 1$
Potential vs. time	$E(p) = V(dE/dy)$
Impedance of insertion reaction electrode	$Z(p) = Q/Da$

where

$Q = \frac{V(dE/dy)}{nFs}$

$a = (p/D)^{1/2}$

dE/dy = slope of the coulometric titration curve

y = composition parameter

n = stoichiometric coefficient

F = Faraday constant

s = surface area

p = complex frequency variable

x = positional coordinate

ID="Par80">V = molar volume

$q(t)$ = charge accumulated in electrode

$i(t)$ = instantaneous current

$F(t)$ = instantaneous electrode potential

1. Upon the imposition of a step in potential F_0, the time dependence of the current $i(t)$ is given by

$$i(t) = \frac{F_0}{Q}\left(\frac{D}{pt}\right)^{1/2} \qquad (7.14)$$

2. The time dependence of the charge accumulated (or produced) $q(t)$ is

$$q(t) = \frac{2F_0}{Q}\left(\frac{t}{p}\right)^{1/2} \qquad (7.15)$$

3. For the case of a step in current i_0, the time dependence of the electrode potential $F(t)$ is given by

$$F(t) = 2Q\left(\frac{t}{pD}\right)^{1/2} \qquad (7.16)$$

4. The time dependence of the current after the imposition of a step potential of F_0 for the more complicated case of a resistance in series with an insertion reaction electrode is found to be

Fig. 7.8 The charge accumulated in an insertion reaction electrode as a function of time for various values of series resistance

$$i(t) = (F_0/R)\exp\left[\left(\frac{Q}{R}\right)^2 t\right]erfc\left[\frac{Qt^{1/2}}{R}\right] \tag{7.17}$$

5. The charge accumulated (or produced) in the case of this series combination is found to be

$$q(t) = \frac{F_0 R}{Q^{1/2}}\left[\exp\left[\left(\frac{Q}{R}\right)^2 t\right]erfc\left[\frac{Qt^{1/2}}{R}\right] - 1\right] + \frac{2F_0}{Q}\left(\frac{t}{p}\right)^{1/2} \tag{7.18}$$

The influence of the value of the series resistance can readily be seen in Fig. 7.8 [19].

The parameters used in the calculation illustrated in Fig. 7.8 are

$$Q^{-1} = 6.33 \text{ ohm sec}^{-1/2}$$

derived from $D = 10^{-8}$ cm^2/s, $V_m = 30$ cm^3/mol, $dE/dy = -2$ V, $s = 1$ cm^2 and $n = 1$.

The applied voltage step was 0.5 V.

Examination of the solutions for the behavior of single components under the first two sets of conditions obtained by this method shows that they are equivalent to those for the analytical solution of the diffusion equation under equivalent experimental conditions.

The impedance of an insertion reaction electrode alone under an AC driving force has also been described in [19].

This Laplace transform procedure is thus an alternative to the more conventional analytical approach. But the real value of the Laplace transform approach becomes clearer, however, under more complex conditions, such as when more than one component is present, and in which the normal procedures become quite cumbersome.

7.8 Energy Storage in Magnetic Systems

The energy storage capability of electromagnets can be much greater than that of capacitors of comparable size. Especially interesting is the possibility of the use of superconductor alloys to carry current in such devices. But before that is discussed, it is necessary to consider the basic aspects of energy storage in magnetic systems.

7.8.1 Energy in a Material in a Magnetic Field

It was shown earlier in this chapter that the energy stored in a parallel plate capacitor with spacing d and area A when a voltage V is applied across it can be written as

$$W_C = \frac{1}{2}\varepsilon A \frac{V^2}{d} = \frac{1}{2}CV^2 \tag{7.19}$$

ε is the *permittivity*, a measure of the polarization of the material between the plates by the electric field, and C the *capacitance*.

Energy can also be stored in magnetic materials and systems. The analogous relation is

$$W_M = \frac{1}{2}\mu H^2 \tag{7.20}$$

where H is the intensity of a *magnetic field*, and μ is the *permeability*, a constant that is dependant upon the material within the field, analogous to the *permittivity*. The magnetic field H is sometimes called the *magnetizing field*, or the *magnetizing force*.

There is also a relation equivalent to that in Eq. (7.3):

$$\mu = \mu_r \mu_0 \tag{7.21}$$

where μ_r is called the *relative permeability* of the material present in the magnetic field, and μ_0 the *permeability of vacuum*, 1.257×10^{-6} Henries per meter.

When a material is placed in a magnetic field, an internal magnetic field will be induced within it whose magnitude depends upon the material's permeability μ. This internal *induced magnetic field, B*, which is sometimes called the *magnetic induction*, or the *magnetic flux density*, is thus related to the external field H by

$$B = \mu H \tag{7.22}$$

Equation (7.20) can therefore be rewritten in terms of the *induced magnetic field B* inside the material instead of the external field H as

$$W_M = \frac{1}{2\mu}B^2 = \frac{1}{2}BH \tag{7.23}$$

Equation (7.22) can be rewritten to show the separate influence of the external field and the internal properties of the material as

$$B = \mu_0 H + \mu_0 M \tag{7.24}$$

M is called the *magnetization*, and $\mu_0 M$ is the additional induced magnetic field due to the properties of the solid. The *magnetization* can also be expressed as

$$M = \frac{(\mu - \mu_0)H}{\mu_0} = \frac{\mu H}{\mu_0} - 1 = \mu_r H - 1 \tag{7.25}$$

The magnetic properties of the solid can also be expressed in terms of the *susceptibility, X*, which is dimensionless, where

$$X = \frac{M}{H} = \frac{\mu - \mu_0}{\mu_0} = \frac{\mu}{\mu_0} - 1 = \mu_r - 1 \tag{7.26}$$

and

$$B = \mu_0(H + M) \tag{7.27}$$

One way to generate a magnetic field H is to pass current through a nearby electrical conductor. In the case of a wire shaped into a spiral, or helix, the value of the H field inside it is

$$H = 4\pi n l \tag{7.28}$$

where n is the number of turns per unit length of the spiral, and I is the magnitude of the current. The direction of this field is parallel to the length of the spiral. This is shown schematically in Fig. 7.9.

The H field is continuous through the material, as well as in its external environment, whereas the B field is only within the solid.

Since magnetic units and their dimensions may not be as familiar to many people as electrostatic units, a list of some of them is given in Table 7.7.

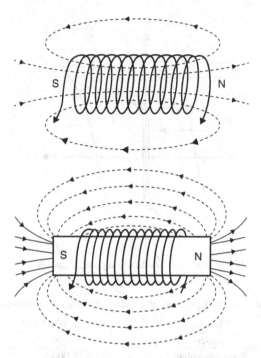

Fig. 7.9 Schematic representation of the H field in the vicinity of a linear helix. It is seen that the presence of the material with a high value of susceptibility amplifies the external H field

Table 7.7 Magnetic quantities, units, and dimensions

Quantity	Unit	Symbol	Dimensions
Magnetic field	Henry	H	A/m
Magnetic induction	Tesla	B	$Wb/m^2 = Vs/Am^2$
Permeability		μ	Vs/Am
Energy product		BH	kJ/m^3
Magnetization per unit volume		M_v	A/m
Magnetization per unit mass		M_m	Am^2/kg
Magnetic flux	Weber	Wb	Vs
Inductance	Henry	L	$Wb/A = Vs/A$

The unit of permeability in the SI system is the Henry. One Henry has the value of 1 Wb per Amp, and is the *inductance* that produces one volt when the current in a circuit varies at a uniform rate of one ampere per second.

It was shown in eqn. 7.23 that the energy in an electromagnet is proportional to the product of B and H. These quantities are related to each other by the value of the magnetization M, or the relative permeability, μ_r. There are two general classes of magnetic materials, one group are generally described as *soft magnetic materials*, and the other are called *hard magnetic materials*. Samples of the latter

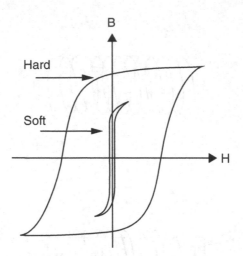

Fig. 7.10 Schematic B–H curves for soft and hard magnetic materials

Table 7.8 B–H products of several hard magnetic materials

Material	(BH)max/Wb-A/m3
Alnico	36,000
Platinum–cobalt	70,000
Samarium–cobalt	120,000

are also often named *permanent magnets*. Their characteristics are shown schematically in Fig. 7.10.

It can be seen that these are very different. The major difference is the large area in the case of the hard material. If such a material is magnetized by increasing H, and then H is reduced to zero, B, and thus the magnetization, remains at a high value. In order to reduce B, or the magnetization, to zero, a large negative value of H is required. Thus the material tends to remain magnetized, i.e., is "hard." The magnitude of this reversed H field that is necessary to demagnetize this material is called the *coercivity*. Thus such a material stores a lot of magnetic energy, but it is very difficult to get it out. The area inside the B–H curve, a measure of hysteresis, represents the energy lost each time the magnetization is reversed. Data on several hard magnetic materials are included in Table 7.8.

It can be seen that the area within the curve for the soft magnetic material in Fig. 7.10 is much smaller than that for the hard magnetic material. It is soft, rather than hard, magnetic materials that are used to reversibly store energy in electromagnetic systems under transient conditions. For this type of application the energy loss due to hysteresis (the area inside the B–H curve) should be as small as possible. Data on the magnetic properties of several types of soft magnetic materials are shown in Table 7.9.

Table 7.9 Data on some soft magnetic materials

Material	Composition/wt%	Relative permeability	Resistivity/Ω-m
Commercial cast iron	99.95 Fe	150	1×10^{-7}
Oriented silicon iron alloy	97 Fe, 3 Si	1,400	4.7×10^{-7}
Permalloy	55 Fe, 45 Ni	2,500	4.5×10^{-7}
Supermalloy	79 Ni, 15 Fe, 5 Mo	75,000	6×10^{-7}
Ferroxcube A	48 MnFe2O4, 52 ZnFe2O4	1,400	2,000
Ferroxcube B	64 ZnFe2O4, 36 NiFe2O4,	650	107

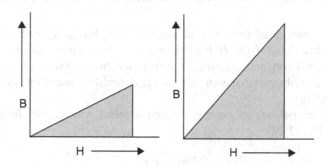

Fig. 7.11 Areas indicating the amount of stored energy for two materials with different values of relative permeability up to the same value of H field

Because they have a large influence over the efficiency, a considerable amount of work has been done over many years to optimize soft magnetic materials for their different uses. An important step forward was the development of iron–silicon alloys with relatively high electronic resistivity, thus reducing the hysteresis losses due to induced *eddy currents*. Processing these materials so that they develop a preferred crystallographic orientation also increases their permeability appreciably, as does annealing them in moist hydrogen to reduce the amount of carbon impurity present. These alloys are now commonly used as insulated laminated sheets in transformers for the transmission of moderate to large amount of electrical power.

Another class of soft magnetic materials includes the nickel–iron *permalloy* type of alloys. These materials have very large values of permeability, and are generally used in very low power applications that require large changes in magnetization with relatively small applied fields.

A third class of soft magnetic materials includes transition metal oxide ceramics called ferrites. Because they have very high values of electronic resistivity, and thus have no appreciable eddy current loss, they can be used in electronic equipment at very high frequencies.

According to Eq. (7.23) the energy stored in a magnetic material is one half the product of B and H, and therefore the area under a plot of B versus H. From Eq. (7.22) the slope of such a curve is the material's permeability. This is shown in Fig. 7.11 for two soft magnetic materials with different values of relative permeability in the same H field.

In this case, it is assumed that there is very little hysteresis, so that the data measured when the H field is increased are essentially the same as those when the H field is decreased.

This is analogous to the energy under a stress/strain curve in mechanical materials systems, and the energy stored under the voltage/composition curve in electrode materials in electrochemical systems.

7.8.2 Energy Storage in Superconducting Magnetic Systems

The magnetic energy of materials in external H fields is dependent upon the intensity of that field. If the H field is produced by current passing through a surrounding spiral conductor, its magnitude is proportional to the current according to Eq. (7.28). It is obvious that high currents are desirable if one wants to store large amounts of energy.

However, the passage of current through a metal wire causes Joule heating according to Eq. (7.29):

$$\text{Heat} = I^2 R \tag{7.29}$$

Thus it is desirable to consider the use of a superconducting material that has essentially no resistance to carry the current. Such systems are generally designed with the high permeability soft magnetic material within a superconductor coil in the shape of a toroid.

Energy can be fed into such a system by use of a DC power supply. Once the current is established in the superconductor, the power supply can be disconnected. The energy is then stored in the magnetic material inside the superconducting coil, where it can be maintained as long as desired without the need for further input.

The transmission of energy to and from the DC superconductor electromagnetic storage system requires special high power AC/DC conversion rectifier, inverter, and control systems. Such a power conditioning system typically causes a 2–3 % energy loss in each direction.

An additional feature that must be taken into account is the generation of large mechanical forces acting on the materials by the large magnetic fields present. The mitigation of this can add considerably to the cost of the whole system.

Superconductor materials have to be maintained below a so-called material-specific *critical temperature*. The maintenance of the required low temperature by the use of a cryostatic refrigerator requires energy, of course. There is another complication in that superconducting materials lose their superconductive property if the value of the surrounding H field is above a critical value, called the *critical field*. Since the field is caused by the current in the superconductor, another way of looking at this limitation is in terms of a *critical current*, rather than a *critical field*.

Such systems are generally used for short-term energy storage, such as improving the power quality and stability of a transmission distribution system, where the rapid response and high short-term power available can be a distinct advantage.

There is a serious potential danger if either the temperature or the field becomes too high, so that the material is no longer superconducting. Its resistance then becomes "normal." This can result in very large, and dangerous, amounts of Joule heating. Safety considerations related to superconducting energy storage devices of any appreciable magnitude generally involve their being placed in caverns deep underground.

7.8.3   Superconductive Materials

The phenomenon of superconductivity was discovered in 1911 by H. Kammerlingh Onnes [20]. He found that the electrical resistance of solid Hg disappeared below about 4 K. Research and development activities aimed at finding materials that remain superconducting to higher temperatures has been vigorously pursued over many years. Several groups of metals and alloys with higher critical temperatures were gradually found. Intermetallic compounds containing niobium have been shown to have attractive properties. The gradual improvement in the superconducting transition temperature resulting from the development of different alloys is shown in Fig. 7.12.

During the 1970s and 1980s it was generally concluded from both experimental and theoretical work that the maximum possible temperature for superconductivity was about 24 K. Then, in 1986, J.G. Bednorz and K.A. Müller showed that superconductivity could remain present in a ceramic oxide material, lanthanum barium yttrium oxide, up to 35 K [21]. A discussion of their work leading to this discovery can be found in [22].

A frantic effort was undertaken in many laboratories to confirm this surprising result, and investigate the possibility that superconductivity could be extended to even higher temperatures in other non-metallic materials. An important step was

Fig. 7.12 The time-dependent increase in the superconducting transition temperature resulting from the development of new materials

Table 7.10 Values of the critical temperature and critical field of some superconducting materials

Material	Critical temperature/K	Critical field/Tesla
Nb-Ti	10	15
Nb_3Al	18	?
Nb_3Ge	23.2	37
Nb_3Sn	18.3	30
NbTi	10	15
MgB_2	39	74
$YBa_2Cu_3O_7$	92	?

the discovery that the phase yttrium barium copper oxide, $YBa_2Cu_3O_7$, remains superconducting up to 93 K [23]. This discovery was particularly important, for this temperature is above the boiling point of relatively inexpensive liquid nitrogen, 77 K. The details of the synthesis of this material have a large influence upon its properties. Particularly important is the oxygen content [24].

Another quite different material that has recently been found to be superconducting is magnesium diboride, MgB_2 [25]. Although its critical temperature is only 39 K, it may become important for some applications, for it is quite inexpensive and relatively easy to make into wires or other shapes.

Data on the properties of some of the important superconductive materials are included in Table 7.10.

All of the higher temperature oxide superconductors are inherently brittle. It has proved to be difficult to find a way to make these materials into the long lengths necessary to make magnetic coils. As a result, the wires used in electromagnets are now typically made from the alloys that have lower critical temperatures. Nb–Ti alloys, that have a ductile BCC crystal structure, and can be formed into wires and made into coils are commonly used for this purpose. Fibers of this material can be imbedded in an aluminum or copper matrix for structural purposes.

However, they cannot withstand as high a magnetic field intensity as the more expensive Nb_3Sn, which is harder to use. Nb_3Sn is very brittle, and cannot be drawn into a wire shape. To overcome this problem, a composite microstructure containing ductile precursor phases can be formed, with separate alloys of Nb, Cu, and Sn. After the material is made into wire and worked into the final shape, it is heat treated, during which the Sn reacts with the Nb to form the Nb_3Sn phase. Because of its high critical field, this material is preferred for the production of high-power magnets and electrical machinery.

$YBa_2Cu_3O_7$, is also brittle, and cannot readily be formed into wires and other shapes. It is expected that this problem can also be circumvented, however, by forming it in situ within other materials that can be readily shaped, and a considerable amount of development effort is currently aimed in this direction.

References

1. Conway BE (1991) J Electrochem Soc 138:1539
2. Conway BE (1993) In: Barnett BM, Dowgiallo E, Halpert G, Matsuda Y, Takeharas Z (eds) Proceedings of Symposium on New Sealed Rechargeable Batteries and Supercapacitors. The Electrochemical Society. p 15
3. Conway BE (1999) Electrochemical Supercapacitors: Scientific Fundamentals and Technological Applications. Plenum, New York, NY
4. Nishino A (1993) In: Barnett BM, Dowgiallo E, Halpert G, Matsuda Y, Takehara Z (eds) Proceedings of Symposium on New Sealed Rechargeable Batteries and Supercapacitors. The Electrochemical Society. p 1
5. Burke LD, Murphy OJ, O'Neill JF, Venkatesan S (1977) J Chem Soc Faraday Trans 1:73
6. Arikado T, Iwakura C, Tamura H (1977) Electrochim Acta 22:513
7. Michell D, Rand DAJ, Woods R (1978) J Electroanal Chem 89:11
8. Trasatti S, Lodi G (1980) Electrodes of Conductive Metallic Oxides-Part A. In: Trasatti S (ed). Elsevier. p 301
9. Craig DR (1985) Canadian patent No.1,196,683
10. Craig DR (1983) Electric Energy Storage Devices. European Patent Application 82,109,061.0, submitted 30 September, 1982, Publication No. 0 078 404, 11 May 1983
11. Raistrick ID, Sherman RJ (1987) In: Srinivasan S, Wagner S, Wroblowa H (eds) Proceedings of Symposium on Electrode Materials and Processes for Energy Conversion and Storage. Electrochem Soc. p 582
12. Jow TR, Zheng JP (1998) J Electrochem Soc 145:49
13. Weston JE, Steele BCH (1980) J Appl Electrochem 10:49
14. Zheng JP, Jow TR, Jia QX, Wu XD (1996) J Electrochem Soc 143:1068
15. Zheng JP, Jow TR (1995) J Electrochem Soc 142:L6
16. Zheng JP, Cygan PJ, Jow TR (1995) J Electrochem Soc 142:2699
17. Boukamp BA, Lesh GC, Huggins RA (1981) J Electrochem Soc 128:725
18. Raistrick ID, Huggins RA (1982) Solid State Ionics 7:213
19. Ho C, Raistrick ID, Huggins RA (1980) J Electrochem Soc 127:343
20. Kammerlingh Onnes H (1911) Commun Phys Lab Univ Leiden 12:120
21. Bednorz JG, Müller KA (1986) Z Phys B 64:189
22. Bednorz JG, Müller KA (1988) Rev Mod Phys 60:585
23. Wu MK, Ashburn JR, Torng CJ, Hor PH, Meng RL, Gao L, Huang ZJ, Wang YQ, Chu CW (1987) Phys Rev Lett 58:908
24. Beyers R, Engler M, Ahn BT, Gür TM, Huggins RA (1988) In: Brodsky MB, Dynes RC, Kitazawa K Tuller HL (eds) High-Temperature Superconductors, vol. 99. Materials Research Society. p 77
25. Nagamatsu J, Nakagawa N, Muranaka T, Zenitani Y, Akimitsu J (2001) Nature 410:63

Chapter 8
Hydrogen Storage

8.1 Introduction

Hydrogen is an important *energy carrier*, and when used as a fuel can be considered to be an alternate to the major fossil fuels, coal, crude oil and natural gas, and their derivatives. It has the potential to be a clean, reliable and affordable energy source, and has the major advantage that the product of its combustion with oxygen is water, rather than CO and CO_2, which contain carbon and are considered greenhouse gases. It is expected to play a major role in future energy systems.

It has been shown that hydrogen can be used directly in internal reciprocating combustion engines, requiring relatively minor modifications, if it is raised to a moderately high pressure, as well as in turbines and process heaters.

It can also be used in hydrogen/oxygen fuel cells to directly produce electricity. Again, the only product is water. The energy efficiency of fuel cells can be as high as 60 %. Fossil fuel systems, on the other hand, are typically about 34 % efficient. When hydrogen is used in elevated temperature fuel cells, it is possible to obtain electricity and also to use the heat generated in the fuel cell, related to its inefficiency, for heating purposes. This is called *co-generation*, and in this way it is possible to obtain total energy efficiencies up to 80 %.

Electrically-powered vehicles have the advantage that electric motors can have energy efficiencies of about 90 %, whereas typical internal combustion engines are about 25 % efficient. On the other hand, fuel cells now cost about 100 times as much as equivalent internal combustion engines of comparable power. One can, however, expect some reduction in cost from further development and larger scale production of fuel cells.

Because of these attractive features, a number of people have long advocated the concept of a simple *hydrogen economy*, in which hydrogen is used as the major fuel.

It reacts with oxygen, either by combustion or in fuel cells, to give energy, and the only product is water. It can be regenerated directly from water by electrolysis.

© Springer International Publishing Switzerland 2016
R.A. Huggins, *Energy Storage*, DOI 10.1007/978-3-319-21239-5_8

Fig. 8.1 The general concept of the ideal hydrogen economy

This is a closed chemical cycle. No chemical compounds are created or destroyed, but there is a net flow of energy.

This concept is illustrated schematically in Fig. 8.1. It is important to recognize that a full hydrogen economy is not necessary. Even partial implementation would be desirable as a way to reduce environmental pollution problems.

The major use of hydrogen at the present time, however, is not as an energy carrier, but as a reactant in a number of important large-scale chemical processes.

8.2 The Production of Hydrogen

As discussed in the Preface, the major, and least expensive, way to obtain hydrogen is to extract it from natural gas, which is primarily methane. The most common method for the conversion of methane to hydrogen involves the use of the *steam reforming*, followed by the *water–gas shift reaction*. It now provides some 95 % of all the hydrogen produced in the USA. This process, as well as several other methods for making hydrogen, will be briefly discussed in the following sections of this chapter.

8.2.1 The Steam Reforming Process

The first step in this procedure is the elimination of impurities, such as sulfur, from the methane-rich natural gas. The methane is then reacted with steam at a relatively high temperature, using nickel oxide as a catalyst. This process is called *steam reforming*, and was already discussed in Chap. 4. It can be written as

$$CH_4 + H_2O = CO + 3H_2 \tag{8.1}$$

This can be followed by a second step in which air is added to convert any residual methane that did not react during the steam reforming.

$$2CH_4 + O_2 = 2CO + 4H_2 \tag{8.2}$$

This is then followed by the *water–gas shift reaction* at a somewhat lower temperature that produces more hydrogen from the CO and steam:

$$CO + H_2O = CO_2 + H_2 \tag{8.3}$$

As discussed already, the driving force for any reaction is the standard Gibbs free energy change, ΔG_r^0 that occurs as the result. This is the difference between the sum of the standard Gibbs free energies of formation of the products and the sum of the standard Gibbs free energies of formation of the reactants. In this case, this can be expressed as

$$\Delta G_r^0 = \Delta G_f^0(CO) + 3\Delta G_f^0(H_2) - \Delta G_f^0(CH_4) - \Delta G_f^0(H_2O) \tag{8.4}$$

Values of the standard Gibbs free energy of formation of the relevant species for three different temperatures are given in Table 8.1.

From these data it is possible to obtain the standard Gibbs free energy of relevant reactions in this system at those temperatures. These are shown in Table 8.2, and the results for Eqs. (8.1) and (8.3) are plotted in Fig. 8.2.

It can be seen that the steam reforming reaction will only go forward if the temperature is *above* about 900 K. Likewise, the subsequent water–gas shift reaction will only proceed if the temperature is *below* about 1025 K.

Table 8.1 Temperature dependence of the standard Gibbs free energies of formation of species in reaction 1

Species	ΔG_f^0 (400 K) (kJ mol^{-1})	ΔG_f^0 (800 K) (kJ mol^{-1})	ΔG_f^0 (1200 K) (kJ mol^{-1})
CO	−146.4	−182.5	−217.8
H$_2$O	−224.0	−203.6	−181.6
CO$_2$	−394.6	−395.5	−396.0
CH$_4$	−42.0	−2.1	+41.6

Table 8.2 Standard Gibbs free energies of reaction at several temperatures

Reaction	ΔG_r^0 (400 K) (kJ mol^{-1})	ΔG_r^0 (800 K) (kJ mol^{-1})	ΔG_r^0 (1200 K) (kJ mol^{-1})
CH$_4$ + H$_2$O = CO + 3 H$_2$	+119.6	+23.2	−77.8
2 CH$_4$ + O2 = 2 CO + 4 H$_2$	−208.8	−360.8	−518.8
CO + H$_2$O = CO$_2$ + H$_2$	−24.2	−9.4	+3.4
CH$_4$ + 2 O$_2$ = CO$_2$ + 2 H$_2$O	−800.6	−800.6	−800.8
C + H$_2$O = CO + H$_2$	+77.6	+21.1	−36.2

Fig. 8.2 Temperature dependence of the standard Gibbs free energy of reactions involved in extraction of hydrogen from methane

8.2.2   The Reaction of Steam with Carbon

A number of countries, especially the USA and China, have large amounts of coal that they can use as an energy source. It is possible to react steam with solid carbon instead of with methane. In this case, the product is *syngas*, a mixture of CO and hydrogen. This reaction is included in Table 8.2. The resultant CO can then be reacted with steam in the water–gas shift reaction, just as is done in the case of steam reforming of methane.

The temperature dependence of the standard Gibbs free energies of these two reactions is shown in Fig. 8.3. As was the case with the reaction of steam with methane, the standard Gibbs free energy of the reaction of steam with carbon also decreases as the temperature increases. It can be seen that this reaction will only go forward if the temperature is above about 940 K. Again, the water–gas shift reaction proceeds only at lower temperatures.

The power plants that generate hydrogen from coal by this two-step process have overall efficiencies of about 34 %. But if they capture the effluent CO_2 from the water–gas reaction, the efficiencies can rise to above 40 %.

The thermal behavior of these various reactions can also be determined from data on the standard enthalpies of the species in these reactions. Such data for three temperatures are included in Table 8.3.

From these data for the various species the enthalpy (heat) effects of these reactions can be calculated. The results are included in Table 8.4, and plotted in Fig. 8.4.

It is seen that the water–gas reaction is endothermic, whereas the other reactions are both exothermic.

Fig. 8.3 Temperature dependence of the standard Gibbs free energies of reactions involved in reaction of steam with carbon

Table 8.3 Temperature dependence of the standard enthalpies of relevant species

Species	ΔH_f^0 (400 K) (kJ mol^{-1})	ΔH_f^0 (800 K) (kJ mol^{-1})	ΔH_f^0 (1200 K) (kJ mol^{-1})
CO	−110.1	−110.9	−113.2
H_2O	−242.8	−246.4	−249.0
CO_2	−393.6	−394.2	−395.0
CH_4	−78.0	−87.3	−91.5

Table 8.4 Standard enthalpies of reactions

Reaction	ΔH_r^0 (400 K) (kJ mol^{-1})	ΔH_r^0 (800 K) (kJ mol^{-1})	ΔH_r^0 (1200 K) (kJ mol^{-1})
$C + H_2O = CO + H_2$	+132.7	+135.5	+135.8
$CO + H_2O = CO_2 + H_2$	−40.7	−36.9	−32.8

If the hydrogen is to be used in a low temperature fuel cell, the gas mixture resulting from the water–gas reaction also generally undergoes a further step, called *methanation*, in which the remaining CO is converted back into methane, which is recycled. This is necessary because CO poisons the platinum catalysts that are typically used in such fuel cells.

It should be noted that all of these gas-phase reactions produce products that consist of mixtures of gases. The separation of hydrogen from the other gas

Fig. 8.4 Temperature dependence of the standard enthalpies of reactions

components involves the use of membranes that are permeable to hydrogen, but not to the other species. There is a need for the development of better selective membranes for this purpose.

8.2.3 Electrolytic Production of Hydrogen

The second major method for the production of hydrogen involves the electrolysis of water by imposing a voltage between two electrodes within it that exceeds its thermodynamic stability range. The result is the evolution of hydrogen gas at the negative electrode, and oxygen gas at the positive electrode. Both of these gases have a significant commercial value. It is easy to design systems to collect them separately.

Relatively pure hydrogen can be produced by the electrolysis of water, and this method appears to be attractive in the long run. There is an enormous amount of water on the earth, with the potential to supply an almost limitless supply of hydrogen. About 4 % of the hydrogen currently used in the world is produced by electrolysis.

The problem with this apparently attractive scenario is that the electrolytic decomposition of water to produce hydrogen is currently quite expensive. Thus it is only a feasible large-scale option where the cost of electricity is relatively low.

An advantage of water electrolysis to produce hydrogen is that, instead of requiring large central facilities, there can be distributed generation using smaller units. They can be located at places near where the hydrogen will be used in order to reduce storage and transportation costs.

The equilibrium (zero current) cell voltage E_{eq} required to decompose water can be found from the value of the standard Gibbs free energy of the formation of water, ΔG_f^0, a thermodynamic quantity. Data of this type for many materials over a range of temperatures can be found in [1]:

$$E = {}^{-\Delta G_f^0}/_{2F} \tag{8.5}$$

F is the Faraday constant, or 96,500 J per volt equivalent. The value 2 is present because the formation of water from hydrogen and oxygen involves two electronic charges (two equivalents):

$$H_2 + {}^1\!/_2 O_2 = H_2O \tag{8.6}$$

The value of ΔG_f^0 is -237.1 kJ/mol for water at 298 K, or 25 °C. Thus the voltage needed to exceed the stability range of water is above 1.23 V at that temperature. If there is a bit of a salt present to provide ionic conductivity, the application of sufficiently higher voltages causes current to flow, and the hydrogen and oxygen bubble off at the electrodes.

The reactions at the electrodes when water is being electrolyzed can be written as

$$H_2O + e^- = {}^1\!/_2 H_2 + OH^- \tag{8.7}$$

on the negative side, and

$$H_2O = 2e^- + {}^1\!/_2 O_2 + 2H^+ \tag{8.8}$$

on the positive side of the cell. Inside the water the result is the formation of OH^- ions at the negative electrode, and H^+ ions at the positive electrode.

But it must be realized that in order for these reactions to occur there must be ionic transport through the water. This would normally be either by the transport of H^+ ions to the other electrode in acid solutions, or the transport of OH^- ions in alkaline solutions.

But where does the water actually disappear? It cannot go away at both electrodes. If the water is acidic, it disappears from the solution at the positive electrode. The protons formed on that side, along with the O_2 gas, are transported to the negative side. When they reach the negative electrode, they react with the OH^- ions that are generated there to form H_2O. Thus there is a net loss of H_2O at the negative side in this case, not at the positive side.

The total cell voltage E_t when current is flowing is

$$E_t = E_{eq} + E_{neg} + E_{pos} + iR_{electrolyte} \tag{8.9}$$

where E_{neg} is the voltage loss at the negative electrode, E_{pos} is the voltage loss at the positive electrode, and $iR_{electrolyte}$ is the voltage loss across the water electrolyte.

This can also be written as

$$E_t = E_{eq} + iZ_{neg} + iZ_{pos} + iR_{electrolyte} \tag{8.10}$$

where i is the current, and the Z values are the impedances at the two electrodes. These can be thought of as resistances whose values depend upon the value of the current. On the other hand, the resistance of the electrolyte, $R_{electrolyte}$, is essentially current-independent.

The energy consumed during the electrolysis of the water is

$$\text{Energy} = iE, \tag{8.11}$$

and the energy efficiency is

$$\text{Efficiency} = {}^{iE_{eq}}\!/_{iE_t} \tag{8.12}$$

As mentioned already, although electrolytic production of hydrogen is significantly more expensive than obtaining it from natural gas, it has the advantage that the resulting gas can be of significantly greater purity. This can be especially important when the hydrogen is used in low-temperature fuel cells with polymeric solid electrolyte membranes. Even minor amounts of impurity species, such as CO, can cause problems by adsorbing on the surfaces of the platinum catalysts that are typically used to assist the conversion of H_2 molecules to H^+ ions and electrons at the negative electrode.

This requires additional treatment, and results in higher costs. The degree of CO adsorption on the catalyst surface decreases as the temperature is raised, so that this type of poisoning is not present in high-temperature fuel cells. That is one of the reasons for interest in the further development of this type of fuel cell.

Current low-temperature fuel cells actually operate at temperatures somewhat above ambient temperature in order to increase the overall kinetics.

As with any fuel, in addition to its acquisition, there must also be methods to transport hydrogen to the locations at which it will be used. Because such matters are typically not fully synchronized, there must also be methods for its storage.

Large commercial electrolyzers now produce hydrogen at a pressure of about 30 bar and a temperature of 80 °C, and have energy efficiencies of 80–90 %. The major source of loss is connected with the processes that take place at the positive electrode, where oxygen is evolved.

The decomposition of water requires that the oxygen species change from oxide ions in the water, that carry a charge of −2, to oxygen molecules, O_2, in which the charge on the oxide ions is effectively zero. An intermediate state, with peroxide O^{-1} ions, must be present on or near the catalyst surface.

High-surface-area nickel or nickel alloy electrodes are typically used on the oxygen side of water electrolysis cells. They have a surface layer of nickel oxy-hydroxide, NiOOH. The properties of this material, which plays an important role in a number of battery systems, will be discussed in a later chapter.

For kinetic reasons, electrolysis cells generally operate at about 2 V, which is substantially greater than the open circuit thermodynamic value for the decomposition of pure water of 1.23 V, and there is a continual effort to reduce the magnitude of this excess voltage.

8.2.4   Thermal Decomposition of Water to Produce Hydrogen

An additional method that can be used to produce hydrogen from water is to thermally decompose it by heating to a very high temperature. The stability of water relative to its two components, hydrogen and oxygen, is expressed in terms of its standard Gibbs free energy of the formation, ΔG_f^0, as mentioned above. This value is temperature-dependent, and can be expressed as

$$\Delta G_f^0 = \Delta H_f^0 - T\Delta S_f^0 \tag{8.13}$$

where ΔH_f^0 and ΔS_f^0 are the enthalpy and entropy values for the formation of water, respectively. The values of both of these terms are relatively independent of the temperature. Using values from [1], the temperature dependence of the standard Gibbs free energy of formation of water is shown in Fig. 8.5.

It is seen that water does not become thermodynamically unstable until the temperature reaches about 4300 K. Thus this method is hardly a practical alternative for splitting water into hydrogen and oxygen, for the required temperature is very high.

Fig. 8.5 Temperature dependence of the standard Gibbs free energy of formation of water

8.2.5   Chemical Extraction of Hydrogen from Water

It is also possible to produce hydrogen by chemically decomposing water. Many species form oxides when in contact with water. In general, however, the oxide that forms produces a protective surface layer that prevents further reaction with the water. There are a few exceptions in which a non-protective product is formed. One of these is lithium, which forms soluble LiOH. Another is aluminum, that can form soluble aluminum hydroxide, or hydrated oxide, instead of the simple oxide Al_2O_3, in basic solutions. This has led to the development of the so-called aluminum/air battery, which utilizes air as the reactant in the positive electrode.

The chemical reaction of aluminum with water can also be used to produce hydrogen, as was first shown by Cuomo and Woodall at the IBM Laboratory in 1968 [2]. This is accomplished by the use of a solution of aluminum in a low-melting-point liquid alloy based on gallium. The aluminum-gallium phase diagram is shown in Fig. 8.6. It can be seen that the melting point of gallium is 30 °C, and slightly below that, there is an eutectic reaction at 26 °C. There is a gradual increase in the solubility of aluminum in the liquid phase as the temperature increases, reaching complete miscibility at the melting point of aluminum at 660 °C. There is also a modest amount of solubility of gallium in solid aluminum all the way up to the melting point.

When this alloy is in contact with water at slightly above room temperature the dissolved aluminum will not form a protective skin, but reacts vigorously to form

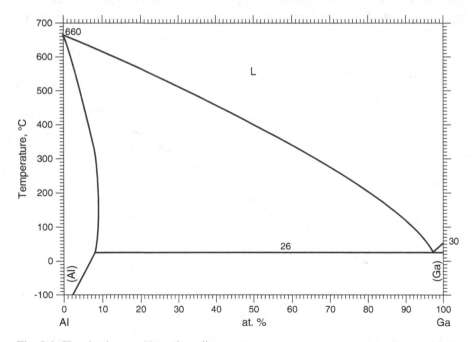

Fig. 8.6   The aluminum-gallium phase diagram

Fig. 8.7 Ternary phase stability diagram for the H–Al–O system

Table 8.5 Thermodynamic data related to the formation of Gibbsite

Species	ΔG_f^0 at 300 K (kJ mol)	ΔH_f^0 at 300 K (kJ mol)
H_2O	−236.8	−285.8
Al_2O_3	−1581.7	−1675.7
$Al_2O_3 \cdot 3\,H_2O$	−2308.4	−2586.6

hydrogen and finely dispersed aluminum oxide, which floats on the top of the liquid gallium alloy, according to the equation

$$Al + 3H_2O = 3/2\,H_2 + Al(OH)_3 + heat \tag{8.14}$$

Aluminum hydroxide is generally described as a hydrated version of Al_2O_3, called Gibbsite, $Al_2O_3 \cdot 3\,H_2O$. There is also a version with a slightly different crystal structure, called Bayerite.

The behavior of this ternary system can be understood by use of the ternary phase stability diagram for the H–Al–O system shown in Fig. 8.7. The principles and methods that can be used to draw and quantitatively interpret such figures will be described in Chap. 12.

The driving force for this reaction, assuming that all species are at unit activity can be calculated from the values of their Gibbs free energy of formation. Using the data in Table 8.5, this is found to be −443.8 kJ per mol of aluminum at 300 K. This value will vary somewhat with temperature, of course.

Similarly, the magnitude of the heat produced by this reaction at 300 °C can be calculated from the respective values of their standard enthalpies. This is found to be 435.9 kJ per mol of aluminum.

In addition to the heat released in this reaction, the hydrogen product can also be subsequently oxidized to produce further heat. This can be done either at the same location or elsewhere, of course.

This second reaction, the combustion of the hydrogen formed, is simply

$$^3/_2H_2 + {}^3/_4O_2 = {}^3/_2H_2O + \text{heat} \tag{8.15}$$

From the enthalpy data in Table 8.5 it can be found that the heat released in this second reaction is 428.7 kJ for the 1.5 mol of hydrogen produced at 300 °C. Thus the overall amount of heat produced from the aluminum is 864.6 kJ per mol of aluminum.

It is useful to consider why it is possible to produce hydrogen and Gibbsite by the reaction of aluminum with water, rather than forming Al_2O_3, which would block further reaction. This can be seen by looking at the ternary phase stability diagram in Fig. 8.7. As will be discussed in Chap. 12, the overall composition during the reaction of aluminum with water moves along the dashed line in Fig. 8.7, starting at the point representing the composition of water.

As aluminum is added, the overall composition first moves into the composition triangle that has H, H_2O and $Al(OH)_3$ at its corners. In this composition regime the activity of the aluminum is low. When the overall composition crosses the line between $Al(OH)_3$ and hydrogen and enters the triangle whose corners are Al_2O_3, $Al(OH)_3$ and H, the aluminum activity is much greater. Upon crossing into the triangle Al, Al_2O_3, and H, the aluminum activity becomes unity. The phase Al_2O_3 is present in either of the latter triangles, and will form, blocking further reaction.

Thus the condition for the absence of a protective layer of Al_2O_3 is that the aluminum activity must be smaller than a critical value. This is accomplished by dissolving it into the liquid alloy.

The use of this very powerful approach to understanding reactions in ternary phases will be discussed much more completely in Chap. 12, and used in both understanding and predicting the potentials and compositional ranges of electrodes in battery systems in subsequent chapters.

Since aluminum from the alloy is consumed in the reaction to form hydrogen and Gibbsite, it must be replenished in order for the reaction to continue. This can easily be done merely by putting solid aluminum in contact with the alloy. The protective Al_2O_3 skin is decomposed, and aluminum dissolved into the liquid alloy.

Subsequent to the discovery that dilute low temperature liquid aluminum alloys can react with water to produce hydrogen gas, it has been found that solid aluminum alloys containing some 5 % of a ternary Ga–In–Sn alloy will also react with water to form hydrogen and finely divided Gibbsite [3–6].

Although gallium dissolves somewhat in solid aluminum, this is essentially not true in the cases of indium and tin. On the other hand, aluminum dissolves slightly in a liquid solution containing all three of these elements. By use of compositions in this ternary system it is possible to obtain aluminum-containing liquids at ambient temperature.

When used as components in alloys that are mostly aluminum, this low-melting phase resides at the aluminum grain boundaries. This grain boundary region becomes liquid at temperatures not far above ambient. When such alloys are in contact with water, it is believed that aluminum in the liquid grain boundary phase reacts with water, forming hydrogen and Gibbsite. As the grain boundary aluminum is reacted, it is replenished by the solution of more of the adjacent solid aluminum phase of the alloy, allowing this reaction to proceed further.

The use of such solid aluminum alloys, instead of aluminum-containing low-melting metals, has significant practical advantages. Such alloys can be readily stored and transported, so that hydrogen and heat can be generated at any location where water is available. The weight of the water is about twice that of the alloy per unit of hydrogen produced.

There are a number of aspects of this method for the production of hydrogen that are quite favorable. It is not necessary that either the aluminum or the alloy constituents be of high purity. No additional materials or electrical power are required. It can also be used as a reserve system to generate hydrogen (and heat) only when needed.

Aluminum is relatively abundant and inexpensive. It is used in many products. This results in the accumulation of a large amount of scrap aluminum. This scrap is generally not reprocessed, for it is generally cheaper to produce more aluminum from the alumina (aluminum oxide) that is extracted from bauxite ore. The result is that there is an immense, and growing, amount of inexpensive scrap aluminum in the world. One estimate was that this currently amounts to about 400 billion kilograms [6].

The Al_2O_3 that is formed by this process can be recycled by electrolysis in the same way that aluminum is produced from its natural oxide ore, bauxite. The Ga-In-Sn alloy is not consumed in this reaction, so that it is completely recoverable.

The specific energy of this system is attractive, 1170 Wh/kg, counting the alloy and water weights, but not considering the weight of the container and any other system components.

However, because of the weights of the aluminum and water, the weight efficiency of the production of hydrogen is not especially attractive—only 3.6 % hydrogen by weight, so such a system is not interesting for on-board vehicle use.

8.2.6 Additional Approaches to the Production of Hydrogen

Another option that has been pursued somewhat has involved the processing hydrocarbons through an electric arc, whereby they decompose to form carbon and hydrogen at temperatures over 1600 °C.

In addition, there has been a considerable amount of research on the possibility of the use of *photoelectrolysis*, in which solar energy is used to decompose water directly, or *photoelectrochemical cells*, in which solar energy is employed to reduce the necessary applied voltage in electrically-driven cells. Although these are very active areas of investigation [7–11], they will not be discussed here.

8.3   Governmental Promotion of the Use of Hydrogen

In 2003 the President of the USA announced a major Hydrogen Fuel Initiative to accelerate the research and development of technologies needed to support hydrogen-powered fuel cells for use in transportation and electricity generation. The underlying objective of this program was to decrease air pollution resulting from the use of petroleum. This program resulted in a significant increase in the amount of research and development on both hydrogen-based fuel cells and on-vehicle hydrogen storage. The proposed budget was $ 1.2 billion over a span of five years, to be used to develop hydrogen production, delivery, storage and fuel cell technologies to enable the automobile and energy industries to commercialize fuel cell vehicles and the hydrogen fuel infrastructure. The general assumption was that hydrogen-powered vehicles should have performance equal or superior to current gasoline-consuming vehicles.

One aspect of this was the achievement of a practical driving range of at least 300 miles for light-duty vehicles. This resulted in the development of targets for the performance of on-board hydrogen storage systems in terms of weight, volume and cost, as well as operating parameters. The more important targets for 2010 are shown in Table 8.6. It has been assumed that meeting these goals would make it possible for some smaller and lighter vehicles to achieve the desired performance. Even more challenging targets have been proposed for the year 2015 that would be appropriate for the full range of light-duty vehicles in North America.

In arriving at these targets it was assumed that fuel cell power plants would have a factor of 2.5–3 times greater efficiencies than current gasoline-powered vehicles. It was also assumed that 5–13 kg of hydrogen would be necessary for fuel cell-driven vehicles, and that 1 kg of hydrogen can contribute about the same amount of energy as 1 gal of gasoline.

Consideration of the magnitudes of some of these parameters is instructive. For example, if hydrogen is stored on board in metal hydrides instead of in high-pressure tanks, care must be taken when recharging them because the reaction of hydrogen with metal hydride materials is typically highly exothermic.

Table 8.6   US Dept. of Energy hydrogen storage system performance targets for year 2010

System parameters	Year 2010 targets
Specific energy from H_2	2 kWh/kg or 6 wt% H_2
Energy density from H_2	1.5 kW/l
Operating ambient temperature	−30 to 50 °C
H_2 delivery temperature	−40 to 85 °C
Cycle life (25–100 %)	1,000 cycles
Minimum delivery pressure	4 atm (fuel cell) or 35 atm (ICE)
Recharging time (for 5 kg H2)	3 min
Minimum flow rate	0.02 g/s per kW

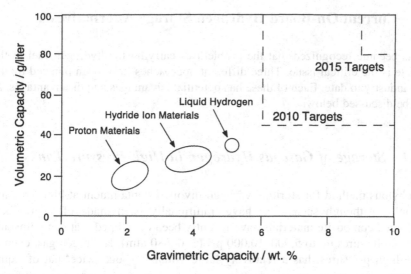

Fig. 8.8 Parameters of current hydrogen storage technologies, related to DOE targets

A material that has a reaction enthalpy of 30 kJ/mol of hydrogen will generate 400 kW of heat if it reacts with 8 kg of hydrogen in 5 min. This heat would obviously have to be removed somehow.

In addition, a fuel cell operating at 80 kW requires a hydrogen flow rate of 1.6 g/s. Therefore the generation of this hydrogen from metal hydrides, which is discussed below, will require an equivalent amount of heat input. As a result, system thermal management can become critical.

These Department of Energy targets are very ambitious, and will require materials with properties significantly better than those presently available, as well as innovative system designs. A rough picture of the current status and future targets is given in Fig. 8.8.

As will be discussed in Chap. 21, it is likely that serious attention will have to be given to compromises that are similar to some of those presently being considered for battery-propelled vehicles, such as plug-in hybrids.

As mentioned above, one of the problems that could limit the rapid expansion of the use of hydrogen-propelled vehicles is the current lack of a large-scale hydrogen support infrastructure. Another is the amount of hydrogen that would be needed if a significant number of vehicles were to be converted to hydrogen propulsion. It was estimated [12] that one million fuel cell-propelled vehicles would consume about 0.4 million tons of hydrogen per year. The magnitude of this becomes evident when it is compared with the fact that the total hydrogen production in the USA is currently about ten million tons. It should also be remembered that the production of hydrogen from natural gas causes about as much pollution as burning it directly.

8.4   Current On-Board Hydrogen Storage Alternatives

It has become recognized that the problem of carrying the hydrogen fuel in the vehicle is the critical issue. Three different approaches have been pursued by the auto industry to date. Each of these has potential advantages and disadvantages, as will be discussed below.

8.4.1   Storage of Gaseous Hydrogen in High-Pressure Tanks

An obvious method for storing hydrogen involves containment at high pressure in tanks. Although such tanks have traditionally been made of steel, fiber-reinforced composite materials have recently been developed that can withstand internal pressures up to 5,000–10,000 psi (340–680 atm). Hydrogen gas, even at these high pressures, has a relatively large volume, about twice that of liquid hydrogen.

This topic was discussed briefly in Chap. 6, where it was pointed out that, according to the ideal gas law, the amount of gas that is stored in a fixed-volume tank is proportional to the pressure. In addition, attention must be given to the thermal effects. Rapid charging of a tank with hydrogen is almost adiabatic (without heat exchange with the surroundings), so that a large amount of heat can be generated, giving rise to high temperatures. If the discharge of the gas from the tank into a fuel cell, for example, is relatively slow, so that heat transfer to the environment can occur, this reverse process is not so adiabatic, and the resulting cooling may not be important.

8.4.2   Storage of Liquid Hydrogen in Insulated Tanks

Hydrogen can be liquefied by cooling. Its boiling point at 1 atm is 20.3 K.

It can be contained and transported in liquid form in thermally-insulated containers.

Hydrogen gas can be readily obtained by applying heat so as to raise the local temperature above the liquid/gas transition. Demonstration vehicles were constructed by the auto firm BMW some years ago in which this method was used for the storage of hydrogen. The hydrogen was combusted in a slightly modified internal combustion engine. A significant disadvantage in this approach is that the process of liquefaction requires some 30–40 % of the final energy content of the hydrogen.

8.4.3   Storage of Hydrogen as Protons in Solids: Metal Hydrides

The third approach is to reversibly absorb hydrogen in solid metal-hydrogen compounds, called *metal hydrides*.

Hydrogen reversibly dissolves in a number of solids, and in some cases to surprisingly high concentrations. The reason why this is possible is that the hydrogen is present in them as protons (H^+) ions, not as hydrogen atoms or hydride (H^{-1}) ions. The electrical charge in solids must always be balanced, so the charged due to the presence of protons is balanced by the presence of an equal number of extra electrons. Thus these materials are always good electronic conductors.

The capacities of such materials depend upon the amount of hydrogen that they can absorb in their crystal structures. This is directly analogous to the absorption of lithium into the electrode materials commonly used in lithium batteries, a topic that is discussed in Chap. 18.

Protons are very small, and easily fit into the intestacies (spaces) between the other atoms present. An example of this is the metal alloy $LaNi_5$. It can absorb up to six hydrogen atoms per mol, to form $LaNi_5H_6$. Another example is the alloy FeTi, which can absorb two hydrogen atoms per mol, forming $FeTiH_2$. The hydrogen densities of these materials, which can absorb up to one hydrogen atom per metal atom, are 5.5×10^{22} and 5.8×10^{22} atoms of hydrogen per cm^3, respectively. These are greater than the density of liquid hydrogen, which is 4.2×10^{22} atoms cm^{-3}. Thus the storage of hydrogen in such materials is very attractive from a volumetric standpoint.

Despite the fact that the hydrogen is present as positively charged ions (protons) such hydrogen-containing metallic alloys are generally called metal hydrides. A number of them are widely used as negative electrode reactants in aqueous electrolyte batteries. The most common examples are the hydride/"nickel" cells that are used in many small applications, as well as the battery component of some hybrid automobiles. One of their especially attractive properties is the high rate at which they can be charged and discharged. This topic will be discussed in some detail in Chap. 16.

Such metal alloy hydrides are certainly candidates for use for hydrogen storage in vehicles. Their major disadvantage, however, is that the presently known materials are too heavy, and do not meet the weight requirements for use in vehicles. As an example, the solid hydride materials that are now employed in the ubiquitous small hydride/"nickel" batteries used in many electronic devices store only about 2–3 % hydrogen by weight. This is far from the 6 % target of the Department of Energy mentioned earlier. There is also some concern about the cost and large-scale availability of some such materials.

Vehicles with hydrogen-powered internal combustion engines, in which the hydrogen was stored in metal hydrides, were demonstrated some years ago by Mercedes Benz. In addition to the matter of cost, the primary problem with hydrogen storage in metal hydrides is the relatively small amount of hydrogen that they can store per unit weight. Brief mention was also made in Chap. 1 of the use of such metal hydrides as hydrogen storage media in fuel cell-propelled submarines. In that case, the volume of the hydrogen-absorbing material, a titanium-manganese alloy, is important, but the weight is not.

8.5 Other Approaches to Hydrogen Storage

In addition to the approaches that have been followed to date in the automobile industry, there are some other approaches to hydrogen storage that might deserve some consideration, for either the vehicle, or other, applications. Two of these are described briefly below.

8.5.1 Hydrogen from the Decomposition of Materials Containing Hydride Anions

Another alternative strategy that is being employed for hydrogen storage involves materials in which the hydrogen is present in the form of hydride (H^-) ions, instead of protons. Whereas protons are very small, and can readily dissolve in a number of metal alloys, hydride ions are large, with an ionic radius of 146 pm, which is close to the size of oxide (O^{-2}) ions. Materials containing hydride ions are quite ionic, rather than metallic, in character, and thus have low values of electronic conductivity. The large hydride ions typically have very low mobilities within their crystal structures.

There are two families of materials containing hydride ions that have received a lot of attention as possible hydrogen storage media. One of these is the *borohydride* family, which can be represented by the general formula $M^+BH_4^-$, where the species M^+ can be Li^+, Na^+, or K^+ or NH_4^+. $LiBH_4$ can theoretically store 13.9 wt% hydrogen, whereas NaBH4 contains 7.9 wt% hydrogen.

There is an analogous family of materials, in which the boron is replaced by aluminum. These materials are called *alanates*. A further group of materials are the boranes, which have the general formula NH_nBH_n, where n can range from 1 to 4. One member of this group is amine borohydride, NH_4BH_4, which has a hydrogen mass ratio of 24 % has received attention [12–14]. This material decomposes in several stages as it is heated, giving off about 6 % of its mass in each step. This is shown in Table 8.7.

Although their hydrogen capacities are quite high, there is a serious practical problem with materials of this general type, for they are not reversible. Hydrogen cannot be simply reacted with them so as to return them to their initial state, as can be done with the common proton-containing metal hydride materials. There is some

Table 8.7 Multi-step decomposition of amine borohydride

Reaction	Wt % H_2 change	Temperature/°C
$NH_4BH_4 = NH_3BH_3 + H_2$	−6.1	<25
$NH_3BH_3 = NH_2BH_2 + H_2$	−6.5	<120
$NH_2BH_2 = NHBH + H_2$	−6.9	<155
$NHBH = BN + H_2$	−7.3	>500

hope that a method might be found for this, however, based on some recent work on titanium-based catalysts [15].

If that cannot be done, they must undergo a chemical reconstitution. This will require their removal from the vehicle and replacement by a new chemical charge. The chemical process to regenerate such hydride-ion materials will probably most effectively be done in large central chemical plants. But then there must also be a system for the transportation of the spent and renewed materials. This requirement for external chemical regeneration constitutes a major disadvantage of this approach.

A similar chemical regeneration requirement led to the demise of the highly touted proposal to use zinc/air batteries for vehicle propulsion some years ago. Although the reaction between zinc and oxygen to form ZnO provides a lot of energy per unit weight, a method for this cell to be electrically recharged was not known. The result was that the discharged battery electrodes had to be shipped to a chemical plant for the conversion of the ZnO product back to elemental zinc. The shipping and chemical processes used to reconstitute the zinc electrodes combined to make this approach unfeasible.

The major problem has traditionally been the ready formation of zinc dendrites when zinc is electrochemically recharged. There has been some recent work aimed at avoiding this problem that involves the use of a solid polymer electrolyte, rather than the conventional aqueous electrolyte.

8.5.2 Ammonia and Related Materials as Hydrogen Storage Media

Another possible approach to consider is the use of ammonia, NH_3, which is 17 % hydrogen by weight. It is possible to thermally decompose ammonia at modest temperatures. This is seen from the data shown in Fig. 8.9, where it can be seen that, at one atmosphere pressure, ammonia decomposes into its elements at about 460 K (187 °C).

Consideration is also being given to the use of more complex materials that can decompose to form NH_3, which can then be further decomposed to provide hydrogen. Metal ammines, in which ammonia is coordinated to a metal ion, are one type of such materials. An example is $Mg(NH_3)_6Cl_2$ [16].

Fig. 8.9 Temperature dependence of the standard Gibbs free energy of formation of ammonia, NH_3

Table 8.8 Standard Gibbs free energy of formation data for phases in the Li–H–N system

Phase	ΔG_f^0 at 298 K (kJ mol^{-1})
Li_3N	−122.2
LiH	−70.0
NH_3	−16.4
Li_2NH	−169.9
$LiNH_2$	−140.6
H_2	0

Another concept is the use of a material containing hydrogen that might react with another material and give off hydrogen as one of the reaction products. One example of this type that has been proposed is lithium amide, $LiNH_2$, with the expectation that it would react with LiH to form lithium imide, Li_2NH and hydrogen.

The feasibility of this concept can be evaluated from thermodynamic data related to the phases concerned. Values of the standard Gibbs free energy of formation for the pertinent phases in the lithium-hydrogen-nitrogen system at 298 K [17] are included in Table 8.8.

The proposed reaction can be written as

$$LiNH_2 + LiH = Li_2NH + H_2 \tag{8.16}$$

The sum of the ΔG_f^0 values of the phases on the left is -210.6 kJ mol^{-1}, and of the phases on the right is -169.9 kJ mol^{-1}. Therefore this reaction will tend to go to the left, not the right. Thus this concept makes no sense at that temperature.

8.5.3   Storage of Hydrogen in Reversible Organic Liquids

In addition to solid materials that absorb hydrogen as protons to become metal hydrides, or already contain hydrogen in the form of hydride anions, it is reasonable to also consider the use of liquids. Many organic liquids have large, and often variable, amounts of hydrogen present in their structures.

If everything else were the same, one potential advantage of the use of liquids as hydrogen-carriers is the assumption that such materials could be inexpensively stored in tanks, and pumping them in and out should be no great problem.

In some countries, such as Switzerland, that have surplus hydropower during part of the year, a large-scale method for the storage of some of this energy for use at other times of the year could be valuable. One possible method for such large-scale and long-term energy storage by the use of organic hydrides was proposed some time ago [18]. This concept involves the electrolytic production of hydrogen when electricity is readily available and inexpensive, and reacting it with a simple organic molecule to produce a product containing more hydrogen. The species with the higher hydrogen content could later be treated such that the hydrogen is released.

The simplest example would be the conversion of benzene (C_6H_6) to form cyclohexane (C_6H_{12}). Benzene is a simple aromatic molecule, in which each of the carbon atoms in the 6-member ring is bonded to one hydrogen atom. In cyclohexane each carbon has two hydrogen neighbors. This possibility cannot be seriously considered for actual use, however, because benzine is considered to be a carcinogen.

Thus it is preferable to use toluene (C_7H_8), and to produce methylcyclohexane (C_7H_{14}) by the reversible addition of six hydrogen atoms per molecule. Both of these liquids can readily be stored in large tanks. They are inexpensive and convenient liquids that are easy to transport and store, with freezing temperatures that are convenient, 178 K and 146.4 K, respectively.

It can readily be seen that the weight of the hydrogen stored in this case is 6.5 % of the weight of its carrier, the toluene. This is substantially better than the comparable values for the solid metal hydrides.

The volumetric density of hydrogen is also important. Methylcyclohexane contains 47.4 g of H_2 per liter. The hydrogen density in gaseous hydrogen varies with the pressure and temperature according to the ideal gas equation, of course. At 200 bar it is only 18 g H_2 per liter.

Both the hydrogen-addition and the hydrogen-deletion processes require catalysts. The conversion of toluene to methylcyclohexane, a well-known large-scale

industrial process, is exothermic, and the reverse reaction, which takes place at
400 °C, is endothermic, so heat has to be supplied at a relatively high temperature
when extracting the hydrogen. This hydrogenation cycle has a round trip efficiency
of about 80 %.

An example of the use of this concept [19] involved installation in a 17-ton truck
in Switzerland. The hydrogen obtained from the onboard catalytic splitting of
methylcyclohexane was use as the fuel for its 150 kW internal combustion engine.
Hydrogen was injected at 10 bar pressure, and resulted in an engine efficiency of
32 %. In an analysis of costs some years ago [20], this scheme for the storage and
use of energy was found to be economically competitive with other carbon-free
large-scale energy storage methods, as well as with the cost of the construction of
additional hydro-power facilities at that time.

Other covalent organic materials with larger molecular structures can also be
considered for this purpose, including perhydrofluorene and several species in the
carbazole family [21].

An additional interesting feature of the reversible liquid hydrogen systems has to
do with their potential use for the transmission of energy over long distances using
simple, low-cost pipelines. This has been pioneered as a method to get energy to
geographically remote areas in Russia and Brazil, and is illustrated schematically in
Fig. 8.10.

Electrical power transmission through wires involves both resistive and capacitive losses that increase linearly with distance, reducing the overall efficiency of
the electrical transmission system. These losses can be appreciable if energy is sent
over long distances. On the other hand, the losses that are involved in the transport
of liquids through pipelines are relatively small.

As a result, there is a critical distance over which energy can be transmitted more
economically be fluids in pipelines than by electrical lines. In one study, the
breakeven point was at 1400 km. At longer distances the use of the reversible
liquid cycle is more effective.

Fig. 8.10 Energy transmission efficiency vs. distance

8.6 The Question of Safety

Almost everyone knows that hydrogen readily burns in air, and with a flame that is almost colorless. Those working with hydrogen are always cautioned to keep it away from open flames. Attention to this potential problem rapidly became widespread after the widely publicized burning of the large Hindenburg zeppelin in New Jersey in 1937.

It is important to give this potential safety problem some attention. The ignition temperature of hydrogen in air is 585 °C at one atmosphere pressure. Thus there is no problem with having a mixture of these gases at ambient temperature. At elevated temperatures the composition range within which flammability is possible is between 4 and 75 % hydrogen by volume.

The density of hydrogen gas is low, only 0.08, so that it rises rapidly in air. This reduces the possibility of reaching the flammability compositional range in open spaces. In enclosed spaces hydrogen tends to accumulate at the highest locations, i.e., near the ceilings of rooms.

The safety question was raised some years ago, when BMW was demonstrating a hydrogen-powered demonstration auto in which the hydrogen was carried as a liquid in an insulated tank. It was contended that any hydrogen that got free would rise so rapidly that its possible combustion would not be a threat.

References

1. Barin I (1995) Thermochemical Data of Pure Substances, 3rd edn. VCH, Weinheim, Published Online 24 Apr 2008. ISBN 9783527619829783527619825
2. Cuomo JJ, Woodall JM (1982) US patent 4,358,291, November 9 1982
3. Woodall JM, Ziebarth J, Allen CR (2007) Proc. 2nd Energy Nanotechnology International Conference. Santa Clara, CA, 5 Sept 2007
4. Woodall JM, Ziebarth JT, Allen CR, Sherman DM, Jeon J, Choi G (2008) Proc Hydrogen 2008, Feb 2008
5. Woodall JM, Ziebarth J, Allen CR, Jeon J, Choi G, Kramer R (2008) Clean Technology. 1 June 2008
6. Woodall JM (2009) Presentation at the Electrochemical Society Meeting, San Francisco, 26 May 2009
7. Graetzl M (2001) Nature 414:15
8. Crabtree GW, Dresselhaus MS, Buchanan MV (2004) Phys Today 57:39
9. Graetzl M (2005) Inorg Chem 44:6841
10. Crabtree GW, Dresselhaus MS (2008) MRS Bull 33:421
11. US Department of Energy (2002) National Hydrogen Energy Roadmap. US Department of Energy, November, 2002. http://www.hydrogen.energy.gov/pdfs/national_h2_roadmap.pdf
12. Gutowska A, Li L, Shin Y, Wang CM, Li XS, Linehan JC, Smith RS, Kay BD, Schmid B, Shaw W, Gutowski M, Autrey T (2005) Angew Chem Int 44, 3578
13. Matus MH, Anderson KD, Camaioni DM, Autrey ST, Dixon DA (2007) J Phys Chem 111:4411
14. Yoon CW, Sneddon LG (2006) J Am Chem Soc 128:13992
15. Muller E, Sutter E, Zahl P, Ciobanu CV, Suttera P (2007) Appl Phys Lett 90:151917

16. Christensen CH, Johannessen T, Soerensen RZ, Norskov JK (2006) Catal Today 111:140
17. Boukamp BA, Huggins RA (1979) Phys Lett 72A:464
18. Taube M, Taube P (1980) Proc of 3rd World Hydrogen Energy Conference, Tokyo
19. Taube M, Rippin D, Knecht W, Milisavijevic B, Hakimifard D (1984) Hydrogen Energy
    Progress, 5th edn. In: Veziroglu TN, Taylor JB. Pergamon. p 1341
20. Scherer GWH, Newson E, Wokaun A (1999) J Hydr Energy 24:1157
21. Cooper A, Scott A, Fowler D, Wilhelm F, Monk V, Cheng H, Pez G (2008) Presentation at
    2008 DOE Hydrogen Program Meeting, June, 2008. http://www.hydrogen.energy.gov/pdfs/
    review08/stp_25_Cooper.pdf

Chapter 9
Introduction to Electrochemical Energy Storage

9.1 Introduction

Among the various methods that can be used for the storage of energy that are discussed in this text, electrochemical methods, involving what are generally called *batteries*, deserve the most attention. They can be used for a very wide range of applications, from assisting the very large scale electrical grid down to tiny portable devices used for many purposes. Battery-powered computers, phones, music players, etc. are everywhere, and one of the currently hot topics involves the use of batteries in the propulsion of vehicles, hybrid autos, plug-in hybrids, and fully electric types.

Many students are put off from discussions of electrochemical systems because of unfamiliarity with electrochemistry. It will be shown here that one can understand the major phenomena and issues in electrochemical systems without considering their truly electrochemical features in detail. As an example, it will be shown that the driving forces of electrochemical cells are related to the driving forces between the electrically neutral components in the electrodes. Electrochemical considerations only come into play in certain features of their mechanisms.

Electrochemical energy storage involves the conversion, or *transduction*, of chemical energy into electrical energy, and vice versa. In order to understand how this works, it is first necessary to consider the *driving forces* that cause electrochemical transduction in electrochemical cells as well as the major types of *reaction mechanisms* that can occur.

This is followed by a brief description of the *important practical parameters* that are used to describe the behavior of electrochemical cells, and how the basic properties of such electrochemical systems can be modeled by the use of *simple equivalent electrical circuits*.

Also included in this chapter is a brief discussion of the principles that determine the major properties of electrochemical cells, their voltages, and their capacities.

© Springer International Publishing Switzerland 2016
R.A. Huggins, *Energy Storage*, DOI 10.1007/978-3-319-21239-5_9

9.2   Simple Chemical and Electrochemical Reactions

Consider a simple *chemical reaction* between two metallic materials A and B, which react to form an electronically conducting product AB. As discussed in Chap. 4, this can be represented simply by the relation

$$A + B = AB \tag{9.1}$$

The driving force for this reaction is the difference in the values of the *standard Gibbs free energy* of the products, only AB in this case, and the standard Gibbs free energies of the reactants, A and B.

If A and B are simple elements, this is called a *formation reaction*, and since the standard Gibbs free energy of formation of elements is zero, the value of the Gibbs free energy change that results per mol of the reaction is simply the *Gibbs free energy of formation* per mol of AB, that is:

$$\Delta G_r^{\circ} = \Delta G_f^{\circ}(AB) \tag{9.2}$$

Values of this parameter for many materials can be found in a number of sources, e.g. [1].

While the morphology of such a reaction can take a number of forms, consider a simple one-dimension case in which the reactants are placed in direct contact and the product phase AB forms between them. The time sequence of the *evolution of the microstructure* during such a reaction is shown schematically in Fig. 9.1. Later times are at the bottom.

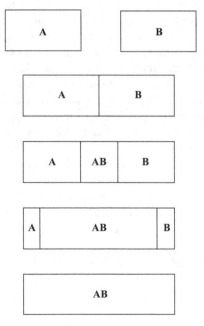

Fig. 9.1 Simple schematic model of the chemical reaction of A and B to form AB, indicating how the microstructure of the system varies with time

It is obvious that in order for the reaction product phase AB to grow, atoms of either A or B must move (*diffuse*) through it to reach its other side to come into contact with the other reactant. If, for example, A moves through the AB phase to the B side, additional AB will form at the AB/B interface. Since some B is consumed, the AB/B interface will move to the right. Also, since the amount of A on the A side has decreased, the A/AB interface will likewise move to the left. The AB will grow in width in the middle. One should note that the same thing will happen in the case that the species B, rather than the species A, moves through the AB phase in this process. There are experimental ways in which one can determine the identity of the moving species in this type of reaction, but it is not necessary to be concerned with them here.

Now suppose that this process occurs by an *electrochemical mechanism*. The time dependence of the microstructure in this case is shown schematically in Fig. 9.2. As in the chemical reaction case, the product AB must form as the result of a reaction between the reactants A and B. But there is an additional phase present in the system, an electrolyte.

Fig. 9.2 Simple schematic model of the time evolution of the microstructure during the electro-chemical reaction of A and B to form AB, a mixed conductor. In this case it is assumed that A^+ ions are the predominant ionic species in the electrolyte. To simplify the figure, the external electronic path is shown only at the start of the reaction

The *function of the electrolyte* is to *act as a filter* that allows the passage of ionic, but not electronic species. This means that the electrolyte contains ions of either A or B, or both, and is an *electronic insulator*.

But the reaction between A and B involves *electrically neutral atoms*, not just ions. This means that in order for the reaction to proceed there must be a path whereby electrons can also move through the system. This is typically an external electrical circuit that connects A and B. In the case that it is A that is transported in the system, and that the electrolyte contains A^+ ions, negatively charged electrons, e^-, must pass through the external circuit in equal numbers, or at an equal rate, to match the charge flux due to the passage of A^+ ions through the electrolyte to the other side.

During an electrochemical discharge reaction of the type illustrated in Fig. 9.2, the reaction at the interface between the phase A and the electrolyte can be written as

$$A = A^+ + e^- \tag{9.3}$$

with the A^+ ions moving into the electrolyte phase and the electrons entering the external circuit through a *current collector*. At the same time there will be a corresponding reaction on the other side of the electrolyte

$$A^+ + e^- = A \tag{9.4}$$

with ions arriving at the interface from the electrolyte and electrons coming to the interface from the external circuit through the electronic current collector. The result is to deposit A atoms onto the adjacent solid phase AB. The result is that the A/electrolyte interface and the electrolyte/AB interface both move incrementally to the left in Fig. 9.2. There must be *interdiffusion* of A and B atoms within the phase AB so that its surface does not have only A atoms. In addition, this phase must be an electronic conductor.

The fact that the overall reaction is between neutral species, and that this requires the concurrent motion of either A or B ions through the electrolyte, and electrons through external circuit, has several important consequences. One is that if flow in either the electronic path or the ionic path is impeded, the whole reaction must stop. For example, if the external electrical circuit is opened so that no electrons can flow through it, no ions can flow through the electrolyte, and the reaction halts. Likewise, if the flow of ions in the electrolyte is impeded—for example, by the presence of some material with a very high resistance for the moving ionic species, or a loss of contact between the electrolyte and the two materials on its sides—there will be no electronic current in the external circuit.

When the electronic circuit is open, and there is no current flowing, there must be a force balance operating upon the electrically charged ions in the electrolyte. A *chemical driving force* upon the mobile ionic species within the electrolyte in one direction is simply balanced by *an electrostatic driving force* in the opposite direction.

The *chemical driving force* across the cell is due to the difference in the chemical potentials of its two electrodes. It can be expressed as the *standard Gibbs free energy change per mol of reaction*, ΔG_r°. This is determined by the difference in the *standard Gibbs free energies of formation* of the products and the reactants in the *virtual chemical reaction* that would occur if the *electrically neutral* materials in the two electrodes were to react chemically. It makes no difference that the reaction actually happens by the transport of ions and electrons across the electrochemical system from one electrode to the other.

The electrostatic energy per mol of an electrically charged species is $-zFE$, where E is the voltage between the electrodes, and z is the *charge number* of the mobile ionic species involved in the virtual reaction. The charge number is the number of elementary charges that they transport. F is the *Faraday constant* (96,500 C per equivalent). An *equivalent* is *Avogadro's number* (1 mol) of electronic charges.

The balance between the chemical and electrical forces upon the ions under open circuit conditions can thus be simply expressed as a chemical energy–electrostatic energy balance

$$\Delta G_r^\circ = -zFE \qquad\qquad\qquad (9.5)$$

Here the value of ΔG_r° is in Joules per mol of reaction, as 1 J is the product of 1 C and 1 V.

Thus this is an interesting situation in which a chemical reaction between neutral species in the electrodes determines the forces upon charged particles in the electrolyte in the interior of an electrochemical system.

If it is assumed that the electrodes on the two sides of the electrolyte are good electronic conductors, there is an externally measurable voltage E between the points where the external electronic circuit contacts the two electrodes. As the result of this voltage, electrical work can be done by the passage of electrons in an external electric circuit if ionic current travels through the electrolyte inside the cell.

Thus this simple electrochemical cell can act as a *transducer* between chemical and electrical quantities; forces, fluxes, and energy. In the ideal case, the chemical energy reduction due to the chemical reaction that takes place between A and B to form mixed-conducting AB is just compensated by the electrical energy transferred to the external electronic circuit.

The flow of both internal ionic species and external electrons can be reversed if a voltage is imposed in the electronic path in the opposite direction that is larger than the voltage that is the result of the driving force of the chemical reaction. Since this causes current to flow in the reverse direction, electrical energy will be consumed and the chemical energy inside the system will increase. This is what occurs when an electrochemical system is recharged.

From these considerations it is obvious that it is not important whether the ionic species are related to element A or to element B. However, the answer to this question will influence the configuration of the cell. The example illustrated schematically in Fig. 9.2 deals with the case in which there are predominantly A^+ ions in the electrolyte. The chemical reaction proceeds by the transport of A^+ ions across the electrolyte and electrons in the external circuit from the left (A) side of the cell to the right-hand side. This involves two electrochemical reactions. On the left side, A atoms are converted to A^+ ions and electrons at the A/electrolyte interface. The electrons travel back through the metallic A and go out into the external electronic circuit. The reverse electrochemical reaction takes place on the other side of the cell. A^+ ions from the electrolyte combine with electrons that have come through the external circuit to form neutral A at the electrolyte/AB interface.

As before, the physical locations of the interfaces, the A/electrolyte interface, the electrolyte/AB interface and the AB/B interface, will move with time as the amounts of the various species in the different phases vary with the extent of the reaction.

It must be recognized that the reaction product AB will not form unless there is a mechanism that allows the newly-arrived A to react with B atoms to form AB. Thus the transport of either A or B atoms within the AB product phase is necessary in this case, as it was in the chemical reaction case illustrated in Fig. 9.1 above. If this did not happen, pure A would be deposited at the right hand electrolyte interface. The chemical composition on both sides of the electrolyte would then be the same, and there would be no driving force to cause further transport of ionic species through the electrolyte, and thus no external voltage.

If B^+ ions, rather than A^+ ions, are present in the electrolyte, so that B species can flow from right to left, the direction of electron flow, and thus the voltage polarity, in the external circuit will be opposite from that discussed above, and the reaction product will form on the left side, rather than on the right side.

It is also possible, of course, that the ions in the electrolyte are negatively charged. In that case, the direction of electron flow in the external circuit will be in the opposite direction.

In any case, it is important to realize that the basic driving force in an electrochemical cell is a *chemical reaction of neutral species* to form an *electrically neutral product*. This is why one can use standard chemical thermodynamic data to understand the equilibrium (no current, or open circuit) potentials and voltages in electrochemical cells.

For any given chemical reaction, the open circuit voltage is independent of the identity of the species in the electrolyte and the details of the reactions that take place at the electrode/electrolyte interfaces.

The situation becomes different if one considers the kinetic behavior of electrochemical cells, for then one has to be concerned with phenomena at all of the interfaces, as well as in the electrodes, the electrolyte, and the external circuit. Such matters will be discussed in some detail later.

9.3 Major Types of Reaction Mechanisms in Electrochemical Cells

As discussed above, the operation of electrochemical cells involves the transport of neutral chemical species into and out of the electrodes, their ionic parts move through the electrolyte, and the charge-balancing electrons move through the external electrical circuit. In many, but not all, cases, this results in changes in the chemical constitution of electrodes, i.e., the amounts and chemical compositions of the phases present.

The result is that the microstructure of one or more of the electrode materials gets significantly changed, or *reconstituted*. There are a number of important chemical, and thus possible electrochemical, reactions in which some phases grow and others disappear.

Reactions in which there is a change in the identity or amounts of the phases present are designated as *reconstitution reactions*.

Phase diagrams are useful *thinking tools* to help understand these types of phenomena. As discussed in Chap. 4, they are graphical representations that indicate the phases and their compositions that are present in a materials system under equilibrium conditions, and were often called *constitution diagrams* in the past. In Chap. 4 the discussion was focused upon phenomena that occur as the result of changes in the temperature. Electrochemical systems, on the other hand, generally operate at a constant temperature, i.e., isothermally. This involves consideration of what occurs as the composition moves horizontally, rather than vertically, across phase diagrams.

Two major types of reconstitution reactions that are relevant to electrochemical systems will be briefly mentioned here, *formation reactions* and *displacement reactions*. This will be followed by an introduction to *insertion reactions*, which play a major role in the operation of electrodes in a number of important modern battery systems.

9.3.1 Formation Reactions

The simple example that was discussed earlier in this chapter, represented by the equation

$$A + B = AB \tag{9.6}$$

is a *formation reaction*, in which a new phase AB is formed in one of the electrodes from its atomic constituents. This can result from the transport of one of the elements, e.g., A, passing across an electrochemical cell through the electrolyte from one electrode to react with the other component in the other electrode. Since

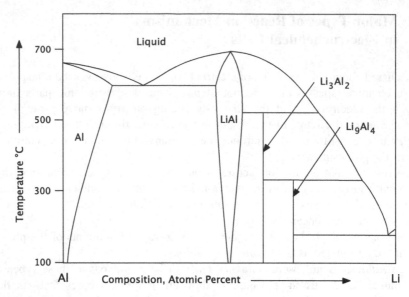

Fig. 9.3 Equilibrium phase diagram of the aluminum–lithium system

this modifies the microstructure, it is an example of one type of *reconstitution* reaction.

There are many examples of this type of formation reaction. There can also be subsequent additional formation reactions whereby other phases can be formed by further reaction of an original product.

As an example, consider the reaction of lithium with aluminum. Lithium–aluminum alloys were explored for use as electrodes in high temperature lithium batteries some time ago [2, 3], and their critical thermodynamic and kinetic properties were studied by the use of molten salt electrolyte electrochemical cells [4–6].

The reactions in this alloy system can be understood by use of the *lithium–aluminum system* phase diagram, as shown in Fig. 9.3.

Assume that the negative electrode is lithium and the positive electrode is initially pure aluminum. Upon the imposition of current by making lithium positively, and aluminum negatively charged, lithium ions pass across the cell and react with the aluminum in the positive electrode, changing its chemical composition. If this were to happen at 100 °C, the lowest temperature in Fig. 9.3, and it could be assumed that equilibrium conditions can readily be attained, it can be seen that a solid solution is initially formed, in which a small amount of lithium dissolves into the aluminum.

As more lithium is passed across the cell the solubility limit of lithium in aluminum is reached, and the composition moves into a region of the phase diagram in which both the lithium-saturated aluminum phase Al_{sat} and a new phase "LiAl" are present in the positive electrode. The quotation marks are used here, for the composition of the phase is not exactly 1:1 Li/Al. Thus in this part of the phase diagram the formation reaction

$$Li + Al_{sat} = \text{``LiAl''} \tag{9.7}$$

takes place. As more lithium passes across the electrochemical cell, the overall composition traverses the two-phase $Al_{sat} + \text{``LiAl''}$ region, more "LiAl" is formed, and the amount of Al_{sat} decreases. By the time the overall composition reaches the low-Li boundary of the "LiAl" region, there is no more of the Al_{sat} present.

The addition of more lithium causes the overall composition to go across the range of the "LiAl" phase. Thereafter, there is another two-phase formation reaction

$$Li + \text{``LiAl''} = Li_3Al_2 \tag{9.8}$$

that is later followed by the reaction

$$Li + Li_3Al_2 = Li_9Al_4 \tag{9.9}$$

as more lithium reacts with the structure that results from reaction (9.8).

The electrical potential varies with the chemical composition of electrodes in the Li-Al alloy system, as well as others that exhibit either ranges of solid solution or multi-phase reactions. This will be discussed in substantial detail in later chapters.

It is also not necessary that both reactants in formation reactions are solids or liquids, of course. For example, the phase LiCl can result from the reaction of lithium with chlorine gas, and ZnO can form as the result of the reaction of zinc with oxygen in the air. Zn/O_2 cells, in which ZnO is formed, are commonly used as the power source in hearing aids.

9.3.2 Displacement Reactions

As discussed in Chap. 4, another type of *reconstitution reaction* involves a *displacement* process, which can be simply represented as

$$A + BX = AX + B \tag{9.10}$$

in which species A displaces species B in the simple binary phase BX, to form AX instead. A new phase consisting of elemental B will be formed in addition. There will be a driving force causing this reaction to tend to occur if phase AX has a greater stability, i.e., has a more negative value of ΔG_f°, than the phase BX. An example of this type that was discussed in Chap. 4 was

$$Li + Cu_2O = Li_2O + Cu \tag{9.11}$$

in which the reaction of lithium with Cu_2O results in the formation of two new phases, Li_2O and elemental copper.

A change in the chemical state in the electrode results in a change in its electrical potential, of course. The relation between the chemical driving forces for such reactions, and the related electrical potentials, will be discussed for this case in later chapters.

9.3.3   Insertion Reactions

Again, as mentioned in Chap. 4, a quite different type of reaction mechanism can also occur in materials in chemical and electrochemical systems. This involves the *insertion* of guest species into normally unoccupied interstitial sites in the crystal structure of an existing stable host material. Although the chemical composition of the host phase initially present can be changed substantially, this type of reaction does not result in a change in the identity, the basic crystal structure, or amounts of the phases in the microstructure. However, in most cases the addition of interstitial species into previously unoccupied locations in the structure causes a change in volume. This involves mechanical stresses, and mechanical energy. The mechanical energy related to the insertion and extraction of interstitial species plays a significant role in the hysteresis, and thus energy loss, observed in a number of reversible battery electrode reactions.

In the particular case of the insertion of species into materials with layer-type crystal structures, insertion reactions are sometimes called *intercalation reactions*. Such reactions, in which the composition of an existing phase is changed by the incorporation of guest species, can also be thought of as a solution of the guest into the host material. Therefore, such processes are also sometimes called *solid solution reactions*.

Generally, the incorporation of such guest species occurs *topotactically*. This means that the guest species tend to be present at specific (low energy) locations inside the crystal structure of the host species, rather than being randomly distributed.

A simple reaction of this type might be the reaction of an amount x of species A with a phase BX to produce the product A_xBX. This can be written as

$$xA + BX = A_xBX \tag{9.12}$$

for such a case. The solid solution phase can have a range of composition, i.e., a range of values of x. As an example, the incorporation of lithium into TiS_2 produces a product in which the value of x can extend from 0 to 1. This was an important early example of this type of insertion reaction [7], and it can be simply represented as

$$x\text{Li} + \text{TiS}_2 = \text{Li}_x\text{TiS}_2 \qquad (9.13)$$

It is also possible to have a *displacement reaction* occur by the replacement of one interstitial species by another inside a stable host material. In this case, only one additional phase is formed, the material that is displaced. The term *extrusion* is sometimes used to describe this process.

In some cases, the new element or phase that is formed by such an *interstitial displacement process* is *crystalline*, whereas in other cases, it can be *amorphous*.

9.4 Important Practical Parameters

When considering the use of electrochemical energy storage systems in various applications, it is important to be aware of the properties that might be relevant, for they are not always the same in every case.

The energy and power available per unit weight, called the *specific energy* and *specific power*, are of great importance in some applications, such as vehicle propulsion.

On the other hand, the amount of energy that can be stored per unit volume, called the *energy density*, can be more important in some other areas of application. This is often the case when such devices are being considered as power sources in portable electronic devices, such as cellular telephones, portable computers, and video camcorders.

The power per unit volume, called the *power density*, can also be especially important for some uses, such as cordless power tools, whereas in others the *cycle life*—the number of times that a device can be effectively recharged before its performance, e.g., its capacity, or perhaps its output kinetics, is degraded too far—is critical. In addition, cost is always of concern, and sometimes can be of overriding importance, even at the expense of reduced performance.

Methods will be described later that allow the determination of the maximum theoretical values of some of these parameters, based upon the properties of the materials in the electrodes alone. However, practical systems never achieve these maximum theoretical values, but instead, often provide much lower performance. One obvious reason is that a practical battery has a number of passive components that are not involved in the basic chemical reaction that acts as the energy storage mechanism. These include the electrolyte, a separator that mechanically prevents the electrodes from coming into contact, the current connectors that transport electrical current to and from the interior of the cell, and the container. In addition, the effective utilization of the active components in the chemical reaction is often less than optimal. Electrode reactant materials can become electronically disconnected, or shielded from the electrolyte. When that happens they cannot participate in the electrochemical reaction, and have to be considered passive. They add to the weight and volume, but do not contribute to the transduction between electrical and chemical energy.

Table 9.1 Approximate values of the practical specific energy and energy density of some common battery systems

System	Specific energy (Wh/kg)	Energy density (Wh/l)
Pb/PbO$_2$	40	90
Cd/Ni	60	130
Hydride/Ni	80	215
Li-Ion	135	320

A rule of thumb that was used for a number of the conventional aqueous electrolyte battery systems was that a practical cell could only produce about 1/5 to 1/4 of its maximum theoretical specific energy. Optimization of a number of factors has made it now possible to exceed such values in a number of cases. In addition, the maximum theoretical values of some of the newer electrochemical systems are considerably higher than those that were available earlier.

Some rough values of the practical *energy density* (Wh/l) and *specific energy* (Wh/kg) of several of the common rechargeable battery systems are listed in Table 9.1. These particular values should not be taken as definitive, for they depend upon a number of operating factors and vary with the designs of different manufacturers. Nevertheless, they indicate the wide range of these parameters available commercially from different technologies.

In addition to their energy capacity, another important parameter relating to the practical use of batteries is the amount of power that they can supply. This is often expressed as specific power, the amount of power per unit weight, and it is very dependent upon the details of the design of the cell, as well as the characteristics of the reactive components. Therefore, values vary over a wide range.

The characteristics of batteries are often graphically illustrated by the use of *Ragone plots*, in which the specific power is plotted versus the specific energy. This type of presentation was named after David V. Ragone, who was the chairman of a governmental committee that wrote a report on the relative properties of different battery systems many years ago. Such a plot, including very approximate data on three current battery systems, is shown in Fig. 9.4.

9.4.1 The Operating Voltage and the Concept of Energy Quality

In addition to the amount of energy stored, another important parameter of a battery system is the voltage at which it operates, both during discharge, when it supplies electrical energy and power, but also when it is being recharged.

As discussed earlier in this chapter, the open circuit, or equilibrium, cell voltage is primarily determined by the thermodynamics of the chemical reaction between the components in the electrodes, for this reaction determines the driving force for the transport of ions through the electrolyte, and electrons through the external

Fig. 9.4 Ragone plot showing approximate practical values of specific power and specific energy of three common battery systems

circuit. During actual use, however, the operating voltage will vary from these theoretical values, depending upon various kinetic factors. These will be discussed extensively later in this text.

In discussing electrochemical energy storage it is useful to consider another parameter, its *quality*, and how it matches the expected applications. The concept of *energy quality* is analogous to the concept of *heat quality* that is well known in engineering thermodynamics.

It is widely recognized that high temperature heat is often more useful (e.g., has higher quality) than low temperature heat in many applications. Similarly, the usefulness of electrical energy is often related to the voltage at which it is available. High voltage energy is often more useful (has higher quality) than low voltage energy. For example, in simple resistive applications the electrical power P is related to the practical (not just theoretical) voltage E and the resistance R by

$$P = E^2 / R \qquad (9.14)$$

Thus the utility of an electrochemical cell in powering a light source or driving an electric motor is particularly voltage-sensitive. Because of the square relation, high voltage stored energy has a much higher quality for such applications than low voltage stored energy.

Rough energy quality rankings can be tentatively assigned to electrochemical cells on the basis of their output voltages as follows:

3.0–5.5 V	High quality energy
1.5–3.0 V	Medium quality energy
0–1.5 V	Low quality energy

There are a number of applications in which a high voltage is required. One example is the electrical system used to propel either hybrid- or all-electric vehicles. Auto manufacturers typically wish to operate such systems at over 200 V. For this type of high voltage application it is desirable that individual cells produce the highest possible voltage, for the greater the voltage of each individual cell, the fewer cells are necessary. There is also the movement toward the use of 36 or 42 V systems, instead of the current 12 V batteries, for the starter, lighting and ignition systems in normal internal combustion engine automobiles, as mentioned earlier.

Despite the implications of this matter of energy quality, it is important that the voltage characteristics of electrochemical energy storage systems *match the requirements of the intended application*. It is not always best to have the highest possible cell voltage, for it can be wasteful if it is too high in some applications.

A further matter that can become especially important in some applications is safety, and this can be a potential problem with some high potential electrode materials. As a result, development efforts aimed at large batteries for vehicle traction applications have been investigating materials that sacrifice some cell voltage to obtain greater safety.

9.4.2   The Charge Capacity

The *energy contained* in an electrochemical system is the integral of the voltage multiplied by the *charge capacity*, i.e., the amount of charge available. That is,

$$\text{Energy} = \int E \, dq \tag{9.15}$$

where E is the output voltage, which can vary with the state of charge as well as kinetic parameters, and q is the amount of electronic charge that can be supplied to the external circuit.

Thus it is important to know the maximum capacity, the amount of charge that can theoretically be stored in a battery. As in the case of the voltage, the maximum amount of charge available under ideal conditions is also a thermodynamic quantity, but it is of a different type. Whereas voltage is an *intensive quantity*, independent of the amount of material present, charge capacity is an *extensive quantity*. The amount of charge that can be stored in an electrode depends upon the amount of material in it. Therefore, capacity is always stated in terms of a measure such as the number of Coulombs per mol of material, per gram of electrode weight, or per ml of electrode volume.

The *state of charge* is the current value of the fraction of the maximum capacity that is still available to be supplied.

9.4.3   The Maximum Theoretical Specific Energy (MTSE)

Consider a simple insertion or formation reaction that can be represented as

$$x\text{A} + \text{R} = \text{A}_x\text{R} \tag{9.16}$$

where x is the number of moles of A that reacts per mol of R. It is also the number of elementary charges per mol of R. If E is the average voltage of this reaction, the theoretical energy involved in this reaction follows directly from Eq. (9.15). If the energy is expressed in Joules, it is the product of the voltage in volts and the charge capacity, in Coulombs, involved in the reaction.

If W_t is the sum of the molecular weights of the reactants engaged in the reaction, the maximum theoretical specific energy (MTSE), the energy per unit weight, is simply

$$\text{MTSE} = (xE/W_t)F \tag{9.17}$$

MTSE is in J/g, or kJ/kg, when x is in equivalents per mol, E is in volts, and W_t is in g/mol. F is the Faraday constant, 96,500 C per equivalent.

Since 1 W is 1 J per second, 1 Wh is 3.6 kJ, and the value of the MTSE can be expressed in Wh/kg as

$$\text{MTSE} = 26,805(xE/W_t) \tag{9.18}$$

9.4.4   Variation of the Voltage as Batteries Are Discharged and Recharged

Looking into the literature, it is seen that the voltage of most—but not all—electrochemical cells varies as their chemical energy is deleted. That is, as they are discharged. Likewise, it changes in the reverse direction when they are recharged. That may not be a surprise. However, not only the voltage ranges but also the characteristics of these state of charge-dependent changes vary widely between different electrochemical systems. It is important to understand what causes these variations.

A characteristic way to present this information is in terms of *discharge curves* and *charge curves*, in which the cell voltage is plotted as a function of the state of charge. These relationships can vary greatly, depending upon the rate at which the energy is extracted from, or added to, the cell.

It is useful to consider the maximum values, the relation between the cell voltage, and the state of charge under equilibrium or near-equilibrium conditions. In this case, a very useful experimental technique, known as *Coulometric titration*, can provide a lot of information. This will be described in a later section.

Fig. 9.5 Examples of battery discharge curves, showing the variation of the voltage shown as a function of the fraction of their available capacity

Some examples of discharge curves under low current, or near-equilibrium, conditions are shown in Fig. 9.5. These are presented here to show the cell voltage as a function of the state of charge parameter. However, different battery systems have different capacities. Thus one has to be careful to not compare the energies stored in different systems in this manner.

The reason for presenting the near-equilibrium properties of these different cells in this way is to show that there are significant differences in the *types* of their behavior, as indicated by the shapes of their curves. It can be seen that some of these discharge curves are essentially flat. Some have more than one flat region, and others have a slanted and stretched S-shape, sometimes with an appreciable slope, and sometimes not. These variants can be simplified into three basic types of discharge curve shapes, as depicted in Fig. 9.6. The reasons behind their general characteristics will be discussed later.

Fig. 9.6 Schematic representation of different types of discharge curves

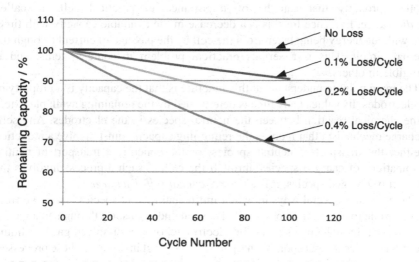

Fig. 9.7 Influence of Coulombic efficiency upon available capacity during cycling

9.4.5   Cycling Behavior

In many applications a battery is expected to maintain its major properties over
many discharge–charge cycles. This can be a serious practical challenge and is
often given a lot of attention during the development and optimization of batteries.
Figure 9.7 shows how the initial capacity is reduced during cycling, assuming three
different values of the *Coulombic efficiency*—the fraction of the prior charge
capacity that is available during the following discharge. This depends upon a
number of factors, especially the current density and the depth of discharge in each
cycle.

It is seen that even a minor amount of inefficiency per cycle can have important consequences. For example, 0.1 % loss per cycle causes the available capacity to drop to only 90 % of the original value after 100 cycles. The situation is worse if the Coulombic efficiency is lower.

Applications that involve many cycles of operation require that cells are designed and constructed such that the capacity loss per cycle is extremely low. This typically means that compromises must be made in other properties. *Supercapacitors* are expected to be used over a very large number of cycles, and they typically have much lower values of specific energy than electrochemical cells which are used for applications in which the amount of energy stored is paramount.

9.4.6 Self-Discharge

Another property that can be of importance in practical cells is called *self-discharge*. Evidence for this is a decrease in the available capacity with time, even without energy being taken from the cell by the passage of current through the external circuit. This is a serious practical problem in some systems, and is negligible in others.

The main point to understand at this juncture is that the capacity is a property of the electrodes. Its value at any time is determined by the remaining available extent of the chemical reaction between the neutral species in the electrodes. Any self-discharge mechanism that reduces the remaining capacity must involve a reduction in either the transport of neutral species, or the concurrent transport of neutral combinations of charged species, through the cell. If such a process involves the transport of charged species, it is *electrochemical self-discharge*.

There are also several ways in which individual neutral species can move from one electrode to the other across a cell. These include transport through an adjacent vapor phase, through cracks in a solid electrolyte, or as a dissolved gas in a liquid electrolyte. Since the transport of charged species is not involved in these processes. They produce *chemical self-discharge*.

It is also possible that impurities can react with constituents in the electrodes or the electrolyte so as to reduce the available capacity with time.

9.5   General Equivalent Circuit of an Electrochemical Cell

It is often useful to devise electrical circuits whose electrical characteristics are analogous to the behavior of important phenomena in physical systems. By examination of the influence of changes in the parameters in such *equivalent circuits*, they can be used as *thinking tools* to obtain useful insight into the significance of particular phenomena to the observable properties of complex physical systems. By use of this approach, the techniques of circuit analysis that have been developed for

Fig. 9.8 Simplified physical model of electrochemical cell

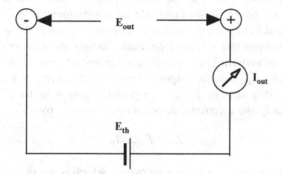

Fig. 9.9 Simple equivalent circuit model of an ideal electrochemical cell

use in various branches of electrical engineering can be very helpful in the analysis of interdependent physical phenomena.

This procedure has proven to be very useful in some areas of electrochemistry, and will be utilized later in this text for a number of purposes. At this point, however, it will only be considered for the case of an ideal electrochemical cell. It will show what happens if the electrolyte is not a perfect filter, and also allows the flow of some electronic current in addition to the expected ionic current through the electrolyte. An electrochemical cell can be simply modeled as shown in Fig. 9.8, and the basic equivalent circuit as shown in Fig. 9.9.

The value of the electrical equivalent of the theoretical chemical driving force is E_{th}, which is given by

$$E_{th} = -\Delta G_r^{\circ}/zF \tag{9.19}$$

as the result of the balance between the chemical and electrical forces acting upon the ionic species in the electrolyte, as mentioned earlier. If there are no impedances or other loss mechanisms, the externally measurable cell voltage E_{out} is simply equal to E_{th}.

9.5.1   Influence of Impedances to the Transport of Ionic and Atomic Species Within the Cell

In practical electrochemical cells E_{out} is not always equal to E_{th}. There can be several possible reasons for this disparity. There will always be some impedance to the transport of the electroactive ions and the related atomic species across the cell. One source is the resistance of the electrolyte to ionic transport. There may also be significant impeding effects at one or both of the two electrolyte/electrode interfaces. Furthermore, there can be a further impedance to the progress of the cell reaction in some cases related to the time-dependent solid state diffusion of the atomic species into, or out of, the electrode microstructure.

Note that *impedances* are used in this discussion instead of resistances, because they can be time-dependent if time-dependent changes in structure or composition are occurring in the system. The impedance is the instantaneous ratio of the applied force (e.g., voltage) E_{appl} and the response (e.g., current) across any circuit element. As an example, if a voltage E_{appl} is imposed across a material that conducts electronic current I_e, the electronic impedance Z_e is given by

$$Z_e = E_{appl}/I_e \qquad (9.20)$$

The inverse of the impedance is the *admittance*, which is the ratio current/voltage. Under steady-state (time-independent) DC conditions, the impedance and resistance of a circuit element are equivalent.

If current is flowing through the cell, there will be a voltage drop related to each of the impedances to the flow of ionic current within the cell. Thus if the sum of these internal impedances is Z_i the output voltage can be written as

$$E_{out} = E_{th} - I_{out}Z_i \qquad (9.21)$$

This relationship can be modeled by the simple circuit in Fig. 9.10.

9.5.2   Influence of Electronic Leakage Through the Electrolyte

The output voltage E_{out} can also be different from the theoretical electrical equivalent of the thermodynamic driving force of the reaction between the neutral species in the electrodes E_{th} even if there is no external current I_{out} flowing. This

Fig. 9.10 Simple equivalent circuit for a battery or fuel cell indicating the effect of the internal ionic impedance Z_i upon the output voltage

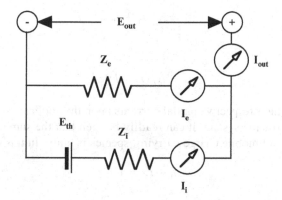

Fig. 9.11 Modified circuit including electronic leakage through the electrolyte

can be the result of electronic leakage through the electrolyte that acts to short-circuit the cell. This effect can be added to the previous equivalent circuit to give the circuit shown in Fig. 9.11.

It is evident that, even with no external current, there is an internal current related to the transport of the electronic species through the electrolyte I_e. Since the current must be the same everywhere in the lower loop, there must be a current through the electrolyte I_i with the same magnitude as the electronic current. There must be *charge flux balance* so that there is no net charge buildup at the electrodes.

The current through the internal ionic impedance Z_i generates a voltage drop, reducing the output voltage E_{out} by the product $I_i Z_i$, which is equal to $I_e Z_e$.

$$E_{out} = E_{th} - I_i Z_i \qquad (9.22)$$

In addition, the fact that both ionic and electronic species flow through the cell means that this is a mechanism of *self-discharge*. This results in a decrease of the available charge capacity of the cell.

9.5.3 Transference Numbers of Individual Species in an Electrochemical Cell

If more than one species can carry charge in an electrolyte, it is often of interest to know something about the relative conductivities or impedances of different species. The parameter that is used to describe the contributions of individual species to the transport of charge when an electrical potential difference (voltage) is applied across an electrolyte is the transference number. This is defined as the fraction of the total current that passes through the system that is carried by a particular species.

In the simple case that electrons and one type of ion can move through the electrochemical cell, we can define the transference number of ions as t_i, and electrons as t_e, where

$$t_i = I_i/(I_i + I_e) \qquad (9.23)$$

and

$$t_e = i_e/(I_i + I_e) \qquad (9.24)$$

and I_i and I_e are their respective partial currents upon the application of an external voltage E_{appl} across the system. It can readily be seen that the sum of the transference numbers of all mobile charge-carrying species is unity. In this case:

$$t_i + t_e = 1 \qquad (9.25)$$

Instead of expressing transference numbers in terms of currents, they can also be written in terms of impedances. For the case of these two species, the transport of charge by the motion of the ions under the influence of an applied voltage E_{appl},

$$t_i = \left(E_{appl}/Z_i\right)/\left[\left(E_{appl}/Z_i\right) + \left(E_{appl}/Z_e\right)\right] = Z_e/(Z_i + Z_e) \qquad (9.26)$$

and likewise for electrons:

$$t_e = Z_i/(Z_i + Z_e) \qquad (9.27)$$

Whereas these parameters are often thought of as properties of the electrolyte, in actual experiments they can also be influenced by what happens at the interfaces between the electrolyte and the electrodes, and thus are properties of the whole electrode-electrolyte system. They are only properties of the electrolyte alone if there is no impedance to the transfer of either ions or electrons across the electrolyte/electrode interface or atomic and electronic species within the electrodes.

Fig. 9.12 Different representation of general equivalent circuit

9.5.4   Relation Between the Output Voltage and the Values of the Ionic and Electronic Transference Numbers

Making the simplifying assumption that the internal impedance is primarily due to the behavior of the ions, the general equivalent circuit of Fig. 9.10 can be rearranged to look like that in Fig. 9.12.

When drawn this way, it can be readily seen that the series combination of Z_i and Z_e acts as a simple voltage divider.

If no current passes out of the system, i.e., under open circuit conditions, the output voltage is equal to the product of E_{th} and the ratio $Z_e/(Z_i + Z_e)$.

$$E_{out} = E_{th}Z_e/(Z_i + Z_e) \tag{9.28}$$

Introducing Eq. (9.26), the output voltage can then be expressed as

$$E_{out} = E_{th}t_i \tag{9.29}$$

or

$$E_{out} = E_{th}(1 - t_e) \tag{9.30}$$

These are well-known relations that can be derived in other ways, as will be shown later. It is clear that the output voltage is optimized when t_i is as close to unity as possible.

9.5.5   Joule Heating Due to Self-Discharge in Electrochemical Cells

Electrochemical self-discharge causes heat generation, often called *Joule heating*, due to the transport of charged species through the cell. The *thermal power* P_{th} caused by the passage of a current through a simple resistance R is given by

$$P_{th} = I^2 R \tag{9.31}$$

But as shown earlier, if self-discharge results from the leakage of electrons through the electrolyte there must be both electronic and ionic current, and they must have equal values. Thus the thermal power due to this type of self-discharge is:

$$P_{th} = I_i^2 Z_i + I_e^2 Z_e = I_e^2 (Z_i + Z_e) \tag{9.32}$$

Measurements of the rate of heat generation by Joule heating under open circuit conditions can be used to evaluate the rate of self-discharge in practical cells.

9.5.6   What If Current Is Drawn from the Cell?

If current is drawn from the cell into an external circuit, the normal mode of operation when chemical energy is converted into electrical energy, it flows through the ionic impedance, Z_i. This results in an additional voltage drop of $I_{out} Z_i$, further reducing the output voltage. If there were no electrochemical self-discharge, this can be written as

$$E_{out} = E_{th} t_i - I_{out} Z_i \tag{9.33}$$

The value of the ionic impedance of the system, Z_i, may increase with the value of the output current as the result of current-dependent impedances at the electrolyte/ electrode interfaces. The difference between E_{th} and E_{out} is often called *polarization* in the electrochemical literature.

The result of the presence of current-dependent interfacial impedances to the passage of ionic species that increase Z_i is that the *effective transference number* of the ions t_i is reduced, since $t_i = Z_e/(Z_i + Z_e)$. This causes an additional reduction in the output voltage.

But in addition to a *reduced output voltage*, there will also be additional *heat generation*. The total amount of Joule heating is the sum of that due to the passage of current into the external circuit I_{ext} and that due to electrochemical self-discharge.

$$P_{th} = I_{ext}^2 Z_i + I_e^2 (Z_i + Z_e) \tag{9.34}$$

In most cases, the first term is considerably larger than the second term.

Measured discharge curves vary with the current density as conditions deviate farther and farther from equilibrium. This is shown schematically in Fig. 9.13.

A parameter that is often used to indicate the rate at which a battery is discharged is the so-called *C*-Rate. The discharge rate of a battery is expressed as C/R, where R is the number of hours required to completely discharge its nominal capacity.

Fig. 9.13 Schematic drawing showing the influence of the current density upon the discharge curve

As an example, if a cell has a nominal capacity of 5 Ah, discharge at the rate of $C/10$ means that it would be fully discharged in 10 h. Thus the current is 0.5 A. And if the discharge rate is $C/5$ the discharge current is 1 A.

Although the C-Rate is often specified when either complete cells or individual electrodes are evaluated experimentally, and the current can be specified, this parameter is often not time-independent during real applications. If the electrical load is primarily resistive, for example, the current will decrease as the output voltage falls. This means that the C-Rate drops as the battery is discharged. Nevertheless, it is often important to consider the C-Rate when comparing the behavior of different materials, electrodes, and complete cells.

It is obvious that not only the average voltage but also the charge delivered can vary appreciably with changes in the C-Rate. But in addition, the amount of energy that can be supplied, which will be seen in later chapters to be related to the area under the discharge curve, is strongly C-Rate dependent.

A further point that should be kept in mind is that not all of the stored energy may be useful. If the load is resistive, the output power is proportional to the square of the voltage according to Eq. (9.14), so that the energy that is available at lower voltages may not be of much benefit.

This behavior can be understood in terms of the equivalent circuit of the battery. The internal ionic impedance Z_i, the sum of the impedances in the electrolyte and at the two electrode/electrolyte interfaces, is a function of the local current density in the cell. This impedance typically also varies with the state of charge. The mechanisms responsible for this behavior will be discussed later in the text.

References

1. Barin I (1995) Thermochemical Data of Pure Substances, 3rd edn. VCH, Weinheim, Published Online 24 Apr 2008. ISBN 9783527619825
2. Yao NP, Heredy LA, Saunders RC (1971) J Electrochem Soc 118:1039
3. Gay EC et al (1976) J Electrochem Soc 123:1591
4. Wen CJ, Boukamp BA, Weppner W, Huggins RA (1979) J Electrochem Soc 126:2258
5. Wen CJ, Weppner W, Boukamp BA, Huggins RA (1980) Met Trans 11B:131
6. Wen CJ, Ho C, Boukamp BA, Raistrick ID, Weppner W, Huggins RA (1981) Int Metals Rev 5:253
7. Whittingham MS (1976) Science 192:1126

Chapter 10
Principles Determining the Voltages and Capacities of Electrochemical Cells

10.1 Introduction

In the prior chapter it was shown that the fundamental driving force across an electrochemical cell is the virtual chemical reaction that would occur if the materials in the two electrodes were to react with each other. If the electrolyte is a perfect filter that allows the passage of ionic species, but not electrons, the cell voltage when no current is passing through the system is determined by the difference in the electrically neutral chemical compositions of the electrodes. The identity and properties of the electrolyte and the phenomena that occur at the electrode/electrolyte interfaces play no role. Likewise, it is the properties of the electrodes that determine the capacity of an electrochemical cell.

These general principles will be extended further in this chapter. Emphasis will be placed upon the equilibrium, or near-equilibrium state. This will address the ideal properties of such systems, which provide the upper limits of various important parameters.

Real systems under load deviate from this behavior. As will be shown later, this is primarily because of kinetic factors. Such factors vary from one system to the next, and are highly dependent upon both the details of the materials present, the cell construction, and the experimental conditions. As a result, it is difficult to obtain reproducible and quantitative experimental results. Such matters will appear later in this text. First, the factors that determine the equilibrium, or near-equilibrium, behavior will be discussed.

10.2 Thermodynamic Properties of Individual Species

It was shown in Chap. 9 that the overall driving force across a simple electrochemical cell is determined by the change in Gibbs free energy, ΔG_r° of the virtual chemical reaction that would occur if the materials in the electrodes were to react

© Springer International Publishing Switzerland 2016   145
R.A. Huggins, *Energy Storage*, DOI 10.1007/978-3-319-21239-5_10

with one another. If there is no current flowing, this chemical driving force is just balanced by an electrical driving force in the opposite direction.

Individual species within the electrolyte in the cell will now be considered. Under open circuit conditions (and no electronic leakage) there is no net current flow. Thus there must be a *force balance* acting on all mobile species.

The thermodynamic properties of a material can be related to those of its constituents by using the concept of the *chemical potential* of an individual species. The chemical potential of species i in a phase j is defined as

$$\mu_i = \partial G_j / \partial n_i \tag{10.1}$$

where G_j is the molar Gibbs free energy of phase j, and n_i is the mol fraction of the i species in phase j. In integral form this is

$$\Delta \mu_i = \Delta G_j \tag{10.2}$$

Since the free energy of the phase changes with the amount of species i, it is easy to see that the chemical potential has the same dimension as the free energy. Thus gradients in the chemical potential of species i produce chemical forces causing i to tend to move in the direction of lower μ_i. It was shown in Chap. 1 that when there is no net flux in the electrolyte, this chemical force must be balanced by an electrostatic force due to the voltage between the electrodes. The energy balance in the electrolyte, and thus in the cell, can be written in terms of the single species i:

$$\Delta \mu_i = -z_i F E \tag{10.3}$$

where z_i is the number of elementary charges carried by particles (ions) of species i.

The chemical potential of a given species is related to another thermodynamic quantity, its *activity*, a_i. The defining relation is

$$\mu_i = \mu_i^o + RT \ln a_i \tag{10.4}$$

where μ_i^o is a constant, the value of the chemical potential of species i in its standard state. R is the gas constant (8.315 J/mol degree), and T is the absolute temperature.

The activity of a species can be thought of as its *effective concentration*. If the activity of species i, a_i, is equal to unity, it behaves chemically as though it is pure i. If a_i is 0.5, it behaves chemically as though it is composed of half species i, and half something else that is chemically inert. In the case of a property such as vapor pressure, a material i with an activity of 0.5 will have a vapor pressure half of that of pure i.

Consider an electrochemical cell in which the activity of species i is different in the two electrodes, $a_i(-)$ in the negative electrode, and $a_i(+)$ in the positive electrode. The difference between the chemical potential on the positive side and that on the negative side can be written as

$$\mu_i(+) - \mu_i(-) = RT[\ln a_i(+) - \ln a_i(-)] = RT \ln[a_i(+)/a_i(-)] \tag{10.5}$$

If this chemical potential difference is balanced by the electrostatic energy from Eq. (10.2):

$$E = -(RT/z_i F)\ln[a_i(+)/a_i(-)] \tag{10.6}$$

This relation, which is often called the *Nernst equation*, is very useful, for it relates the measurable cell voltage to the chemical difference across an electrochemical cell. That is, it transduces between the chemical and electrical driving forces. If the activity of species i in one of the electrodes is a standard reference value, the *Nernst equation* provides the relative electrical potential of the other electrode.

10.3   A Simple Example: The Lithium/Iodine Cell

As an initial example, the thermodynamic basis for the voltage of a lithium/iodine cell will be considered. Primary (non-rechargeable) cells based upon this chemical system were invented by Schneider and Moser in 1972 [1, 2], and they are currently widely used to supply the energy in cardiac pacemakers.

The typical configuration of this electrochemical cell employs metallic lithium as the negative electrode and a composite of iodine with about 10 wt% of poly-2-vinylpyridine (P2VP) on the positive side. The composite of iodine and P2VP is a charge transfer complex, with the P2VP acting as an electron donor, and the iodine as an acceptor. The result is that the combination has a high electronic conductivity and the chemical properties are essentially the same as those of pure iodine. Reaction between the Li and the (iodine, P2VP) composite produces a layer of solid LiI. This material acts as a solid electrolyte in which Li^+ ions move from the interface with the negative electrode to the interface with the positive electrode, where they react with iodine to form more LiI. The transport mechanism involves a flux of lithium ion vacancies in the opposite direction. Although LiI has relatively low ionic conductivity, it has negligible electronic transport, meeting the requirements of an electrolyte.

This system can be represented simply as

$$(-)\text{Li/electrolyte/I}_2(+) \tag{10.7}$$

The *virtual reaction* that determines the voltage is thus

$$\text{Li} + 1/2\text{I}_2 = \text{LiI} \tag{10.8}$$

More and more LiI forms between the lithium electrode and the iodine electrode as the reaction progresses. The time evolution of the microstructure during discharge is shown schematically in Fig. 10.1.

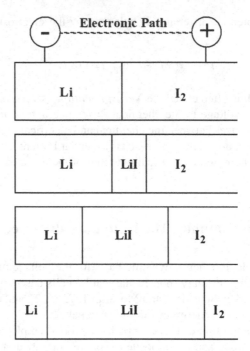

Fig. 10.1 Schematic representation of the microstructure of a Li/I_2 cell at several stages of discharge

The voltage across this cell under open circuit conditions can be readily calculated from the balance between the chemical and electrical driving forces, as shown in Chap. 9:

$$E = -\Delta G_r / z_i F \tag{10.9}$$

where

$$\Delta G_r = \Delta G_f(\text{LiI}) \tag{10.10}$$

and z_i is +1, for the electroactive species are the Li$^+$ ions.

According to the data in Barin [3], the Gibbs free energy of formation of LiI is −269.67 kJ/mol at 25 °C. Since the value of the Faraday constant is 96,500 C per equivalent (mol of electronic charge), the open-circuit voltage can be calculated to be 2.795 V at 25 °C.

Data on the properties of commercial Li/I_2 cells are shown in Fig. 10.2 [4]. It is seen that during most of the life of this battery the voltage corresponds closely to that which was calculated above. It is also seen in this figure that the resistance across the cell increases with the extent of reaction, due to the increasing thickness of the solid electrolyte product that grows as the cell is discharged. Such cells are

Fig. 10.2 Output voltage and internal resistance of a typical Li/I_2 battery of the type used in cardiac pacemakers. After [4]

typically designed to be *positive-electrode limited*. This means that the positive electrode capacity is somewhat less than the negative electrode capacity, and therefore is the part of the cell that determines the overall capacity.

10.3.1   Calculation of the Maximum Theoretical Specific Energy

The value of the maximum theoretical specific energy of a Li/I_2 cell can now be calculated from this information and the weights of the reactants. It was shown in Chap. 9 that the MTSE, in Wh/kg, is given by

$$MTSE = 26,805(xE/W_t) \qquad (10.11)$$

The reactant weight W_t is the weight of a mol of Li (6.94 g) plus half a mol of I_2 (126.9 g), or 133.84 g. The value of x is 1, and E was calculated to be 2.795 V. Thus the value of the MTSE is 559.77 Wh/kg.

This is a large number, about 15 times the value that is typical of the common Pb-acid cells that are so widely used as SLI batteries in automobiles, as well as for a number of other purposes. The lack of rechargeability as well as the cost of the ingredients and the low discharge rate unfortunately limit the range of application of Li/I_2 cells, however.

10.3.2 The Temperature Dependence of the Cell Voltage

As it has been seen, the quantity that determines the voltage is the Gibbs free energy change associated with the virtual cell reaction between the chemical species in the electrodes. That quantity is, however, temperature dependent. This can be seen by dividing the Gibbs free energy into its enthalpy and entropy components:

$$\Delta G_r = \Delta H_r - T\Delta S_r \qquad (10.12)$$

so that

$$d(\Delta G_r)/dT = -\Delta S_r \qquad (10.13)$$

and

$$dE/dT = \Delta S_r/z_i F \qquad (10.14)$$

The value of ΔS for the formation of LiI is given by

$$\Delta S_r(\text{LiI}) = S(\text{LiI}) - S(\text{Li}) - 1/2S(\text{I}_2) \qquad (10.15)$$

Entropy data for these materials, as well as a number of others, are given in Table 10.1. Note that these entropy values are in J/mol deg, whereas Gibbs free energy values are typically in kJ/mol. From these data, the value of ΔS_r for the formation of LiI is -1.38 J/K mol. Thus, from Eq. (10.13), the cell voltage varies only -1.43×10^{-5} V/K. This is very small. As will be seen later, the temperature dependence of the voltage related to many other electrochemical reactions, and thus

Table 10.1 Entropy data for some species at 25 and 225 °C [3]

Species	S (25 °C) (J/K mol)
Li	29.08
Zn	41.63
H_2	130.68
O_2	205.15
Cl_2	304.32
I_2	116.14
LiF	35.66
LiCl	59.30
LiBr	74.06
LiI	85.77
H_2O (liquid)	69.95
ZnO	43.64
H_2	145.74
O_2	220.69
H_2O (gas)	206.66

Fig. 10.3 Theoretical open-circuit voltage of a H_2/O_2 fuel cell as a function of the absolute temperature

of other batteries, is often much greater. An example is the small Zn/O_2 battery that is commonly used in hearing aids, where it is -5.2×10^{-4} V/K.

The data in Table 10.1 show that the entropy values of simple solids are considerably lower than those of liquids, which, it turn, are lower than gases. This is reflected, of course, in the temperature dependence of electrochemical cells.

An interesting example is the H_2/O_2 fuel cell. In that case the voltage varies -1.7×10^{-3} V/K near room temperature where water, the product of the reaction, is a liquid. But at 225 °C, where the product of the cell reaction is a gas, steam, the variation is only -0.5×10^{-3} V/K. The resultant variation of the cell voltage with temperature from about room temperature to the operating temperature of high temperature oxide-electrolyte fuel cells is shown in Fig. 10.3. Operation at a high temperature results in a significantly lower voltage. The theoretical open-circuit voltage is 1.23 V at 25 °C, but only 0.91 V at 1025 °C.

10.4 The Shape of Discharge Curves and the Gibbs Phase Rule

It was shown earlier that the voltage of batteries often varies with the state of charge, and it was pointed out that their discharge curves typically have one of three general shapes. Some are relatively flat, others have more than one relatively flat portion, and others have a slanted or stretched-S shape, sometimes with a relatively large slope. The data in Fig. 10.2 show that the Li/I_2 cell falls in the first category.

To understand how the voltage across an electrochemical cell varies with the state of charge, and why it is essentially flat in the case of the Li/I_2 cell, it is useful to consider the application of the *Gibbs phase rule* to such systems.

The *Gibbs phase rule* is often written as

$$F = C - P + 2 \tag{10.16}$$

in which C is the *number of components* (e.g., elements) present, and P is the *number of phases present* in this materials system in a given experiment. The quantity F may be more difficult to understand. It is the *number of degrees of freedom*; that means the number of *intensive thermodynamic parameters* that must be specified in order to *define the system* and *all of its associated properties*. One of these properties is, of course, the electric potential.

To understand the application of the phase rule to this situation, it must be determined what thermodynamic parameters should be considered. They must be intensive variables, which means that their values are independent of the amount of material present. For this purpose, the most useful thermodynamic parameters are the temperature, the overall pressure, and either the chemical potential or the chemical composition of each of the phases present.

How does this apply to the Li/I_2 cell? Starting with the negative electrode; there is only one phase present, Li, so P is 1. It is a single element, with only one type of atom. Thus the number of components C is also equal to 1. Thus F must be equal to 2.

What is the meaning of $F = 2$? It means that if the values of two intensive thermodynamic parameters, such as the temperature and the overall pressure, are specified, there are no degrees of freedom left over. Thus the *residual value of F* is zero. This means that all of the intensive properties of the negative electrode system are fully defined, e.g., have fixed values.

Thus in the case of the lithium negative electrode the chemical potentials of all species (i.e. the pure lithium), as well as the electrical potential, have fixed values, *regardless of the amount* of lithium present. The amount of lithium in the negative electrode decreases as the cell becomes discharged and the product LiI is formed. That is, the amount of lithium varies with the state of charge of the Li/I_2 battery. But since $F = 2$, and thus the residual value of F, if the temperature and total pressure are held constant, is zero, none of the intensive properties change. This means that the electrical potential of the lithium electrode is independent of the state of charge of the cell. This is shown schematically in Fig. 10.4.

On the other hand, if some iodine *could* dissolve into the lithium, forming a solid solution, *which it does not*, the number of components in the negative electrode would be two. In a solid solution there is only one phase present. Thus $C = 2$, $P = 1$ and $F = 3$.

In this hypothetical case the system would not be fully defined after fixing the temperature and the overall pressure. There would be a residual value of F, i.e., 1. Thus the electrical potential of the lithium-iodine alloy would not be fixed, but would vary, depending upon some other parameter, such as the amount of iodine in the Li-I solid solution. This is shown schematically in Fig. 10.5.

Although it is *not true* in the lithium/iodine cell it is quite common in other electrochemical cells for the electrical potential of electrodes to vary with the

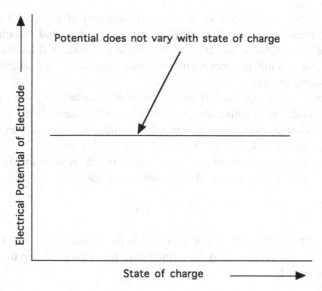

Fig. 10.4 The potential of a pure lithium electrode does not vary with the state of charge of the Li/I$_2$ cell

Fig. 10.5 Schematic representation of the variation of the electrical potential of an electrode as a function of its composition for the case in which the residual value of F is not 0

composition, and thus with the state of charge. A number of examples will be discussed in subsequent chapters.

Now consider the positive electrode. There is only one active component (element) present, iodine. There is also only one electrochemically active phase, iodine. Thus both C and P have values of 1. The number of degrees of freedom is thus again 2. Therefore, the values of all intensive variables and associated properties, such as the electrical potential, of the iodine electrode will be determined if the values of the two independent thermodynamic parameters, the temperature and the total pressure, are fixed.

This means that the potential of the I_2 electrode does not vary with its state of charge. Since both the negative and positive electrode potentials are independent of the state of charge, the voltage across the cell must also be independent of the state of charge of the Li/I_2 battery. This was illustrated in Fig. 10.2.

The earlier discussion showed that the chemical potential of an element depends upon its activity, and for the case of the iodine electrode

$$\mu(I_2) = \mu^\circ(I_2) \tag{10.17}$$

where $\mu^\circ(I_2)$ is the chemical potential of iodine in its standard state, i.e., pure iodine at a pressure of one atmosphere at the temperature in question. When the activity is unity, i.e., for pure I_2,

$$\mu(I_2) = \mu^\circ(I_2) \tag{10.18}$$

Now consider the voltage of the Li/I_2 cell. This is determined by the Gibbs free energy of formation of the LiI phase, as given in Eqs. (10.8) and (10.9). But it is also related to the difference in the chemical potential of iodine at the two electrode/electrolyte interfaces according to the relation

$$E = -\Delta\mu(I_2)/z_i F \tag{10.19}$$

where the value of z_i is -2. Therefore the activity of iodine at the positive side of the electrolyte is unity, but it is very small at the interface on the negative electrode side. Likewise, the cell voltage is related to the difference in the chemical potential of lithium at the two electrode/electrolyte interfaces:

$$E = -\Delta\mu(Li)/z_i F \tag{10.20}$$

in which the value of z_i is $+1$. In this case the activity of lithium is unity at the negative interface, and very small at the positive interface, where the electrolyte is in contact with I_2.

Whereas this discussion has focused on the potential of a single electrode, the shape of the equilibrium discharge curve (voltage versus state of charge) of an electrochemical cell is the result of the change of the potentials of both electrodes as the overall reaction takes place. If the potential of one of the electrodes does not vary, the variation of the cell voltage is obviously the result of the change of the potential of the other electrode as its overall composition changes.

There are a number of materials that are used as electrodes in electrochemical cells in which more than one reaction can occur in sequence as the overall discharge process takes place. In some cases, these reactions are of the same type, whereas in others they are not.

As one example, a *series of multiphase reactions* in which the number of residual degrees of freedom is zero can result in a discharge curve with several constant voltage plateaus. This is illustrated schematically in Fig. 10.6.

It is also possible for an electrode to undergo sequential reactions that are not of the same type. An example of this is the reaction of lithium with a spinel phase in the Li-Ti-O system. Experimental data are shown in Fig. 10.7 [5].

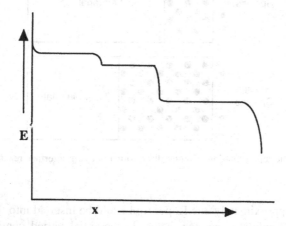

Fig. 10.6 Schematic equilibrium discharge curve of an electrode that undergoes a series of multiphase reactions in which the residual value of F is 0

Fig. 10.7 Equilibrium discharge curve of a material in the Li-Ti-O system that initially had a composition with a spinel type of crystal structure

Fig. 10.8 Schematic representation of a one-dimensional moving interface reaction

In this case approximately one Li per mol could be inserted into the host spinel phase as a *solid solution reaction*. Thus the potential varied continuously as a function of composition.

The introduction of additional lithium causes the nucleation, and subsequent growth, of a second phase that has the rocksalt structure and a composition of approximately two Li per mol of the original host. Thus a *reconstitution reaction* takes place when more than one lithium was added. During a reconstitution reaction there are two regions within the material with different Li contents. As the reaction proceeds the compositions of the two phases do not change, but the relative amount of the phase with the higher Li content increases, and that of the initial solid solution phase is reduced. This occurs by the *movement of the interface* between them. This type of a *moving interface reconstitution reaction* can be schematically represented as shown in Fig. 10.8.

Another example, in which several reactions occur as the overall composition is changed is the Li-Mn-O system. In this case there is a series of three different reactions. This is seen from the shape of the equilibrium discharge curve in Fig. 10.9 [6]. There is a two-phase plateau, followed by a single-phase solid solution region, and then by another two-phase plateau.

This interpretation was reinforced by the results of X-ray diffraction experiments, which are shown in Fig. 10.10 [6]. It is seen that the lattice parameters remain constant within two-phase regions, and vary with the composition within the single-phase solid solution region.

Fig. 10.9 Equilibrium discharge curve for $Li_xMn_2O_4$. After [6]

10.5 The Coulometric Titration Technique

The simple examples that have been discussed so far in this chapter assume that the requisite thermodynamic data are already known. Thus one can calculate the open-circuit voltage of an electrochemical cell from the value of the Gibbs free energy of the appropriate virtual reaction, and the ideal capacity can be determined from the *reaction's stoichiometry*.

It is also possible to do the opposite, using electrochemical measurements to obtain thermodynamic information. A useful tool for this purpose is the *coulometric titration technique*, which was first introduced by Wagner [7] to study the phase $Ag_{2+x}S$, which exists over a relatively narrow range of composition x. Its composition, or stoichiometry (the relative amounts of silver and sulfur) depends upon the value of the activity of silver within it. One can use a simple electrochemical cell to both change the stoichiometry and evaluate the activity of one of the species, e.g., silver in this case.

This method was further developed and applied by Weppner and Huggins [8] to the investigation of poly-phase alloy systems. It was demonstrated that the phase diagram can be determined, as well as the thermodynamic properties of the individual phases within it, by the use of this technique.

Consider the use of the following simple electrochemical cell to investigate the properties of the *vario-stoichiometric* (the stoichiometry can have a range of values) *phase* A_yB. This can be represented schematically as

$$A/\text{Electrolyte that transports } A^+ \text{ ions}/A_yB$$

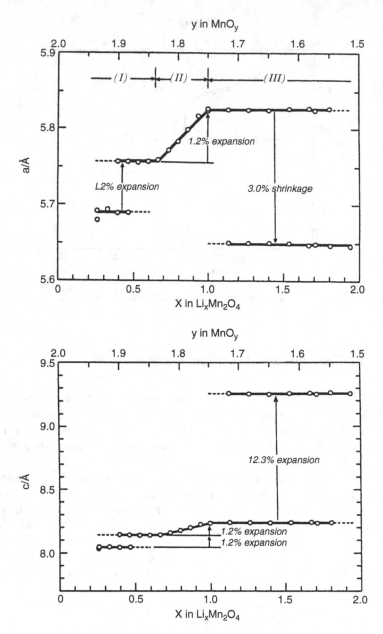

Fig. 10.10 Changes in unit cell dimensions as a function of composition in $Li_xMn_2O_4$ [6]

In this case, the element A acts as both a source and sink for the electroactive species A and a thermodynamic reference for component A. For simplicity, it can be assumed that this electrode is pure A, and thus has an activity of unity. It can also be assumed that both A and A_yB are good electronic conductors, that the ionic transference number in the electrolyte is unity, and that the system is under isobaric and isothermal conditions.

Under these conditions, the open-circuit voltage E is a direct measure of the chemical potential and the activity of A in the phase A_yB according to Eq. (10.5) that appeared earlier. As the electrode of pure A has an activity of unity, this relation can be written as

$$E = -\Delta\mu_A/(z_A,F) = -(RT/z_A,F)\ln a(A) \tag{10.21}$$

where z_{A+} is the charge number of the A^+ ions in the electrolyte, which is 1.

If a positive current is passed through the cell by the use of an electronic source, A^+ ions will be transported through the electrolyte from the left electrode to the right electrode. An equal current of electrons will go through the outer circuit because of the *requirement for charge flux balance*. The result is that the value of y in the A_yB phase will be increased.

For the case that a steady value of current I is applied for a fixed time t, the amount of charge Q that is passed across the cell is simply

$$Q = It \tag{10.22}$$

The number of mols of species A that are transported during this current pulse is

$$\Delta m(A) = Q/z_A,F \tag{10.23}$$

so that the change in the value of y, the mol fraction of species A, is

$$\Delta y = \Delta m(A)/m(B) = Q/(z_A,Fm(B)) \tag{10.24}$$

where $m(B)$ is the number of moles of B present in the electrode.

This method can be used to make *very minute* changes in the composition of the electrode material. One can see how sensitive this procedure is by putting some numbers into this relation.

Suppose that the electrode has a weight of 5 g, and the component B has a molecular weight of 100 g/mol. The value of $m(B)$ is thus 0.05 mol. Now suppose that a current of 0.1 mA is run through the cell for 10 s. The value of Q is then 0.001 C. With $z_{A+} = 1$ equivalent per mol and $F = 96,500$ C per equivalent, then Δy is only about 2×10^{-7}.

This is very small. Thus it is possible to investigate the compositional dependence of the properties of phases with very narrow compositional ranges. It is very difficult to get such a high degree of compositional resolution by other techniques.

By waiting for a sufficiently long time to allow the composition to become homogeneous throughout the electrode material, as evidenced by reaching a steady-state value of open-circuit voltage, information can be obtained about the equilibrium chemical potential and activity of the mobile electroactive species as a function of composition. This technique has been used to investigate a wide variety of materials of potential interest in battery systems, and numerous examples will be discussed in later chapters.

The success of this method depends upon a number of assumptions. One is that the electrolyte is essentially only an ionic conductor, i.e., the ionic transference number is very close to unity. Another is that there can be no appreciable loss of either component from the electrode material A_yB by evaporation, dissolution, or interaction with the electrical lead materials, the so-called *current collectors*. Furthermore, the rate of *compositional equilibration via chemical diffusion* in the electrode material must be sufficiently fast. This means that it may be necessary to use thin samples as electrodes in order to reduce the time necessary for concentration homogenization.

It should also be recognized that this coulometric titration technique gives information about the influence of compositional changes, but not the absolute composition. That must be determined by some other method.

References

1. Moser JR (1972) US Patent 3,660,163
2. Schneider AA, Moser JR (1972) US Patent 3,674,562
3. Barin I (1995) Thermochemical Data of Pure Substances, 3rd edn. VCH, Weinheim, Published Online 24 Apr 2008. ISBN 9783527619825
4. Courtesy of Catalyst Research Corp.
5. Liebert BE, Weppner W, Huggins RA (1977) In: McIntyre JDE, Srinivasan S, Will FG (eds) Proceedings of the Symposium on Electrode Materials and Processes for Energy Conversion and Storage. Electrochemical Society, Princeton, p 821
6. Ohzuku T, Kitagawa M, Hirai T (1990) J Electrochem Soc 137:769
7. Wagner C (1953) J Chem Phys 21:1819
8. Weppner W, Huggins RA (1978) J Electrochem Soc 125:7

Chapter 11
Binary Electrodes Under Equilibrium or Near-Equilibrium Conditions

11.1 Introduction

The theoretical basis for understanding and predicting the composition-dependence of the potentials, as well as the capacities, of both binary (two element) and ternary (three element) alloys has now been established. The relevant principles are discussed for the case of binary systems in this chapter. Ternary systems will be treated in the next chapter.

Under equilibrium and near-equilibrium conditions these important practical parameters are directly related to the thermodynamic properties and compositional ranges of the pertinent phases in the respective *phase diagrams*. Phase diagrams, which were touched upon briefly in both Chaps. 4 and 9, are graphical representations that indicate the phases and their compositions that are present in a materials system under equilibrium conditions. They can be useful *thinking tools* to help understand the fundamental properties of electrodes in electrochemical systems.

One can often understand the behavior under dynamic conditions in terms of simple deviations from the equilibrium conditions assumed in phase diagrams. In other cases, however, *metastable* phases may be present in the microstructure of an electrode whose properties are considerably different from those of the *absolutely-stable* phases. The influence of *metastable* microstructures will be discussed in a later chapter. In addition, it is possible that the compositional changes occurring in an electrode during the operation of an electrochemical cell can cause *amorphization* of its structure. This will also be discussed later.

© Springer International Publishing Switzerland 2016
R.A. Huggins, *Energy Storage*, DOI 10.1007/978-3-319-21239-5_11

11.2   Relationship Between Phase Diagrams and Electrical Potentials in Binary Systems

In order to demonstrate the relationship between phase diagrams and some of the important features of electrodes in electrochemical systems, a schematic phase diagram for a hypothetical binary alloy system A–B is shown in Fig. 11.1. In this case there are 4 one-phase regions. The solid phases are designated as phases a, b, and γ. In addition, there is a liquid phase at higher temperatures. It can be seen in the figure that the single phases are all separated by two-phase regions as the composition moves horizontally (isothermally) across the diagram.

It was shown earlier that according to the Gibbs phase rule all intensive properties, including the electrical potential, vary continually with the composition within single-phase regions in a binary system. Correspondingly, the intensive properties are composition-independent when two phases are present in a binary system. Since the equilibrium electrical potential of such an electrode, E, in an electrochemical cell is determined by the chemical potential or activity of the electroactive species, it also varies with composition within single-phase regions, and is composition-independent when there are two phases present under the equilibrium conditions that are assumed here.

The variation of the electrical potential with the overall composition in this hypothetical system at temperature T_1 is shown in Fig. 11.2. It is seen that it alternately goes through composition regions in which it is constant (*potential plateaus*) and those in which it varies. If B atoms are added to pure element A the overall composition is initially in the solid solution phase α and the electrical

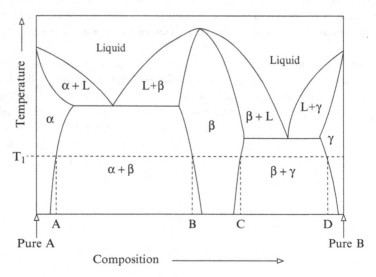

Fig. 11.1 Schematic binary phase diagram with an intermediate phase β, and solid solubility in terminal phases α and γ

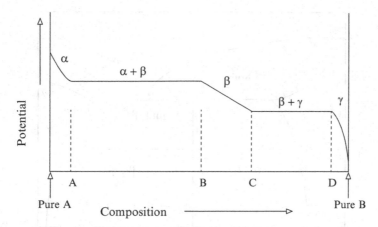

Fig. 11.2 Schematic variation of electrical potential with composition across the binary phase diagram shown in Fig. 11.1

potential varies with composition. When the α solubility limit is reached, indicated as composition A, the addition of more B causes the nucleation and growth of the β phase. Two phases are then present, and the potential maintains a fixed value. When the overall composition reaches composition B, all of the α phase will have been consumed and there will only be phase β present. Upon further compositional change the electrical potential again becomes composition dependent. At composition C, the upper compositional limit of the β phase at that temperature, the overall composition again enters a two-phase (β and γ) range and the potential is again composition-independent. Upon reaching composition D the potential again varies with composition.

It is also possible for the composition ranges of phases to be quite narrow, and then they are sometimes called *line phases*. As an example, a variation upon the phase diagram presented in Fig. 11.1 is shown in Fig. 11.3.

The corresponding variation of the electrical potential with composition is shown in schematically in Fig. 11.4. The potential drops abruptly, rather than gradually, across the narrow β phase in this case.

11.3   A Real Example, the Lithium–Antimony System

As a concrete example to demonstrate these principles consider the Li-Sb system. It has been studied both experimentally and theoretically in some detail [1–3]. The phase diagram is shown in Fig. 11.5.

Below 615 °C there are two intermediate phases between Sb and Li, Li_2Sb, and Li_3Sb. Both have rather narrow ranges of composition and are represented simply as vertical lines in the phase diagram. Thus, if an electrode starts as pure Sb and

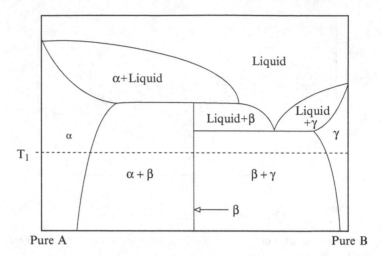

Fig. 11.3 A hypothetical binary-phase diagram in which the intermediate β phase has a small range of composition

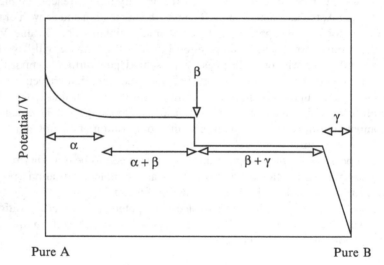

Fig. 11.4 Schematic variation of electrical potential with composition across the binary phase diagram shown in Fig. 11.3

lithium is added it successively goes through two different reactions. The first involves the formation of the phase Li_2Sb, and can be written as

$$2Li + Sb = Li_2Sb \tag{11.1}$$

Upon the addition of more lithium a second reaction will occur that results in the formation of the second intermediate phase from the first. This can be written as

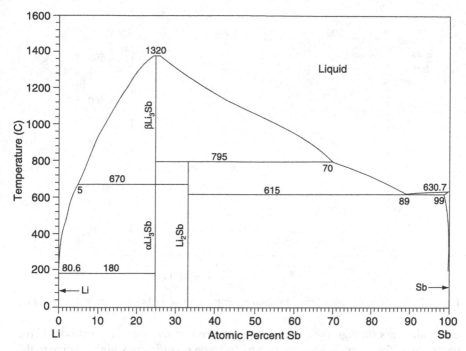

Fig. 11.5 Lithium–antimony phase diagram

Fig. 11.6 Schematic drawing of electrochemical cell to use to study the Li-Sb system

$$Li + Li_2Sb = Li_3Sb \qquad (11.2)$$

This process can be studied experimentally by the use of an electrochemical cell whose initial configuration is similar to that shown schematically in Fig. 11.6.

By driving current through this cell from an external source that causes the voltage between the two electrodes to be reduced from the open circuit value that it has when the positive electrode is pure antimony, lithium will leave the negative

Fig. 11.7 Results from a coulometric titration experiment on the Li-Sb system at 360 °C. After [3]

electrode, pass through the electrolyte, and arrive at the positive electrode. If the chemical diffusion rate within the Li_xSb electrode is sufficiently high relative to the rate at which lithium ions arrive at the positive electrode surface, this lithium will be incorporated into the bulk of the electrode crystal structure, changing its composition. That is, the value of x in the positive electrode material Li_xSb will increase.

If the lithium is either added very slowly, or stepwise, allowing equilibrium to be attained within the positive electrode material after each step, the influence of the lithium concentration in the positive electrode upon its potential under equilibrium or near-equilibrium conditions can be investigated. This is the *coulometric titration technique* discussed earlier. Data from such an experiment at 360 °C are shown in Fig. 11.7.

These results can be understood by consideration of the Gibbs phase rule, which was discussed in Chap. 10.

After an initial, invisibly narrow, range of solid solution, the first plateau in Fig. 11.7 corresponds to compositions in the phase diagram in which both (almost pure) Sb and the phase Li_2Sb are present. Thus it is related to the reaction in Eq. (11.1).

There is also a very narrow composition range in which only one phase, Li_2Sb, is present and the potential varies. Upon the addition of further Li the overall composition moves into the region of the phase diagram in which two phases are again present, in this case Li_2Sb and Li_3Sb, and the potential follows along a second plateau, related to Eq. (11.2).

The potentials of these two plateaus can be calculated from thermodynamic data on the standard Gibbs free energies of formation of the two phases, Li_2Sb and

Li_3Sb. According to [3] these values are -176.0 kJ/mol and -260.1 kJ/mol, respectively, at that temperature.

The standard Gibbs free energy change, $\Delta G_r°$, related to virtual reaction (11.1) is simply the standard Gibbs free energy of formation of the phase Li_2Sb, $\Delta G_f°(Li_2Sb)$. From this the potential of the first plateau can calculated from

$$E - E° = -\Delta G_r°/2F \tag{11.3}$$

where $E°$ is the potential of pure Li. This was found to be 912 mV in the experiment.

The potential of the second plateau is related to virtual reaction (11.2), where

$$\Delta G_r° = \Delta G_f°(Li_3Sb) - \Delta G_f°(Li_2Sb) \tag{11.4}$$

and in this case

$$E - E° = -\Delta G_r°/F \tag{11.5}$$

The result is that the potential of this plateau was experimentally found to be 871 mV vs. pure Li.

The maximum theoretical energy that can be obtained from this alloy system is the sum of the energies involved in the two reactions. These relationships are shown schematically in Fig. 11.8.

The total energy that can be stored is proportional to the total area under the titration curve. The energy released in the first reaction is the product of the voltage of the first plateau times its capacity, i.e., the charge passed through the cell in connection with that reaction. That corresponds to the area inside rectangle A. The energy released in the second discharge reaction step is the product of its voltage

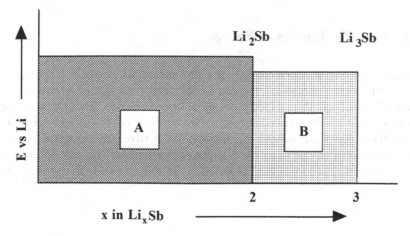

Fig. 11.8 Relation between energy stored and the titration curve in the Li-Sb system

and its capacity, and corresponds to the area inside rectangle B. The total energy is the sum of the two areas, A and B.

These energy values can be converted into *specific energy*, energy per unit weight. In the case of the first plateau the *maximum theoretical specific energy*, MTSE, is simply the standard Gibbs free energy of the reaction divided by the sum of the atomic weights in the product. This was found to be 1298 kJ/kg. This can also be expressed as 360 Wh/kg, since 3.6 kJ is equal to 1 Wh.

The maximum theoretical specific energy can also be calculated if the composition were to only vary between the compositions Li_2Sb and Li_3Sb. In that case the voltage was 871 mV and the capacity was only 1 mol of lithium per mol of original Li_2Sb. When calculating the MTSE the weight of the product is then that of Li_3Sb, 142.57 g/mol. The result is 589 kJ/kg and 164 Wh/kg for a cell operated in this composition range.

However, if the experiment is performed starting with pure Li and pure Sb, and the energy relating to both plateaus is used, the relevant weight for both steps is the final weight, that of Li_3Sb. Thus the MTSE of the first reaction in the two-reaction scheme is less than it would be if it were used alone. Instead of 1298 kJ/kg, it is only 1234 kJ/kg. The total MTSE is then $1234 + 589 = 1823$ kJ/kg.

This is the same result that would be found if it were assumed that the intermediate phase, Li_2Sb, did not form, and that there is only a single voltage plateau between Li and Li_3Sb.

If the electrochemical titration curve were calculated from the experimental value of the total energy, it would have only a single plateau, and at a voltage that is the weighted average of the voltages of the two reactions that actually take place. This is a false result, due to the lack of recognition of the existence of the intermediate phase. Thus one has to be careful to be aware of all of the stable phases when making voltage predictions from thermodynamic data.

11.4   Stability Ranges of Phases

Whereas emphasis has been placed upon the potentials at which reactions take place in this discussion thus far, there is another important type of information that is available from equilibrium electrochemical titration curves. The potential ranges over which the various intermediate phases are stable can be readily obtained. Since intermediate phases are present at compositions between two plateaus, they are stable at all potentials between the two plateau potentials. This can be important information if they are to be used as *mixed conductors*, as will be described later.

11.5 Another Example, the Lithium–Bismuth System

The lithium–bismuth binary system has also been extensively explored by the use of the coulometric titration technique. The phase diagram is shown in Fig. 11.9. Note that this diagram is drawn with lithium on the right hand side, i.e., in the opposite direction from the Li-Sb diagram. This difference is, in principle, not important, however.

The titration curve that resulted from measurements made at 360 °C is shown in Fig. 11.10.

It can be seen that there are three differences from the Li-Sb system. The phase diagram shows that there is a considerable amount of solubility of bismuth in liquid lithium at that temperature. This results in the appearance of a single-phase region in the titration curve. Also, there is a phase LiBi in the Li-Bi case, but Li_2Sb in the Li-Sb case.

In addition, the phase diagram in Fig. 11.9 indicates that the solid phase "Li_3Bi" has an appreciable range of composition. This can also be seen in the titration curve. Because of the very high sensitivity of the coulometric titration technique, the electrochemical properties of this phase could be explored in much more detail. This is shown in Fig. 11.11.

Fig. 11.9 The lithium–bismuth binary phase diagram

Fig. 11.10 Results from a coulometric titration experiment on the Li-Bi system at 360 °C. After [3]

Fig. 11.11 Coulometric titration measurements within the composition range of the phase "Li_3Bi." After [3]

11.6 Coulometric Titration Measurements on Other Binary Systems

Coulometric titration experiments have been made on a number of other binary metallic systems at several temperatures. In order to obtain reliable data, it is important that the experiments be undertaken under conditions such that equilibrium can be reached within a reasonable time. This requirement is fulfilled much more easily at elevated temperatures, but in some cases, equilibrium data can also be obtained at ambient temperatures, albeit with a bit more patience. A number of further examples will be discussed in later chapters.

11.7 Temperature Dependence of the Potential

The early measurements of the equilibrium electrochemical properties of binary lithium alloys and their relationship to the relevant phase diagrams were made at elevated temperatures using a LiCl-KCl molten salt electrolyte. These included experiments on the Li-Al, Li-Bi, Li-Cd, Li-Ga, Li-In, Li-Pb, Li-Sb, and Li-Si, and Li-Sn systems [4–10]. This molten salt electrolyte was being used in research efforts aimed at the development of large scale batteries for electric vehicle propulsion and load leveling applications. Subsequently, measurements were made with lower temperature molten salts, $LiNO_3$-KNO_3 [11], and at ambient temperatures with organic solvent electrolytes. The latter will be discussed later.

It has been found, as expected, that the temperature dependence of the potentials and capacities can be explained in terms of the relevant phase diagrams and thermodynamic data in all of these cases.

To demonstrate the principles involved, experimental results on materials in the Li-Sb and Li-Bi systems over a wide range of temperature will be described. The results are shown in Fig. 11.12.

Each of these systems has two intermediate phases at low temperatures. The temperature dependence of the potentials of the plateaus due to the presence of two-phase equilibria in the Li-Sb system fall upon two straight lines, corresponding to the reactions

$$2Li + Sb = Li_2Sb \tag{11.6}$$

and

$$Li + Li_2Sb = Li_3Sb \tag{11.7}$$

In the Li-Bi case, however, where the comparable reactions are

Fig. 11.12 Temperature dependence of the potentials of the two-phase plateaus in the Li-Sb and Li-Bi systems. After [12]

$$Li + Bi = LiBi \tag{11.8}$$

and

$$2Li + LiBi = Li_3Bi \tag{11.9}$$

the temperature dependence of the plateau potentials is different from the Li-Sb case. There is a change in slope at the *eutectic melting point* (243 °C), and the data for the two plateaus converge at about 420 °C, which corresponds to the fact that the LiBi phase is no longer stable above that temperature. This can be seen in the phase diagram for that system shown in Fig. 11.9. At higher temperatures there is only a single reaction,

$$3Li + Bi = Li_3Bi \tag{11.10}$$

In addition, the potentials of the second reaction fall along two straight line segments, depending upon the temperature range. There is a significant change in slope at about 210 °C, resulting in a negligible temperature dependence of the potential at low temperatures, due to the melting of bismuth.

The potentials are related to the standard Gibbs free energy change $\Delta G_r°$ relating to the relevant reaction, and the temperature dependence of the value of $\Delta G_r°$ can be seen from the relation between the Gibbs free energy, the enthalpy, and the entropy

Table 11.1 Reaction entropies in the lithium–antimony and lithium–bismuth systems

Reaction	Molar entropy of reaction (J/K mol)	Temperature range (°C)
$2\,Li + Sb = Li_2Sb$	-31.9	25–500
$Li + Li_2Sb = Li_3Sb$	-46.5	25–600
$Li + Bi = LiBi$	0	25–200
$2\,Li + LiBi = Li_3Bi$	-36.4	25–400

$$\Delta G_r^\circ = \Delta H_r^\circ - T\Delta S_r^\circ \qquad (11.11)$$

where ΔH_r° is the change in the standard enthalpy and ΔS_r° is the change in the standard entropy resulting from the reaction. Thus it can be seen that

$$d\Delta G_r^\circ / dT = \Delta S_r^\circ \qquad (11.12)$$

From these data, one can obtain value of the standard molar entropy changes involved in these several reactions. They are shown in Table 11.1. Thus the potentials at any temperature within this range can be predicted.

11.8   Application to Oxides and Similar Materials

This discussion thus far has been concerned with binary metallic alloys. However, the same principles can be applied to binary metal–oxygen systems. The Nb-O system will be used as an example. The niobium–oxygen phase diagram is shown in Fig. 11.13.

There are three intermediate phases in this system, and thermodynamic data can be used to calculate the potentials of the various two-phase plateaus at any temperature, as before. In addition, these potentials can be converted into values of the oxygen activity at the respective temperatures. Data on the plateau potentials, which define the limiting values for the stability of each of the phases, are shown for two temperatures in Fig. 11.14. In this case, the potentials are shown as voltages relative to the potential of pure oxygen at one atmosphere at the respective temperatures.

Similar procedures can be employed to the analysis of other metal–gas systems, such as iodides and chlorides.

11.9   Ellingham Diagrams and Difference Diagrams

Another thinking tool that is sometimes used to help understand the behavior of metal–oxygen systems are the so-called *Ellingham diagrams*. These are plots of the Gibbs free energy of formation of oxides as a function of temperature.

Fig. 11.13 Niobium–oxygen phase diagram

Fig. 11.14 Equilibrium potentials of the 3 two-phase plateaus in the Nb-O system at two temperatures

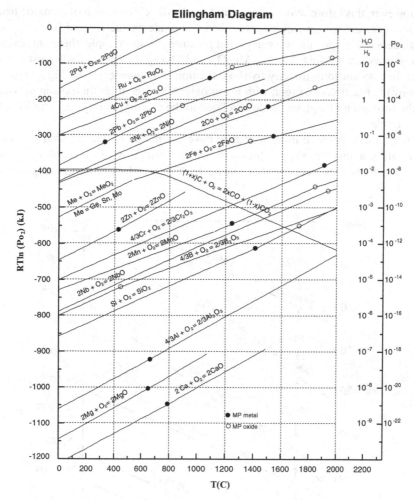

Fig. 11.15 Integral Ellingham diagram, courtesy of Prof. A. Petric, McMaster Univ

There are two different ways in which this information can be presented. The Ellingham diagram shown in Fig. 11.15 is of the type that is most often seen in textbooks. It is of the *integral type*, and shows the temperature dependence of the Gibbs free energy necessary for the formation of a particular oxide from its component elements. The formation reactions are generally written on a "per mol oxygen" basis, so that the lines relating to different oxides are roughly parallel, as the entropy of gaseous oxygen makes the major contribution to the entropy of the formation reaction.

An equilibrium oxygen pressure scale is generally added on the right side to provide a simple graphical means to determine the oxygen partial pressure of the oxides as a function of temperature [13]. In some cases, as in Fig. 11.15, a scale indicating values of the ratio H_2O/H_2 is also included.

However, this information is only valid for the direct formation of an oxide from its elements, and in many cases more than one oxide can be formed from a given metal, depending upon the oxygen partial pressure. As an example, there are several manganese oxides: MnO, Mn_3O_4, Mn_2O_3, and MnO_2. Except for the lowest oxide, MnO they do not form directly by the oxidation of manganese. The higher oxides form by the reaction of oxygen with lower oxides, and the relevant oxygen pressure for the formation of a given oxide is related to the reaction of oxygen with its next lower oxide.

Therefore, it is much more useful to have a *difference diagram* that provides information about the oxygen pressure for the formation of given phases from their neighbors. An example that shows both types of information is given in Fig. 11.16.

Fig. 11.16 Diagram that shows both integral and difference data for the manganese oxide system. After [14]

11.10   Comments on Mechanisms and Terminology

It has been shown that the incorporation of species into solid electrodes can involve either a change in the composition of a single phase that is already present in the microstructure, or the nucleation and growth of an additional phase. When that species is deleted, the same two types of phenomena can occur, but in the opposite sense. Consider how this can happen. If, for example, lithium is added to an existing phase it forms a solid solution, and the composition will change, becoming more lithium-rich. If this involves more than merely the surface layer it must involve the diffusive motion of lithium atoms or ions into the crystal structure.

In many metallic alloys and ceramic materials the inserted material occupies the same type of lattice positions as the host material. This is called a *substitutional solid solution*. In order for the composition to change there must be a mechanism that allows atomic, or ionic, motion through the crystal structure. In substitutional solid solutions this typically involves the presence and motion of empty lattice sites, *vacancies*. Atoms can jump into these vacant lattice sites from adjacent positions in the structure. The result of a single jump is the effective motion of the vacancy in one direction, and the atom in the opposite direction. Compositional changes occur by this vacancy mechanism if one type of atom has a greater probability of jumping into adjacent vacancies than the other type and a gradient in the composition is present.

Another type of diffusion mechanism is often present in crystal structures in which one type of atom or ion is appreciably smaller than the other types present. This is often the case in lithium and hydrogen systems, as these species are quite small. The smaller atoms can occupy interstitial sites between the other atoms in the crystal lattice. They can move about by jumping from one interstitial site to the next. Diffusion by this *interstitial mechanism* does not require the motion of either the other atoms or vacancies, and it is typically very much faster than the vacancy diffusion mechanism. That is, interstitial *diffusion coefficients*, or *diffusivities*, are typically greater than vacancy mechanism diffusion coefficients, or diffusivities.

Because of the large concentration of interstitial positions in most crystal structures, it is structurally possible for a large number of guest species to be inserted into the host crystal structure, if other factors, such as the electronic energy spectrum, are favorable.

If the basic structure of the host material is not appreciably altered by the *insertion* of additional guest species, so that there is a definite relation between the initial crystal structure and the structure that results, the reaction is called *topotactic*. This is often the case in materials of interest as electrode reactants in lithium battery systems.

Topotaxy implies a three-dimensional relation between the structures of the *parent* and *product* structures, whereas the term *epitaxy* is used to describe a two-dimensional correspondence between two structures. Likewise, an *insertion reaction* that has a two-dimensional character is often called *intercalation*.

From this structural viewpoint it can be readily understood that there can be a limit to the concentration of interstitial guest species that can be inserted into a host crystal structure. This limit can be due to either crystallographic or electronic factors, and will not be discussed further here.

If there is a thermodynamic driving force for the incorporation of *additional* guest species than can be accommodated interstitially, this must occur by a different mechanism. A *second phase*, with a different crystal structure as well as a higher solute concentration, must be nucleated. As more and more of the guest species atoms arrive, the extent of this second phase increases, gradually replacing the interstitial phase that was initially present. This change in the microstructure, in which one phase is gradually replaced by another phase, is an example of a *reconstitution* reaction.

When a reconstitution reaction is taking place the initial and product phases are both present in the microstructure. This is sometimes called a *heterophase structure*, in contrast to a *homophase* structure, in which only a single phase is present. Thus this range of compositions must be in a two-phase region of the corresponding phase diagram.

Phase diagrams are expected to provide information about the *absolutely stable* phases that tend to be present in a chemical system as a function of intensive thermodynamic variables such as temperature and composition. The term *absolutely stable* has been used to describe the *most stable equilibrium structure* possible for a given composition. On the other hand, a phase that is *stable relative to small perturbations*, and thus meets the general requirement for equilibrium, yet is not the most stable variation, is termed *metastable*.

11.11 Summary

Many batteries use binary systems as either negative or positive, or both, electrode reactants today. The theoretical limits of the potentials and capacities of such electrodes can be determined from a combination of thermodynamic data and phase diagrams. This has been demonstrated here for several examples of binary systems.

There are two general types of reactions that can take place, *homophase* reactions, in which guest atoms are inserted into an existing phase, often *topotactically*, and *reconstitution* reactions in which phases nucleate and grow in *heterophase* microstructures. The potential varies with the overall composition of an electrode in the insertion reaction homophase case, but is composition-independent when reconstitution reactions take place in heterophase microstructures.

The electrochemical titration method can be used to investigate the important parameters experimentally. When the composition is within a single-phase region of the relevant phase diagram the potential varies as guest species are inserted or extracted during an electrochemical reaction. On the other hand, when the composition is within two-phase regions of the phase diagram, reconstitution reactions take place, and the potential is independent of composition. Experimental results are now available for a number of systems of each type, both at elevated temperatures and at ambient temperatures.

Under equilibrium, or near-equilibrium, conditions the potentials are directly related to the values of the standard Gibbs free energies of formation of the phases involved. Thus thermodynamic data can be used to predict experimental results. Likewise, experiments can provide thermodynamic data. As an example, the temperature dependence of potential plateaus can be used to determine the standard entropy changes in the relevant reaction. These experimental data also correlate with the stability of phases in the phase diagram. In addition, the maximum theoretical specific energy of an electrochemical system can be determined from the equilibrium electrochemical titration curve and the related thermodynamic data.

These principles are also applicable to metal oxides, as well as liquid binary materials, as illustrated by the Nb-O system.

References

1. Weppner W, Huggins RA (1977) In: McIntyre JDE, Srinivasan S, Will FG (eds) Proceedings of the Symposium on Electrode Materials and Processes for Energy Conversion and Storage. Electrochemical Society, Princeton, p 833
2. Weppner W, Huggins RA (1977) Z Phys Chem N F 108:105
3. Weppner W, Huggins RA (1978) J Electrochem Soc 125:7
4. Wen CJ et al (1979) J Electrochem Soc 126:2258
5. Wen CJ (1980) Ph.D. Dissertation, Stanford University
6. Wen CJ, Huggins RA (1981) J Electrochem Soc 128:1636
7. Wen CJ, Huggins RA (1980) Mater Res Bull 15:1225
8. Saboungi ML et al (1979) J Electrochem Soc 126:322
9. Wen CJ, Huggins RA (1981) J Solid State Chem 37:271
10. Wen CJ, Huggins RA (1981) J Electrochem Soc 128:1181
11. Doench JP, Huggins RA (1982) J Electrochem Soc 129:341C
12. Wang J, Raistrick ID, Huggins RA (1986) J Electrochem Soc 133:457
13. Richardson FD, Jeffes JHE (1948) J Iron Steel Inst 160:261
14. Godshall NA, Raistrick ID, Huggins RA (1984) J Electrochem Soc 131:543

Chapter 12
Ternary Electrodes Under Equilibrium or Near-Equilibrium Conditions

12.1 Introduction

The previous chapter described binary electrodes, in which the microstructure is composed of phases made up of two elements. It was pointed out that there are also cases in which three elements are present, but only partial equilibrium can be obtained in experiments, so the electrode behaves as though it were composed of two, rather than three, components.

This chapter discusses active materials that contain three elements, but have kinetic behavior such that they behave as true ternary systems. As before, it will be seen that phase diagrams and equilibrium electrochemical titration curves are very useful thinking tools in understanding the potentials and capacities of electrodes containing such materials.

It is generally more difficult to obtain complete equilibrium in ternary systems than in binary systems, so that much of the available equilibrium, or near-equilibrium, information stems from experiments at elevated temperatures. Selective, or partial, equilibrium is much more common at ambient temperatures. This will be discussed in another chapter.

12.2 Ternary Phase Diagrams and Phase Stability Diagrams

In order to represent compositions in a three-component system one must have a figure that represents the concentrations of three components. This can be done by using a two-dimensional figure, as will be seen shortly. However, if information about the influence of temperature is also desired, a three-dimensional figure is required. This is often done in metallurgical and ceramic systems in which experiments commonly involve changes in the temperature. Most electrochemical

© Springer International Publishing Switzerland 2016
R.A. Huggins, *Energy Storage*, DOI 10.1007/978-3-319-21239-5_12

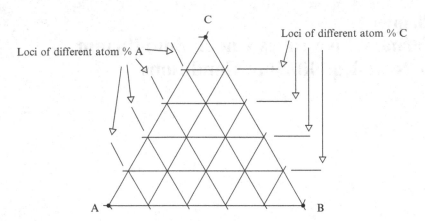

Fig. 12.1 General coordinate scheme used to depict compositions and phase equilibria in ternary systems on isothermal Gibbs triangles

systems operate at or near constant temperatures, so three-dimensional figures are not generally considered necessary.

Compositions in isothermal ternary systems can be represented on paper by using a triangular coordinate system. The method that is commonly employed in materials systems involves the use of *isothermal Gibbs triangles*. This scheme is illustrated in Fig. 12.1.

Compositions are expressed in terms of the atomic percent of each of the three components, indicated as A, B, and C in this case. For the purposes of this discussion it is desirable to have elements as components, so that three elements are placed at the corners, and the atomic percent of an element varies from zero along the opposite side to 100 % at its corner. Thus the position of each point within the triangle represents the atomic fraction of each of the elements present in the system.

Although phases in ternary systems often have ranges of composition, as they do in binary systems, it is often useful to simplify the phase equilibrium situation by assuming that they act as *point phases*. That is, that they have very narrow composition ranges. The term *phase stability diagram* will be used in this discussion to describe this approximation to the actual ternary phase diagram. It will be seen that it is possible to get a large amount of useful information by the use of such an approximate isothermal Gibbs triangle.

If there are phases inside the Gibbs triangle, the influence of the Gibbs phase rule must be considered. It was shown earlier that the Gibbs phase rule can be written as

$$F = C - P + 2 \qquad (12.1)$$

If the temperature and total pressure are kept constant, the number of residual degrees of freedom F will be zero when there are three phases present in a ternary system. Three phases are in equilibrium with each other within triangles inside the overall Gibbs triangle. Two phases are in equilibrium if their compositions are

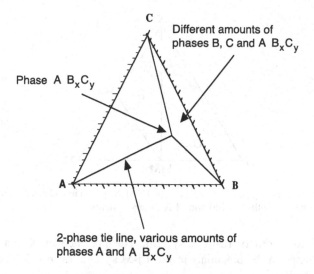

Fig. 12.2 Isothermal phase stability diagram ABC for the case in which there is a single intermediate phase whose composition is A_xB_yC

connected by a line, called a *two-phase tie line*. As shown in Fig. 12.2, if intermediate ternary phases are present, the total area within the Gibbs triangle is divided into *sub-triangles* whose sides are *two-phase tie lines*.

All of the compositions that lie within a given triangle have microstructures that are composed of mixtures of the three phases that are at the corners of that triangle. The overall composition determines the amounts of these different phases present, but not their compositions, for the latter are specified by the locations of the points at the corners of the relevant triangle. Any materials having compositions that fall along one of the sides of a triangle will have microstructures composed of the two phases at the ends of that tie line. The amounts are determined by the position of the composition along the tie line. Points closer to a given end have greater amounts of the phase whose composition is at the end.

Because the compositions of the phases present within triangles are constant, determined by the locations on the corner points, all of the intensive (amount-independent) thermodynamic parameters and properties are the same for all compositions inside the triangle. Important intensive properties include the chemical potentials and activities of each of the components, and the electrical potential.

12.3   Comments on the Influence of Sub-triangle Configurations in Ternary Systems

Binary systems can be changed to ternary systems by the addition of an additional element. As an example, consider a lithium-based binary system Li-M, in which the lithium composition can be varied. The addition of an additional element X

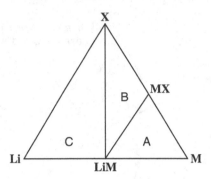

Fig. 12.3 Schematic ternary phase diagram for the Li-M-X system in which there are intermediate phases in the centers of both the Li-M and M-X binary systems

converts this to a ternary Li-M-X system. The presence of X can result in a significant change in the potentials in the Li-M system, even if X does not react with lithium itself.

Consider a simple case. Assume that the thermodynamic properties of this system lead to a ternary phase stability diagram of the type shown in Fig. 12.3, in which it is assumed that there are two stable binary phases, LiM, and MX.

If there is no X present, the composition moves along the Li-M edge of the ternary diagram, which is simply the binary Li-M system, and there will be a constant potential plateau for all compositions between pure M and LiM. The voltage vs. pure lithium in this compositional range, and therefore in triangle A of the ternary system, will be given by

$$E_A = \Delta G^{\circ}_f(LiM)/F \tag{12.2}$$

What happens if lithium reacts with a material that has an original composition containing some X? The overall composition will follow a trajectory that starts at that position along the X-M side of the triangle and goes in the direction of the lithium corner of the ternary diagram. The addition of X to the M will not change the plateau potential for all compositions in triangle A. Therefore, there will be a plateau at that potential. Its length, however, will vary, depending upon the starting composition.

In addition, an additional plateau will appear at higher lithium concentrations as the overall composition enters and traverses triangle B. The potential of all compositions in that triangle will be given by

$$E_B = \left(\Delta G^{\circ}_f(MX) - \Delta G^{\circ}_f(LiM) \right)/F \tag{12.3}$$

As in the case of the binary Li-M system, when the overall composition gets into triangle C the potential will be the same as that of pure lithium. These effects are illustrated in Figs. 12.4 and 12.5.

In Fig. 12.5 the variation of the electrode potential with overall composition is shown schematically for three different starting electrode compositions in Fig. 12.4.

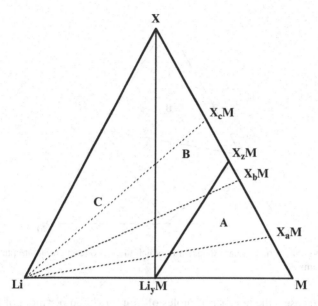

Fig. 12.4 The ternary Li-M-X system shown in Fig. 12.3, showing the loci of the overall composition for three different initial compositions

Fig. 12.5 Variation of the potential as lithium is added to electrodes with the three different starting compositions shown in Fig. 12.4. *Top*: X_aM, *middle*: X_bM, and *bottom*: X_cM

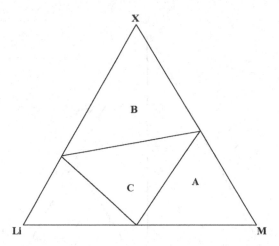

Fig. 12.6 Hypothetic ternary-phase diagram in which there is one intermediate phase in each of the binary systems

In all three cases, the number of moles of lithium stored per mole of M does not change, but the weight of the electrode will change, depending upon the relative weights of M and X. In addition, the average electrode potential becomes closer to that of pure lithium. This can be either advantageous or disadvantageous, depending upon whether the material is used as a negative electrode or as a positive electrode in a lithium-based cell.

Another ternary-phase configuration is shown in Fig. 12.6. In this case, it is assumed that there is also an intermediate phase in the Li-X binary system. The weight of the electrode per mol of Li will be reduced if the weight of Li_xX per mol of Li is less than that of Li per mol of Li_yM.

In practice, a binary system containing several intermediate phases may not be useful over its entire range of lithium composition, due to the change of the potential with composition. Poor diffusion kinetics in one of the intermediate phases or the terminal phase can also be deleterious.

There are many other possible phase diagram configurations in ternary systems, including those containing ternary phases in the interior of the diagram. In screening possible systems for study, however, a logical starting point is to examine systems with known binary and ternary phases.

12.4   An Example: The Sodium/Nickel Chloride "Zebra" System

Some years ago an interesting battery system suddenly appeared that had been initially developed in secret in South Africa and England. It is based upon the use of the solid electrolyte Na β-alumina, as in the Na/Na_xS system, which will be discussed later.

The β-alumina is a ceramic material that is a sodium-aluminum oxide with a nominal composition of $NaAl_{11}O_{17}$. It has a layer-type crystal structure in which the sodium ions have a very high mobility, so that it has the properties of a solid electrolyte.

This novel battery soon became known as the *"Zebra" cell* as the result of its development in South Africa.

It operates at 250–300 °C, and uses liquid sodium as the negative electrode, which is enclosed in a solid β-alumina tube. At this temperature sodium is liquid, and the ionic conductivity of the β-alumina is quite high. When the cell is fully charged, the positive electrode reactant is finely powdered $NiCl_2$, which is present adjacent to the β-alumina inside a solid container. Because the contact between the solid β-alumina tube and the particles of $NiCl_2$ is only at their points of contact, a second (liquid) electrolyte, $NaAlCl_4$, is also present in the outer, positive electrode compartment, part of the cell. Thus the full surface area of the $NiCl_2$ particles acts as the electrochemical interface, which greatly increases the kinetics.

Thus this electrochemical system, when charged, has the configuration:

$$Na/Na\ \beta-alumina/NaAlCl_4/NiCl_2$$

The physical arrangement of this cell is shown schematically in Fig. 12.7.

The electrochemical behavior of a Zebra cell can be understood by consideration of the Na-Ni-Cl ternary phase diagram. Thermodynamic data indicate that there are only two binary phases in this ternary system, $LiCl_2$, and NaCl. They lie on two different sides of the ternary Na-Ni-Cl-phase diagram. Since the total area must be

Fig. 12.7 Schematic view of the "Zebra" cell, which operates at 250–300 °C

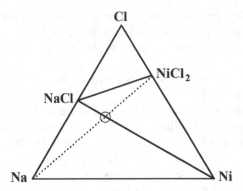

Fig. 12.8 The Na-Ni-Cl ternary phase diagram, showing the locus of the overall composition as Na reacts with NiCl$_2$

divided into triangles, it is evident that there are two possibilities. There is either a tie line from NiCl$_2$ to the Na corner, or there is one from NaCl to the Ni corner. The decision as to which of these is stable can be determined by the direction of the virtual reaction

$$2Na + NiCl_2 = 2NaCl + Ni \qquad (12.4)$$

The Gibbs free energy change in this virtual reaction is given by

$$\Delta G_r^{\circ} = 2\Delta G_f^{\circ}(NaCl) - \Delta G_f^{\circ}(NiCl_2) \qquad (12.5)$$

Values of the standard Gibbs free energies at 275 °C of NaCl and NiCl$_2$ are -360.25 kJ/mol and -221.12 kJ/mol, respectively. Therefore the reaction in Eq. (12.4) will tend to go to the right, and the tie line between NaCl and Ni is more stable than the one between NiCl$_2$ and Na.

As a result, the phase stability diagram must be as shown by the solid lines in Fig. 12.8. As Na reacts with NiCl$_2$ the overall composition of the positive electrode follows the dotted line in that figure. When it reaches the composition indicated by the small circle, all of the NiCl$_2$ will have been consumed, and only NaCl and Ni are present.

So long as the overall composition remains in the NaCl-NiCl$_2$-Ni triangle, the potential is constant. Its value can be calculated from the Gibbs free energy of reaction value corresponding to Eq. (12.5). The voltage of the positive electrode with respect to the pure Na negative electrode is given by

$$\Delta E = -\Delta G_r^{\circ}/zF \qquad (12.6)$$

where $z = 2$, according to the reaction is Eq. (12.4). The result is that the potential of all compositions within that triangle in the ternary diagram, and also across the Zebra cell, is constant, and equal to 2.59 V. This is also what is observed experimentally.

12.5 A Second Example: The Lithium-Copper-Chlorine Ternary System

The Li-Cu-Cl system will be used as a further example to illustrate these principles, and show how useful information can be derived from a combination of a ternary phase diagram and thermodynamic data related to the stable phases within it.

Thermodynamic information shows that there are three stable phases within this system at 298 K, LiCl, CuCl, and $CuCl_2$. Values of their standard Gibbs free energies of formation are given in Table 12.1.

All of these phases lie on the edges of the isothermal Gibbs triangle. If they are assumed to be point phases, with negligible ranges of composition, the phase stability diagram can be constructed by following a few simple rules and procedures.

1. The total area must be divided into triangles. Their edges are tie lines between pairs of phases.
2. No more than three phases can be present within a triangle. Their compositions must be at the corners.
3. Tie lines cannot cross.

The first task is to determine the stable tie lines in this system. This can be done by drawing all the possible tie lines between the stable phases on a trial basis, and then determining which of them are stable. The end result must be that the overall triangle is divided into sub-triangles.

The line between LiCl and $CuCl_2$ must be stable, as there are no other possible lines that could cross it. There are four additional possibilities, lines between Li and $CuCl_2$, Li and CuCl, LiCl and CuCl, and Li and Cu. A method that can be used to determine which of these is actually stable is to write the virtual reactions between the phases at the ends of conflicting (crossing) tie lines. Which of the two pairs of phases are more stable in each case can be determined from the available thermodynamic data.

As an example, consider whether there is a tie line between LiCl and Cu or one between CuCl and Li. Both cannot be stable, for they would cross.

The virtual reaction between the pairs of possible end phases can be written as

$$LiCl + Cu = CuCl + Li \qquad (12.7)$$

Table 12.1 Gibbs free energies of formation of phases in the Li-Cu-Cl system

Phase	ΔG_f° at 298 K (kJ/mol)
LiCl	−384.0
CuCl	−138.7
$CuCl_2$	−173.8

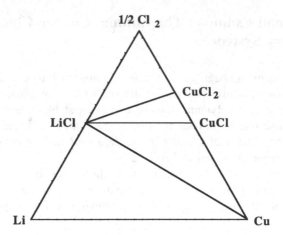

Fig. 12.9 Isothermal phase stability diagram for the Li-Cu-Cl ternary system at 25 °C

As before, the direction in which this virtual reaction would tend to go can be determined from the value of the standard Gibbs free energy of reaction. In this case, it is given by

$$\Delta G_r^\circ = \Delta G_f^\circ(\text{CuCl}) - \Delta G_f^\circ(\text{LiCl}) \tag{12.8}$$

The result is that ΔG_r° is $(-138.7) - (-384.0) = +245.3$ kJ/mol. Thus this reaction would tend to go to the left. This means that the combination of the phases LiCl and Cu is more stable than the combination of CuCl and Li. Thus the tie line between LiCl and Cu is stable in the phase diagram.

This implies that the tie line between LiCl and Cu is also more stable than one between CuCl_2 and Li, and also that a line between LiCl and CuCl exists. These conclusions can be verified by consideration of the virtual reaction between LiCl and Cu, and CuCl_2 and Li. This reaction would be written as

$$2\text{LiCl} + \text{Cu} = \text{CuCl}_2 + 2\text{Li} \tag{12.9}$$

for which the standard Gibbs free energy of reaction is $(-173.8) - 2(-384.0) = +594.2$ kJ/mol. Thus these conclusions were correct.

The resulting isothermal phase stability diagram for this system is shown in Fig. 12.9.

12.5.1   Calculation of the Voltages in this System

From this diagram and the thermodynamic data the voltages and capacities of electrodes in this system can also be calculated. As the first example, consider the

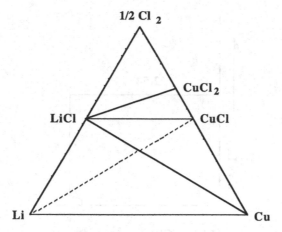

Fig. 12.10 Use of ternary phase diagram to understand the reaction of lithium with CuCl

reaction of lithium with CuCl. This reaction can be understood in terms of the ternary phase diagram as shown in Fig. 12.9.

By the addition of lithium the overall composition moves from the initial composition at the CuCl point along the dotted line toward the Li corner, as shown in Fig. 12.10. In doing so, it moves into and across the LiCl–CuCl–Cu triangle. So long as it is inside this triangle its voltage remains constant.

This voltage can be calculated from the virtual reaction that takes place by the addition of lithium as the overall composition moves into, and through, the LiCl-CuCl-Cu triangle:

$$Li + CuCl = LiCl + Cu \qquad (12.10)$$

The standard Gibbs free energy change as the result of this reaction is $(-384.0) - (-138.7) = -245.3$ kJ/mol. The voltage can be calculated from

$$E = -(-245.3)/[(1)(96.5)] \qquad (12.11)$$

The result is 2.54 V vs. pure Li. This voltage remains constant as long as the overall composition stays in the LiCl-CuCl-Cu triangle. It is obvious from Eq. (12.10) and the phase diagram in Fig. 12.10 that up to 1 mol of Li can participate in this reaction. Thus the equilibrium titration curve, the variation of the voltage of a cell of this type as a function of composition can be drawn as in Fig. 12.11.

If, on the other hand, the positive electrode were to consist of $CuCl_2$ instead of CuCl, the overall composition would move along the dotted line shown in Fig. 12.12.

The overall composition first enters the LiCl-$CuCl_2$-CuCl triangle. The relevant virtual reaction for this triangle is

Fig. 12.11 Variation of the equilibrium voltage of Li/CuCl cell as a function of the extent of reaction

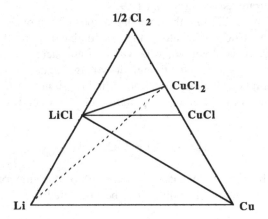

Fig. 12.12 Use of ternary phase stability diagram to understand the reaction of Li with $CuCl_2$

$$Li + CuCl_2 = LiCl + CuCl \tag{12.12}$$

The standard Gibbs free energy change as the result of this reaction is (-384.0) $+(-138.7) - (-173.8) = -348.9$ kJ/mol. The voltage with respect to pure Li can be calculated from

$$E = -\Delta G_r^\circ /zF = 348.9/96.5 \tag{12.13}$$

or 3.615 V vs. Li

Fig. 12.13 Equilibrium titration curve for the reaction of lithium with $CuCl_2$ to form LiCl and CuCl, and then more LiCl and Cu

There will be a plateau at this voltage in the equilibrium titration curve. The LiCl cannot react further with Li. But the CuCl that is formed in this reaction can undergo a further reaction with additional lithium. When this happens the overall composition moves into and across the second triangle, whose corners are at LiCl, CuCl and Cu. Although the reaction path is different, this is the same triangle whose voltage was calculated above for the reaction of lithium with CuCl. Thus the same voltage will be observed, 2.54 V vs. Li, in this second reaction, written in Eq. (12.10). The equilibrium titration curve will therefore have two plateaus, related to the two triangles that the overall composition traverses as lithium reacts with $CuCl_2$. This is shown in Fig. 12.13.

12.5.2   Experimental Arrangement for Lithium/Copper Chloride Cells

Cells based upon the reaction of lithium with either of the copper chloride phases can be constructed at ambient temperature using an electrolyte with a nonaqueous solvent, such as propylene carbonate, containing a lithium salt such as $LiClO_4$. There are a number of alternative solvents, as well as alternative salts, and this topic will be discussed in a later chapter. The important thing at the present time is that water and oxygen must be avoided, and the salt should have a relatively high solubility in the nonaqueous solvent.

12.6   Calculation of the Maximum Theoretical Specific Energies of Li/CuCl and Li/CuCl$_2$ Cells

The maximum values of specific energies that might be obtained from electrochemical cells containing either CuCl or CuCl$_2$ as positive electrode reactants can be calculated from this information.

As shown in an earlier chapter, the general relation for the maximum theoretical specific energy (MTSE) is

$$\text{MTSE} = (xV/W)(F) \text{ kJ/kg} \tag{12.14}$$

where x is the number of mols of Li involved in the reaction, V the average voltage, and W the sum of the atomic weights of the reactants. F is the Faraday constant, 96,500 C/mol.

In the case of a positive electrode that starts as CuCl and undergoes reaction (12.10), the sum of the atomic weights of the reactants is $(7 + 63.55 + 35.45) = 106.0$ g. The value of x is unity, and the average cell voltage is 2.54 V. Thus the MTSE is 2312.4 kJ/kg.

This can be converted to Wh/kg by dividing by 3.6, the number of kJ per Wh. The result is that the MTSE can be written as 642.3 Wh/kg for this reaction.

If the positive electrode starts as CuCl$_2$ and undergoes reaction (12.12) to form LiCl and CuCl the weight of the reactants is $(7 + 63.55 + (2 \times 35.45)) = 141.45$ g. The value of x is again unity, and the cell voltage was calculated to be 3.615 V. This then gives a value of MTSE of 2466.2 kJ/kg. Alternatively, it could be expressed as 685.1 Wh/kg.

If further lithium reacts with the products of this reaction, the voltage will proceed along the lower plateau, as was the case for an electrode whose composition started as CuCl. Thus additional energy is available. However, the total specific energy is not simply the sum of the specific energies that have just been calculated for the two plateau reactions independently. The reason for this is that the weight that must be considered in the calculation for the second reaction is the starting weight before the first reaction in this case.

Then, for the second plateau reaction:

$$\text{MTSE} = (1)(2.54)(96,500)/141.45 = 1732.8 \text{ kJ/kg} \tag{12.15}$$

This is less than for the second plateau alone, starting with CuCl, which was shown above to be 2312 kJ/kg. Alternatively, the specific energy content of the second plateau for an electrode that starts as CuCl$_2$ is 481.3 Wh/kg instead of 642.3 Wh/kg, if it were to start as CuCl.

Thus if the electrode starts out as CuCl$_2$, the total MTSE can be written as

$$\text{MTSE} = 2466.2 + 1732.8 = 4199 \text{ kJ/kg} \tag{12.16}$$

Or alternatively, $685.1 + 481.3 = 1166.4$ Wh/kg.

12.7   Specific Capacity and Capacity Density in Ternary Systems

As mentioned earlier, other parameters that are often important in battery systems are the capacity per unit weight or per unit volume. In the case of ternary systems, the capacity along a constant potential plateau is determined by the length of the path of the overall composition within the corresponding triangle. This is determined by the distance along the composition line between the binary tie lines at the boundaries of the triangles.

12.8   Another Group of Examples: Metal Hydride Systems Containing Magnesium

Binary alloys are often used as negative electrodes in hydrogen-transporting electrochemical cells. When they absorb or react with hydrogen, they are generally called *metal hydrides*. Because of the presence of hydrogen as well as the two metal components, they become ternary systems.

There is a great interest in the storage of hydrogen for a number of purposes related to the desire to reduce the dependence on petroleum. The reversible hydrogen absorption in some metal hydrides is a serious competitor for this purpose.

If the kinetics of hydrogen absorption or reaction are relatively fast, and the motion of the other constituents in the crystal structure is very sluggish, so that no structural reconstitution of the metal constituents in the microstructure takes place in the time scale of interest, such metal hydride systems can be treated as *pseudo-binary systems*, i.e., hydrogen plus the metal alloy. This is the general assumption that is almost always found in the literature on the behavior of metal hydrides.

On the other hand, there are materials in which this is not the case, and the hydrogen-metal hydride combination should be treated as a ternary system. Experiments have shown that the reaction of hydrogen with several binary magnesium alloys provides examples of such ternary systems [1, 2].

The prior examples of the reaction of lithium with the two copper chloride phases were used to illustrate how thermodynamic information can be used to determine the phase diagram and the electrochemical properties. These hydrogen/magnesium-alloy systems will be discussed, however, *as reverse examples*, in which electrochemical methods can be used in order to determine the relevant phase diagrams and thermodynamic properties, as well as to determine the practical parameters of energy and capacity.

Metal hydride systems are typically studied by the use of gas absorption experiments, in which the hydrogen pressure and temperature are the primary external variables. Electrochemical methods can generally also be employed by the use of a suitable electrolyte and cell configuration. Variation of the cell voltage can cause a change in the difference between the effective hydrogen pressure in the two electrodes. If one electrode has a fixed hydrogen activity, the hydrogen activity in the other can be varied by the use of an applied voltage. This then causes either the absorption or desorption of hydrogen. This can be expressed by the Nernst relation:

$$E = (RT/zF)\Delta \ln p(H_2) \tag{12.17}$$

where E is the cell voltage, R the gas constant, T the absolute temperature, z the charge carried by the transporting ion (hydrogen), and F the Faraday constant. The term $\Delta \ln p(H_2)$ is the difference in the natural logarithms of the effective partial pressures, or activities, of hydrogen at the two electrodes.

Electrochemical methods can have several advantages over the traditional pressure-temperature methods. Since no temperature change is necessary for the absorption or desorption, data can be obtained at a constant temperature. If a stable reference is used, variation of the cell voltage determines the hydrogen activity at the surface of the alloy electrode. Large changes in hydrogen activity can be obtained by the use of relatively small differences in cell voltage. Thus the effective pressure can be easily and rapidly changed over several orders of magnitude. The amount of hydrogen added to, or deleted from, an electrode can be readily determined from the amount of current that passes through the cell.

One of the important parameters in the selection of materials for hydrogen storage is the amount of hydrogen that can be stored per unit weight of host material, the specific capacity. This is often expressed as the ratio of the weight of hydrogen absorbed to the weight of the host material. Magnesium-based hydrides are considered to be potentially very favorable in this regard, for the atomic weight of magnesium is quite low, 24.3 g per mol. MgH_2 contains 1 mol of H_2, and the ratio 2/24.3 means 8.23 w/% hydrogen. This can be readily converted to the amount of charge stored per unit weight, i.e. the number of mAh/g. One Faraday is 96,500 C, or 26,800 mAh, per equivalent. The addition of two hydrogens per magnesium means that two equivalents are involved. Thus 2204 mAh of hydrogen can be reacted per gram of magnesium.

On the other hand, one is often interested in the amount of hydrogen that can be obtained by the decomposition of a metal hydride. This means that the weight to be considered is that of the metal plus the hydrogen, rather than just the metal itself. When this is done, it is found that 7.6 w/%, or 2038 mAh/g hydrogen can be obtained from MgH_2.

These values for magnesium hydride are over five times those of the materials that are commonly used as metal hydride electrodes in commercial battery systems. Thus there is continued interest in the possibility of the development of useful alloys based upon magnesium. The practical problem is that magnesium forms a very stable oxide, which acts as a barrier to the passage of hydrogen. It is very difficult to prevent the formation of this oxide on the alloy surface in contact with the aqueous electrolytes commonly used in battery systems containing metal hydrides.

Fig. 12.14 Magnesium–nickel binary phase diagram

One of the strategies that have been explored is to put a material such as nickel, which is stable in these electrolytes, on the surface of the magnesium. It is known that nickel acts as a mixed conductor, allowing the passage of hydrogen into the interior of the alloy. However, this surface covering cannot be maintained over many charge/discharge cycles, with the accompanying volume changes.

A different approach is to use an electrolyte in which magnesium is stable, but its oxide is not. This was demonstrated by the use of a novel intermediate temperature alkali organo-aluminate molten salt electrolyte NaAlEt$_4$ [1]. The hydride salt NaH can be dissolved into this melt, providing hydride ions, H$^-$, that can transport hydrogen across the cell. This salt is stable in the presence of pure Na, which can then be used as a reference, as well as a counter, electrode.

This experimental method was used to study hydrogen storage in three ternary systems involving magnesium alloys, the H-Mg-Ni, H-Mg-Cu and H-Mg-Al systems. In order to be above the melting point of this organic anion electrolyte, these experiments were performed somewhat above 140 °C.

The magnesium-nickel binary phase diagram is shown in Fig. 12.14. It shows that there are two intermediate phases, Mg$_2$Ni and MgNi$_2$. It is also known that magnesium forms the dihydride MgH$_2$. These compositions are shown on the H-Mg-Ni ternary diagram shown in Fig. 12.15. Note that the ternary diagram is drawn with hydrogen at the top in this case.

In order to explore this ternary system, an electrochemical cell was used to investigate the reaction of hydrogen with three compositions in the Mg-Ni binary

Fig. 12.15 The H-Mg-Ni ternary diagram showing only the known compositions along the binary edges

Fig. 12.16 Loci of the overall composition as hydrogen reacts with three initial Mg-Ni alloy compositions

alloy system, $MgNi_2$, Mg_2Ni, and $Mg_{2.35}Ni$. Thus the overall compositions of these materials moved along the dashed lines shown in Fig. 12.16 as hydrogen was added.

It was found that the voltage went to zero as soon as hydrogen was added to the phase $MgNi_2$. However, in the other cases, it changed suddenly from one plateau potential to another as certain compositions were reached. These transition compositions are indicated by the circles in Fig. 12.17. The values of the voltage versus the hydrogen potential in the different compositions regions are also shown in that figure.

Fig. 12.17 Plateau voltages found in different composition regions

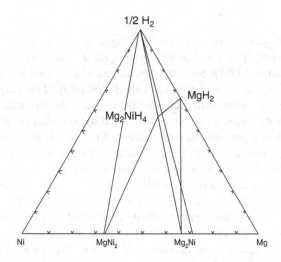

Fig. 12.18 Ternary phase stability diagram for the H-Mg-Ni system at about 140 °C, derived from the compositional variation of the potential as hydrogen was reacted with three different initial binary alloy compositions

This information can be used to construct the ternary equilibrium diagram for this system. As described earlier, constant potential plateaus are found for compositions in three-phase triangles, and potential jumps occur when the composition crosses two-phase tie lines. The result is that there are no phases between $MgNi_2$ and pure hydrogen, but there must be a ternary phase with the composition Mg_2NiH_4. The resulting H-Mg-Ni ternary diagram at this temperature is shown in Fig. 12.18.

Fig. 12.19 Variation of the potential as hydrogen is added to alloy with an initial composition $Mg_{2.35}Ni$

The phase Mg_2Ni reacts with four hydrogen atoms to form Mg_2NiH_4 at a constant potential of 79 mV versus pure hydrogen. The weight of the Mg_2Ni host is 107.33 g, which is 26.83 g per mol of hydrogen atoms. This amounts to 3.73 % hydrogen atoms stored per unit weight of the initial alloy. This is quite attractive, and is considerably more than the specific capacity of the materials that are currently used in the negative electrodes of metal hydride/nickel batteries.

On the other hand, pure magnesium reacts to form MgH_2 at a constant potential of 107 mV versus pure hydrogen. Because of the lighter weight of magnesium than nickel, this amounts to 8.23 % hydrogen atoms per unit weight of the initial magnesium, or 7.6 w% relative to MgH_2. Thus the use of magnesium, and its conversion to MgH_2, is very attractive for hydrogen storage. There is a practical problem, however, due to the great sensitivity of magnesium to the presence of even small amounts of oxygen or water vapor in its environment.

If the initial composition is between Mg_2Ni and Mg, as is the case for the composition $Mg_{2.35}Ni$ that has been discussed above, there will be two potential plateaus, and their respective lengths, as well as the total amount of hydrogen stored per unit weight of the electrode, will have intermediate values, varying with the initial composition. As an example, the variation of the potential with the amount of hydrogen added to the $Mg_{2.35}Ni$ is shown in Fig. 12.19.

Similar experiments were carried out on the reaction of hydrogen with two other magnesium alloy systems, the H-Mg-Cu and H-Mg-Al systems [1]. Their ternary equilibrium diagrams were determined by using analogous methods. They are shown in Figs. 12.20 and 12.21.

Fig. 12.20 Ternary phase stability diagram for the H-Mg-Cu system at about 140 °C, derived from the compositional variation of the potential as hydrogen was reacted with different initial binary alloy compositions using organic anion molten salt electrolyte

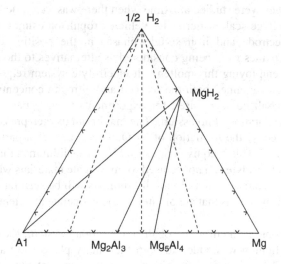

Fig. 12.21 Ternary phase stability diagram for the H-Mg-Al system at about 140 °C, derived from the compositional variation of the potential as hydrogen was reacted with different initial binary alloy compositions using organic anion molten salt electrolyte

12.9   Further Ternary Examples: Lithium-Transition Metal Oxides

These same concepts and techniques have been used to investigate several lithium-transition metal oxide systems [3, 4]. They will be discussed briefly here. These examples are different from those that have been discussed thus far, for in a number of cases the initial compositions are, themselves, ternary phases, not just binary phases.

They further illustrate how electrochemical measurements on selected compositions can be used to determine the relevant phase diagrams. This makes it possible to predict the potentials and capacities of other materials within the same ternary system without having to measure them individually.

In addition, it will be seen that one can obtain a correlation between the activity of lithium, and thus the potential, and the equilibrium oxygen partial pressure, of phases and phase combinations in some cases. This provides the opportunity to predict the potentials of a number of binary and ternary materials with respect to lithium from information on the properties of relevant oxide phases alone.

Data on the ternary lithium–transition metal oxide systems that will be presented here were obtained by the use of the LiCl–KCl eutectic molten salt as electrolyte at about 400 °C. They were studied at a time when there was a significant effort in the USA to develop large-scale batteries for vehicle propulsion using lithium alloys in the negative electrode and iron sulfide phases in the positive electrode. The transition metal oxides were being considered as alternatives to the sulfides.

Experiments employing this molten salt electrolyte system required the use of glove boxes that maintained both the oxygen and nitrogen concentrations at very low levels. This salt could be used for experiments to very negative potentials, limited by the evaporation of potassium. The maximum oxygen pressure that can be tolerated is limited by the formation of Li_2O. This occurs at a partial pressure of 10^{-15} atm at 400 °C. This is equivalent to 1.82 V versus lithium at that temperature. As a result, this electrolyte can not be used to investigate materials whose potentials are above 1.82 V relative to that of pure lithium. As will be seen later, many of the positive electrode materials that are of interest today operate at potentials above this limit.

The first example is the lithium-cobalt oxide ternary system. Experiments were made in which lithium was added to both the binary phase CoO and the ternary phase $LiCoO_2$. The variations of the observed equilibrium potentials as lithium was added to these phases are indicated in Fig. 12.22. It is seen that there were sudden drops from 1.807 to 1.636 V, and then to zero in the case of CoO. Starting with $LiCoO_2$, however, only one voltage jump was observed, from 1.636 to 0. Since these jumps occur when the composition crosses binary tie lines in such diagrams, it was very easy to plot the ternary figure in this case. The result is shown in Fig. 12.23, in which the values of the potential (voltage versus pure lithium), lithium activity, cobalt activity and oxygen partial pressure for the two relevant compositional triangles are indicated. As mentioned earlier, it was not possible to

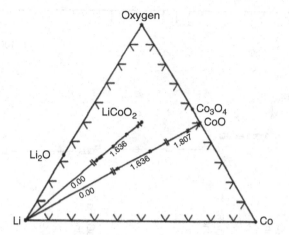

Fig. 12.22 Results of coulometric titration experiments on two compositions in the lithium-cobalt oxide system. After [4]

Fig. 12.23 Ternary phase stability diagram derived from the coulometric titration experiments shown in Fig. 12.22. After [4]

investigate the higher potential regions that are being used in positive electrodes today.

A further example is the Li-Fe-O system. Fig. 12.24 shows the variation of the equilibrium potential as lithium was added to Fe_3O_4 under near-equilibrium conditions. It is seen that this is a more complex case, for after a small initial solid solution region there are three jumps in the potential.

Similar experiments were undertaken on materials with two other initial compositions, $LiFe_5O_8$ and $LiFeO_2$. From these data it was possible to plot out the whole ternary system within the accessible potential range, as shown in Fig. 12.25.

Fig. 12.24 Results of a coulometric titration experiment on a sample with an initial composition Fe_3O_4. After [4]

Fig. 12.25 Ternary phase stability diagram derived from coulometric titration measurements on materials in the Li-Fe-O ternary system. After [4]

Investigation of the Li-Mn-O system produced results that were somewhat different from those in the Li-Co-O and Li-Fe-O systems. The variation of the potential as lithium was added to samples with initial compositions MnO, Mn_3O_4, $LiMnO_2$ and Li_2MnO_3 is shown in Fig. 12.26. The ternary equilibrium diagram that resulted in shown in Fig. 12.27. It is seen that all of the two-phase tie lines do not go

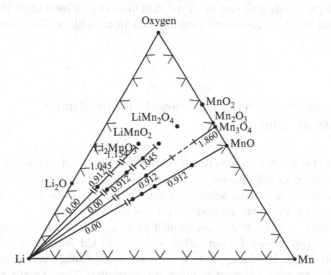

Fig. 12.26 Results of coulometric titration experiments on several phases in the Li-Mn-O ternary system. After [4]

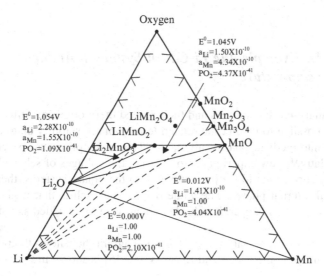

Fig. 12.27 Ternary phase stability diagram that resulted from the coulometric titration results shown in Fig. 12.26. After [4]

to the transition metal corner in this case. Instead, three of them lead to the composition Li_2O. Nevertheless, the principles and the experimental methods are the same.

It will be shown later, in Chap. 19, that some materials of this type behave quite differently at ambient temperature and higher potentials. In some cases lithium can be extracted from individual phases, which then act as insertion-extraction electrodes, with potentials that vary with the stoichiometry of individual phases. The principles involved in insertion-extraction reactions will be discussed later, in Chap. 13.

12.10   Ternary Systems Composed of Two Binary Metal Alloys

In addition to the ternary systems that involve a nonmetal component that have been discussed thus far in this chapter, it is also possible to have ternaries in which all three components are metals. Some such materials are possible candidates for use as reactants in the negative electrode of lithium battery systems.

One example will be briefly mentioned here, the Li-Cd-Sn system, which is composed of two binary lithium alloy systems, Li-Cd and Li-Sn. As will be described in Chap. 18, these, as well as a number of other binary metal alloy systems, have been investigated at ambient temperature. Their kinetic behavior is sufficiently fast that they can be used at these low temperatures. This system, as well as others, will be discussed in connection with the important mixed-conductor matrix concept.

12.10.1   An Example, the Li-Cd-Sn System at Ambient Temperature

If the two binary-phase diagrams and their related thermodynamic information are known, it is possible to predict the related ternary-phase stability diagram, assuming that no intermediate phases are stable. This assumption can be checked by making a relatively few experiments to measure the voltages of selected compositions. If they correspond to the predictions from the binary systems, there must be no additional internal phases. The value of this approach is that it gives a quick picture of what would happen if a third element were to be added as a dopant to a binary alloy.

As an example, the ternary-phase stability diagram that shows the potentials of all possible alloys in the Li-Cd and Li-Sn system [5] is shown in Fig. 12.28.

Fig. 12.28 Ternary phase stability diagram for the Li-Cd-Sn system. The numbers are the values of the voltage of all compositions in the various sub-triangles relative to pure lithium. After [5]

12.11 What About the Presence of Additional Components?

Practical materials often include additional elements, either as deliberately added dopants, or as impurities. If these elements are in solid solution in the major phases present in the ternary system, they can generally be considered to cause only minor deviations from the properties of the basic ternary system. Thus it is not generally necessary to consider systems with more than three components.

12.12 Summary

This rather long chapter has shown that the ideal electrochemical behavior of ternary systems, in which the components can be solids, liquids or gases, can be understood by the use of phase stability diagrams and theoretical electrochemical titration curves. The characteristics of phase stability diagrams can be determined from thermodynamic information, and from them the related theoretical electrochemical titration curves can be determined. Important properties, such as the maximum theoretical specific energy, can then be calculated from this information. A number of examples have been discussed that illustrate the range of application of this powerful method.

References

1. Luedecke CM, Deublein G, Huggins RA (1984) In: Veziroglu TN, Taylor JB (eds) Hydrogen Energy Progress V. Pergamon Press, New York, p 1421
2. Luedecke CM, Deublein G, Huggins RA (1985) J Electrochem Soc 132:52
3. Godshall NA, Raistrick ID, Huggins RA (1980) Mater Res Bull 15:561
4. Godshall NA, Raistrick ID, Huggins RA (1984) J Electrochem Soc 131:543
5. Anani AA, Crouch-Baker S, Huggins RA (1988) J Electrochem Soc 135:2103

Chapter 13
Potentials

13.1 Introduction

Potentials and potential gradients are important in battery systems. The difference in the potentials of the two electrodes determines the voltage of electrochemical cells, being larger when they are charged, and smaller when they are discharged. On the other hand, potential gradients are the driving forces for the transport of species within electrodes.

All potentials (potential energies) are relative, rather than having absolute values.

Since they cannot be measured on an absolute scale, it is desirable to establish useful references against which they can be measured. This is not the case in electrochemistry alone, but is true for all disciplines. For example, when dealing with the potential energies of electrons in solids, the solid-state physics community uses two different references, depending upon the problem being addressed. One is the potential energy of an electron at the bottom of the *valence band* in a solid, and the other is the so-called *vacuum level*, the energy of an isolated electron at an infinite distance from the solid in question. There is no universal relation between these two reference potentials, as the first is dependent upon the identity of the material involved, while the latter is not.

In electrochemical systems potential differences are measured electrically as voltages between some reference electrode system and an electrode of interest. The voltage that is measured is a measure of the difference in the electrochemical potentials of the electrons in the two electrodes.

The approaches to this matter are different between the conventional electrochemical community, whose interests have traditionally been mostly concerned with aqueous systems, and the solid-state electrochemical community, many of whose members have come from a solid-state materials background. This is despite

© Springer International Publishing Switzerland 2016   209
R.A. Huggins, *Energy Storage*, DOI 10.1007/978-3-319-21239-5_13

the fact that some of the electrochemical systems of interest to the latter group also often include liquid electrolytes. It will be seen that one difference is the focus on the properties of neutral species in the solid-state electrochemical community, and upon ionic species in the aqueous electrochemical community.

The matter of the distribution of the different electrical and chemical potentials within electrochemical cells is often misunderstood. It will be seen that this often depends upon the experimental conditions.

13.2   Terminology

The term "potential" is often used for both a single potential and a potential difference. The standard practice in electrochemistry is to use certain reactions to provide reference electrode potentials against which other potentials can be measured. In aqueous systems a standard procedure is to use the reaction

$$2H^+ + 2e^- = H_2 \tag{13.1}$$

as the reference potential, and electrodes which involve this reaction are often called standard hydrogen electrodes (SHE), as discussed in Sect. 13A.9. On the same scale the potential of a lithium electrode at which the reaction is

$$Li^+ + e^- = Li \tag{13.2}$$

occurs at a potential that is -3.045 V with regard to the potential of the SHE. Further, a fluorine electrode operating at a pressure of 1 atm of fluorine gas and for which the reaction can be written as

$$2F^- - 2e^- = F_2 \tag{13.3}$$

has a potential of $+2.87$ V relative to the SHE at ambient temperature.

These potentials will be modified somewhat in other electrolytes because of differences in the *solvation energies*. If the solvation energy is not considered, the difference in electrode potential is always equal to that of the Nernst equation voltage for neutral species outside the electrolyte and one can always write

$$\Delta G_r^0 = -zFE \tag{13.4}$$

where ΔG_r^0 is the Gibbs free energy change of the relevant reaction, z is the number of electrons transferred in the reaction to which ΔG_r^0 refers, F is the Faraday constant, and E is the cell voltage, which is equal to the difference in the electrode potentials on the two sides.

13.3   Potential Scales

Another alternative way of looking at electrode potentials involves the use of a general potential scale based upon a particular reaction equilibrium. In molten salts, for example, it may be useful to use the chlorine or fluorine evolution electrode reaction as a reference against which other electrode potentials are measured.

In the subsequent chapters, reference will be made to chemical and electrostatic *macropotential differences* across a solid, as well as to gradients in those potentials within the solid. The chemical and electrostatic potentials are only two of a number of thermodynamic potentials. Since there is often a considerable amount of confusion relating to the different types of potentials inside solids and near their surfaces, this question will now be considered.

Understanding of the spatial distribution of the various thermodynamic potentials within a solid is important because of the relationship between the values of specific potentials and the local structure. Since many of the properties of solids are dependent upon the local structure, variations in properties with position are both possible and commonplace. Furthermore, under proper conditions, they can be controlled to advantage.

This relationship between local potentials, structure, and properties leads to two general types of application. Local values of some potentials may be experimentally observed by the proper use of appropriate probe techniques. This can lead to valuable information about the structure and can therefore be used as an analytical technique for a host of purposes. In addition, however, the situation can be reversed, and specific values of certain potentials can be imposed upon a material in order to change or control its structure and properties.

The total thermodynamic potential for a given species i at any point can arbitrarily be composed of several factors. Each of these factors has the dimensions of energy, as does the total thermodynamic potential. The total thermodynamic potential of a particular species has the properties of potential energy, and gradients in it produce forces tending to cause the superposition of a long range drift motion upon the local random motion of that species within the solid.

13.4   Electrical, Chemical, and Electrochemical Potentials in Metals

First, consider the matter of electrical potentials and the various related electrostatic potentials of individual species. In order to compare electrical potentials as well as the electrostatic energies of charged particles within and near different solids, it must be recognized that neither of these quantities has an absolute value. Therefore, it is desirable to establish some sort of reference level electric potential. For this purpose it is useful to compare the thermodynamic potential of a charged particle i within a solid phase with the potential of an isolated particle of the same chemical composition in a vacuum at an infinite distance from all other charges. The value of the electrical potential at this charge-free infinity is defined arbitrarily as zero. This fixed reference value is called the *reference vacuum level*, E^∞. Unfortunately, the term *vacuum level* is also used in some of the current semiconductor literature for a different potential, as will be described later.

Consider a hypothetical experiment in which this charged particle is transferred from infinity to a position inside a solid. In the absence of any other potential gradients, the work that would be done can be divided into two parts. One of these is due to the interaction of the particle with the other particles within the bulk solid phase. This will typically include electrostatic, polarization, and repulsive interactions, and thus is dependent upon the identity of the particle, as well as the constitution of the bulk solid. It represents the chemical binding energy of the species in the solid, and is called the *chemical potential* of particle i, μ_i.

The second part of the work involved in transferring the particle from infinity to the interior of the solid is purely electrostatic, and is thus $z_i q(\Phi - E^\infty)$, where z_i is the charge number of the particle (and represents both the sign and the number of elementary charges carried), q is the magnitude of the elementary charge (the value of the charge of the proton), and Φ is the local value of the *electrostatic macropotential* within the solid, which is called the *inner potential*.

The total work involved in this hypothetical experiment is called *the electrochemical potential* of the particle of species i, η_i, within the solid, and since $E^\infty = 0$, can thus be written as

$$\eta_i = \mu_i + z_i q \Phi \qquad (13.5)$$

If the interior of the solid can be considered to be compositionally homogeneous, it will thus have a uniform value of the inner potential Φ. However, the solid phase must have an exterior surface that separates it from its surroundings whether vacuum, gas, liquid, or another solid phase.

For simplicity, consider the case of an isolated solid phase surrounded by vacuum. Because of the structural discontinuity at the surface, there must be local redistributions of both particles and electrical charge compared to the configurational structure within the bulk solid. There can also be differences in the concentrations of intrinsic species between the surface and the interior, as well as adsorbed foreign species upon the surface.

It is useful to utilize a simple model in which the solid is divided into two parts, a uniform interior and a separate surface region. The latter is sometimes called the *selvedge*, the term used for the edge region of a piece of cloth, which is often woven differently from the interior to prevent it from unraveling. The selvedge thus contains all of the various local redistributions and compositional effects, which can be described as producing an electrical double layer. In addition, this surface region contains all the excess charge if the solid has a net electrostatic charge different from E^∞.

Therefore, the value of the inner potential Φ can be divided into two terms, one called the *surface potential X*, which is related to the dipolar effects of the electrical double layer in the selvedge. The second is known as the *outer potential* Ψ, and is the net externally measurable electrostatic potential of the solid.

The value of outer potential Ψ is dependent upon the amount of excess charge Q and the dimensions of the solid. For the case of a sphere of radius a,

$$\Psi = Q/a \tag{13.6}$$

The surface potential can be interpreted in terms of a simple model consisting of a uniform distribution of dipoles of moment M perpendicular to the surface with a concentration of N per cm^2 within the selvedge. If the positively charged ends of the dipoles are on the outside,

$$X = -4\pi NM \tag{13.7}$$

The relationship between these potentials for the case of two chemically identical solids with different amounts of excess charge, and thus different values of the outer potential, is shown in Fig. 13.1.

Changes in the charge on a solid body are actually accomplished by the transfer of charged particles, e.g., electrons, so that the composition is actually slightly changed when the net charge is varied. However, this involves such minor changes in the concentrations of the particles present in the solid that they can be neglected.

Now consider the question of the *energies* (also sometimes called *potentials*) of charged species. As mentioned already, the electrostatic part of the total potential energy of a particle i of charge $z_i q$ inside a solid with an inner potential Φ is $z_i q\Phi$. In

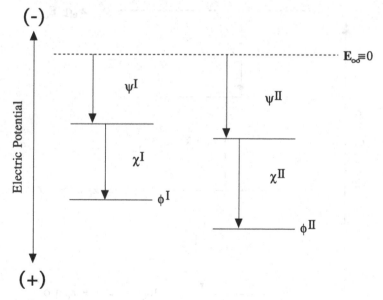

Fig. 13.1 Relationship between potentials related to two chemically identical solids with different values of outer potential

the case of electrons, $z_i = -1$. Thus this part of the total potential energy of an electron has an absolute magnitude that is greater and more positive the lower the value of Φ, since Φ is negative relative to the zero reference E^∞,

Therefore, the total energy of an electron within a solid is its electrochemical potential $\mu_{e'}$, which has two components. One is related to the fact that the electron is within the solid and has the characteristics of a *chemical binding energy*, and the other is *purely electrostatic*. Thus

$$n_{e^-} = \mu_{e^-} + z_{e^-}q\Phi \tag{13.8}$$

and

$$\eta_{e^-} = \mu_{e^-} + RT\ln a_{e^-} + z_{e^-}q(\Psi + X) \tag{13.9}$$

These energy relations for a single electron in the interior of a metal are shown in Fig. 13.2. Note that the value of the chemical potential for the electron is also negative.

To reinforce the understanding of these matters, a hypothetical experiment can be considered in which a single electron exists between two parallel plates of the same metal which are maintained in a vacuum, but are connected to the opposite terminals of a battery. This is illustrated in Fig. 13.3, in which the right-hand plate is connected to the positive pole, and the left-hand plate to the negative pole of the battery.

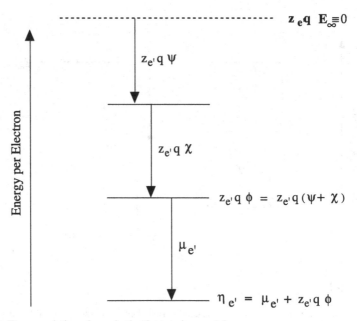

Fig. 13.2 Energy relations for a single electron in a metal

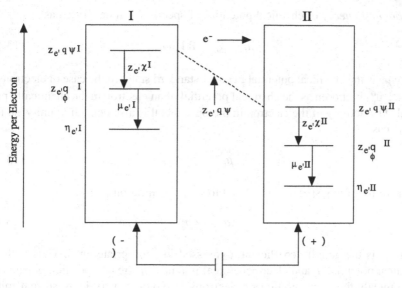

Fig. 13.3 Relationship between the potentials in two identical materials connected to the terminals of a battery

There will be a force acting upon the electron that is proportional to the negative value of the gradient in its potential energy in the vacuum $z_{e-}q \, d\Psi/dx$. This will cause it to accelerate from left to right. This is consistent with our expectation from general electrostatics that a negatively charged particle will be attracted to a positively charged electrode.

It can be seen from this example that the values of $z_i q\Psi$, μ_i and η_i all change as the externally measurable electrostatic potential of the solid Ψ is varied. However, if the chemical constitution is not altered, the values of X and μ_i remain the same, so that the relative values of $z_i q\Psi$, $z_i q\Phi$ *and* η_i do not change.

It will be seen later that the quantity Ψ can be varied externally, as in the above example, and also experimentally measured. It has therefore been found useful to define another quantity, the *real potential* α_i, which is independent of the value of Ψ. This is given as

$$a_i = \mu_i + z_i qX \tag{13.10}$$

For the case of an uncharged solid, where $\Psi = 0$, $\alpha_i = \mu_i$ and α_i, which generally has a negative value, is the work done in transferring a particle of species i from infinity to the interior of the uncharged solid.

The real potential of an electron $\alpha_{e'}$ thus has the same magnitude, but opposite sign, as the *electronic work function* $W_{e'}$, which is defined as the Gibbs free energy necessary to extract an electron from an uncharged solid into an exterior vacuum (at infinite distance). That is,

$$\alpha_{e-} = W_{e-} \tag{13.11}$$

The *binding energy* or chemical potential of species i can be written as

$$\mu_i = \mu_i^0 + RT\ln a_i \tag{13.12}$$

where μ_i^0 is the chemical potential in some standard state. In the case of electrons in a metal, $\mu_{e^-}^0$ is chosen as the chemical potential of an electron in the chemically pure metal. The activity of the electron in the pure metal is also defined as unity, so that in this case

$$\mu_{e^-} = \mu_{e^-}^0 \tag{13.13}$$

The activity of any species i is related to its concentration $[i]$ by

$$a_i = \gamma_i[i] \tag{13.14}$$

where γ_i is the activity coefficient, expressed in appropriate units. Thus both the chemical potential μ_i and electrochemical potential η_i are composition-dependent.

In metals the concentration of electrons is typically very high, so that minor changes in composition due to impurities or doping produce negligible effects upon μ_{e^-} and η_{e^-}. However, this factor should be taken into consideration in more heavily alloyed metals as well as in non-metals. The two contributions to μ_i, and thus to η_i are shown in Fig. 13.4.

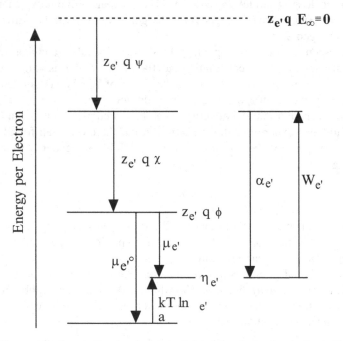

Fig. 13.4 The contributions to the chemical potential and the electrochemical potential

Table 13.1 Values of the electronic work function measured on polycrystals

Metal	Work function/eV
Ag	4.33
Ba	2.39
Be	3.92
Ca	2.71
Co	4.41
Cs	1.87
Fe	4.48
K	2.26
Li	2.28
Mg	3.67
Mo	4.20
Na	2.28
Ni	4.61
Rb	2.16
Ta	4.19
U	3.27
W	4.49
Zn	4.28

Table 13.2 Crystallographic orientation dependence of the work function: single crystals

Metal	Normal to Surface	Work Function/eV
Cu	111	4.39
Cu	100	5.64
Ag	111	4.75
Ag	100	4.81
W	111	4.39
W	112	4.69
W	001	4.56
W	110	4.68

Because the real potential and the work function include the term $z_{e^-}qX$ that relates to the electrical double layer effects in the selvedge, these values are dependent upon the details of the structure in that region. Experiments have shown that this includes both the crystal face from which electrons are emitted and the presence of any impurities upon the surface. Some experimental values are given in Table 13.1 for polycrystalline metals. Table 13.2 shows the variation with crystal face on single crystals of several metals.

Fig. 13.5 Simple Band model of a metal

13.5 Relation to the Band Model of Electrons in Solids

A combination of knowledge about the variation of the density of energy states available for electron occupation with energy and the state occupation probability, which is expressed in terms of the Fermi-Dirac relation, provides information about the distribution of the electrons within a solid among their allowed energy states. This type of information is often displayed in simplified form by use of an *energy band model*, in which the energy per electron is plotted versus distance. An example of a simple band model of a metal is shown in Fig. 13.5. This figure also shows the relationship to the various thermodynamic potentials discussed here.

13.6 Potentials in Semiconductors

While the discussion thus far has centered upon metals, similar conclusions are found for semiconductors. In (undoped) intrinsic semiconductors, $\mu_{e-} = \mu_{e-}{}^0$ and the electrochemical potential is the same as the *Fermi level* E_F The Fermi level is midway between the electron energies at the top of the valence band E_V and the bottom of the conduction band E_C, regardless of the temperature. This situation is shown in Fig. 13.6.

Fig. 13.6 Relation between thermodynamic potentials and potentials commonly used in the energy band model of an intrinsic semiconductor. E_G is the band gap (E_C–E_V). The potential of the Fermi level E_F is equal to the electrochemical potential of the electrons, η_{e-}

An important difference between metals and semiconductors, however, is that the electrostatic contribution to the total energy of an electron $z_{e-}q\Phi$ is generally not the same throughout the solid in semiconductors. It often increases or decreases significantly near the surface, or at locations where the chemical composition varies, such as at *p-n junctions*.

What if the semiconductor is doped with an *altervalent*, or *aliovalent*, species, an atom that carries a different amount of electrical charge from those that are normally present. The *electroneutrality requirement* causes the ratio of conduction electrons to holes to change to compensate for the charge of this *foreign species*, or *dopant*. As an example, if *donors* are present the concentration of itinerant conduction band electrons is increased. The activity of the electrons is thus greater in such an *extrinsic* semiconductor than in the corresponding *intrinsic* (undoped) material. This raises the values of kT ln a_{e-}, thus reducing μ_{e-} (which is negative), and raising η_{e-}. In semiconductor band model language this raises E_F toward E_C. This also, of course, decreases the work function W_{e-} since one can assume that X is not changed.

13.7   Interactions Between Different Materials

Many applications of these concepts involve interactions between the various potentials and energies that have been discussed here. In order to understand such matters a simple case will be considered.

Fig. 13.7 Potentials of two chemically different metals separated by a vacuum, and not in equilibrium with each other. The relative positions of the Fermi levels (electrochemical potentials) is arbitrary, depending upon prior history

Consider two chemically different metals. If they are physically separated and not in equilibrium with each other, their potentials can be portrayed as illustrated in Fig. 13.7.

13.8   Junctions Between Two Metals

If these two metals are brought into contact, so that thermal equilibrium can be established between them, electrons will flow from one to the other until the total energy per electron is the same in both. This means that after equilibrium is attained, $\mu_{e'}^{I}$ must be equal to $\mu_{e'}^{II}$, or in band model language, $E_{F}^{I} = E_{F}^{II}$. The question is how this is achieved. The values of $\mu_{e'}^{\infty I}$ and $\mu_{e'}^{\infty II}$ are chemically binding energies of electrons in the lowest levels of the respective conduction bands. These are thus fixed by the chemical compositions of the two metals. The values of $kT \ln a_{e'}$ are determined by the electron concentrations in the metals. Since these concentrations are typically very high in metals, the relatively small number of electrons that pass from one metal to the other upon contact will make relatively small changes in the activities of the electrons. Thus this term will also not change significantly.

This means that the equilibration of the Fermi levels occurs primarily by that adjustment of the electrostatic energy term $z_{e'}q\Phi$ or $z_{e'}q(\Psi + X)$.

Since in each case $\mu_{e'} = \mu_{e'} + z_{e'}q\Phi$, upon equilibration of the $\mu_{e'}$ values (Fermi levels) a fixed value of the difference in the internal electrostatic potentials will be established, directly related to the difference in the chemical potentials of electrons in the two metals. That is

$$z_{e'}q\left(\Phi^{I} - \Phi^{II}\right) = \mu_{e'}^{II} - \mu_{e'}^{I} \tag{13.15}$$

The value of $z_{e'}q\left(\Phi^{I} - \Phi^{II}\right)$ is called the *Galvanic voltage*, or *Galvanic potential difference*, and is characteristic of the two metals in question. It cannot be measured, however, because it is not possible to separate the two contributions to the value of $\Delta\Phi$, the differences in the outer potential $\Delta\Psi$ and in the surface double layer potential ΔX. The transfer of only a relatively few electrons from one metal to the other is expected to modify the X values significantly. As a result, $\left(\Phi^{I} - \Phi^{II}\right)$ is not equal to $\left(\Psi^{I} - \Psi^{II}\right)$.

However, it is possible to measure experimentally $\left(\Psi^{I} - \Psi^{II}\right)$, since these are externally observable values. The energy difference $z_{e'}q\left(\Psi^{I} - \Psi^{II}\right)$ is called the *Volta potential difference*. It is also sometimes called the *contact potential*. The use of this latter term is unfortunate, for it actually relates to a difference in electric potential between two free surfaces that are *not* in physical contact with each other. The relations between the various potentials when two chemically different metals are brought into equilibrium are illustrated in Fig. 13.8.

It is seen that the Volta potential difference is equal to the difference in electron work functions

$$z_{e'}q\left(\Psi^{I} - \Psi^{II}\right) = W_{e'}^{I} - W_{e'}^{II} \tag{13.16}$$

or

$$z_{e'}q\left(\Psi^{I} - \Psi^{II}\right) = z_{e'}q\left[\left(\Psi^{I} - \mu^{I}\right) - \left(\Psi^{II} - \mu^{II}\right)\right] \tag{13.17}$$

13.9 Junctions Between Metals and Semiconductors

Similar considerations are important in the case of equilibration between a metal and a semiconductor. Again, the important feature is that the Fermi levels must be equal under equilibrium conditions. This will not be discussed here, however. The principles are the same as have been elucidated in the lasts few pages, and this topic is addressed in many other places in the literature.

Fig. 13.8 Relationship between various potentials when two chemically different metals are brought into electronic equilibrium, so that their Fermi energies become equal. For purposes of illustration, the two metals are shown physically separated

13.10   Selective Equilibrium

This discussion has assumed that local equilibrium can be maintained within solids. In most practical cases this is reasonable at elevated temperatures. However, it is often not true for all species at lower temperatures. Indeed, it is often found that *selective equilibrium* occurs at low (e.g., ambient) temperatures. The concentrations of less-mobile species can be established during processing at high temperatures, and *frozen in* by cooling to lower temperatures, where they may not be in accord with low temperature equilibria. More mobile species can react to the influence of various forces and reach appropriate equilibria at lower temperatures. The frozen-in less-mobile species do, however, influence the local electrostatic charge balance and thus can play a major role in determining the concentrations of the more mobile defects, as all species participate in the charge balance.

13.11 Reference Electrodes

Reference electrodes play an important role in the study of many aspects of electrochemical systems. Experimental work reported in the literature can involve the use of different reference systems, and it is sometimes difficult to translate between measurements made with one from those made using another.

This situation is made even worse by the fact that reference electrodes that are used in solid-state electrochemical systems are based upon the potentials of electrically neutral chemical species that can be understood by the use of normal chemical thermodynamics. On the other hand, the general practice in aqueous electrochemistry is to use reference electrodes that involve the properties of ions, and the pH of the electrolyte becomes important in some cases, but not in others.

These matters are discussed in terms of the Gibbs Phase Rule, showing the difference between zero-degree-of-freedom (ZDF) electrodes, and those in which an additional parameter, such as the electrolyte pH, must be specified.

The interrelationship between these two types will be illustrated using potential-pH plots, or Pourbaix diagrams. Use of this *thinking tool* provides a simple understanding of the glass electrode systems that are used to measure the pH of electrolytes, for example.

It will also be shown that in electrodes with a mixed-conducting matrix and an internal ZDF reaction, the potential is determined by the internal chemical reaction, rather than the external electrochemical reaction.

13.12 Reference Electrodes in Nonaqueous Lithium Systems

Much of the current interest in batteries involves lithium-based systems with nonaqueous electrolytes. Thus attention should first be directed to the matter of reference electrodes in lithium systems.

13.12.1 Use of Elemental Lithium

Pure metallic lithium is typically used as a reference electrode in experimental activities to investigate the properties of individual electrode components, both those of interest as negative electrodes and those that act as positive electrodes, at ambient and near-ambient temperatures.

Because it is so extremely reactive, it is very difficult to maintain the surface of lithium free of oxide or other layers in even the cleanest gaseous and liquid environments. It is also important to realize that the organic electrolytes that are often used with lithium reference electrodes are typically not stable in the presence of elemental lithium. Reaction product layers are commonly present on the surface of the lithium, and separate the lithium from the bulk electrolyte. This topic is discussed in Chaps. 14 and 16.

Therefore, the reaction that takes place at the electrochemical interface is typically not really known. It is also important that the identity of the electrolyte is not important, so long as it acts to transport Li ions, and not electrons. Despite these factors, elemental lithium is a widely used and highly reliable primary potential reference in a wide range of lithium-based electrochemical systems.

13.12.2 Use of Two-Phase Lithium Alloys

Some years ago there were substantial efforts to develop elevated temperature lithium-based batteries. Since they operated above its melting point, metallic lithium could not be used. One of the reference electrodes that was often employed was a two-phase mixture of aluminum and the phase LiAl.

In auxiliary experiments the potential of that electrode could be compared to that of pure lithium, which was considered to be the primary reference, and to which all potentials were referred. Because of the entropy change involved in the formation of LiAl by the reaction of lithium with aluminum, this difference is temperature-dependent. Experiments [1] from 375 °C to 600 °C showed that the potential of a two-phase mixture of lithium and LiAl is more positive than that of pure lithium, and that this potential difference ΔE can be expressed by the following relation:

$$\Delta E = 451 - 0.220T(K) \qquad (13.18)$$

where ΔE is in millivolts.

This is shown in Fig. 13.9. The reason for the use of a two-phase mixture in this case, and why it is suitable, will become obvious from the discussion below.

Because of their high lithium activity, these 2-phase electrodes also can have reaction product layers on their surfaces in the presence of some electrolytes. As in the case of pure lithium reference electrodes, neither the presence of such layers nor the details of the interfacial reactions have any significant influence upon their potential.

13.13 Reference Electrodes in Elevated Temperature Oxide-Based Systems

Electrochemical systems with solid electrolytes are employed in high temperature fuel cells, oxygen sensors, and related applications. In such cases there are also two general types of reference electrodes used.

Fig. 13.9 Temperature dependence of the voltage between two-phase Li-LiAl electrode and pure lithium [1]

13.13.1 Gas Electrodes

An inert metal such as platinum in contact with pure oxygen gas is often used as a primary reference in these systems. Alternatively, air or some other gas with a known oxygen activity can be used. Gases with lower oxygen partial pressures will have less positive potentials. The potential difference between pure oxygen and a gas with a lower oxygen partial pressure is typically expressed in terms of the Nernst equation:

$$\Delta E = -RT/zF\ln p(O_2) \tag{13.19}$$

where $p(O_2)$ is the oxygen partial pressure of the gas in question. R is the gas constant, z the charge carried by the electrons involved in the assumed reaction (-4), and F the Faraday constant. Air is often used as a reference instead of pure oxygen. Using equation 2 it is possible to compare the potential of an air reference with that of pure oxygen. If it is assumed that air has an oxygen partial pressure of 0.79 atm, ΔE is equal to 6.09 mV at 1200 K, or 927 °C. Thus the air reference potential is 6.09 mV lower than that of pure oxygen at that particular temperature.

13.13.2 Polyphase Solid Reference Electrodes

An alternative to the use of a gas reference electrode is the use of solid electrodes. One example is a 2-phase mixture of Ni and NixO. If conditions are such that an equilibrium between Ni and its oxide can be attained, this combination will have a fixed value of oxygen activity, equal to that in NixO at its Ni-rich compositional limit. Thus this two-phase mixture can be used as a secondary reference instead of pure oxygen. The oxygen activity and the potential relative to oxygen can be calculated if the Gibbs free energy of formation of NiO is known. The formation reaction is:

$$Ni + \frac{1}{2}O_2 = NiO \tag{13.20}$$

The potential of this materials combination is also less positive than that of pure oxygen. The difference can be calculated from the simple relation:

$$\Delta E = -\left(\Delta G_r^0 \middle/ zF\right) \tag{13.21}$$

where ΔG_r^0 is the Gibbs free energy change involved in the formation reaction. In this case, $z = -2$, as only one oxygen atom is involved. Because the Gibbs free energy contains an entropy term, the value of ΔE will be temperature-dependent. At 925 °C the value of ΔG_r^0 is -132.16 kJ/mol. Thus the potential of the two-phase Ni, NixO system is 0.685 V less positive than the potential of pure oxygen at that temperature.

One should note that, as was the case in the lithium systems, the potentials and potential differences in these oxide-related cases are independent of the details of the interfacial reactions. The identity of the electrolyte is not important, so long as it effectively transports oxygen ions and has a relatively low electronic conductivity.

The open circuit voltage of an H_2/O_2 fuel cell is also independent of the details of the interfacial electrochemical reactions as well as the identity and detailed properties of the electrolyte. Regardless of whether the electrolyte transports hydrogen ions or oxygen ions, the voltage is always determined by the thermodynamics of the reaction in which water is formed from hydrogen and oxygen. The electrolyte does not need to be solid; it can also be liquid, and either acidic or basic. It can also have a composite structure, such as when a liquid electrolyte is contained within a solid polymer, such as Nafion.

13.14   Relations Between Binary Potential Scales

One can determine the relation between different potential scales if they both refer to a common reference. As an example, consider the relation between a scale based upon the potential of pure lithium, or one based on sodium. Lithium and sodium both react with oxygen to form their respective oxides. If it can be assumed that those reactions were to occur with oxygen at unit activity (1 atm), the difference between the potentials of Li and Na and that of pure oxygen can be calculated from their respective oxide formation reactions.

At 25 °C, the Gibbs free energy of formation values [2] are -562.104 kJ/mol for Li_2O, and -379.090 kJ/mol for Na_2O. Using the relation between the voltages and these Gibbs free energies of formation, it is found that the potential of pure Li is 2.91 volts, and that of sodium 1.96 volts, negative of pure oxygen at 298 K. Those values are the ranges of stability of their respective oxides.

13.15   Potentials in the Ternary Lithium: Hydrogen: Oxygen System

This situation is modified in the case of a ternary system. As an example, if lithium is also present in addition to hydrogen and oxygen, the potential limits of the stability of water are shifted. To understand this, the isothermal Li-H-O ternary phase diagram must be considered.

There are several phases in this system in addition to the elements: Li_2O, LiH, LiOH, $LiOH \cdot H_2O$ and H_2O. The values of the standard Gibbs free energy of formation of these phases are given in Table 13.3.

The locations of these phases are shown in the isothermal ternary Gibbs triangle in Fig. 13.10. Assuming that all of these phases are at unit activity, the potentials of all of the 3-phase sub-triangles can be calculated with respect to each of the elements at the corners of the Gibbs triangle, from thermodynamic information, as discussed earlier. The values shown in that figure are voltages versus pure lithium.

The potential range in which water is stable is bounded by the potentials of two triangles, the three-phase triangles that have H_2O at their corners. It can be seen that their potentials with respect to lithium are 2.23 V and 3.46 V, and 1.23 V apart. The presence of the phases LiOH and $LiOH \cdot H_2O$ caused their potentials to both shift in the positive direction relative to that of Li.

It was pointed out that these calculations relate to a very basic aqueous electrolyte, with a pH value of 14. Conversion of the potential of the triangle that has both hydrogen and water present (2.23 V) to that at pH zero can be done by adding the product of 14 and 0.059 V, the change in potential per pH unit. The result is 3.05 V, which is the value found in electrochemical tables for the potential of the standard hydrogen electrode (SHE).

Table 13.3 Values of the standard Gibbs free energy of formation of phases in the Li-H-O system at 25 °C

Phase	ΔG_f^0/kJ/mol
Li_2O	-562.1
LiOH	-439.0
$LiOH . H_2O$	-689.5
H_2O	-237.1
LiH	-68.5

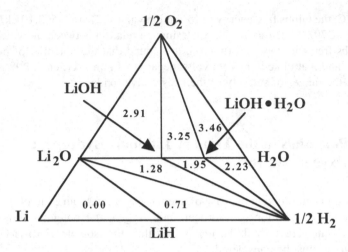

Fig. 13.10 Isothermal Gibbs triangle for the Li-H-O system at 25 °C. The numbers within the sub-triangles are the calculated values of their respective potentials vs. pure lithium

13.16 Lithium Cells in Aqueous Electrolytes

It was pointed out earlier that this relationship between these different potential scales means is that if a material has a potential that is between 2.09 and 3.32 V positive of pure lithium and does not dissolve or otherwise react chemically, it will be stable versus water containing LiOH. Thus one can use electrodes that react with lithium in aqueous electrolytes if Li ions are present in the electrolyte. Lithium-based electrochemical cells can operate in aqueous electrolytes, so long as both electrodes react with lithium and their potentials are within this range. This has been demonstrated experimentally [3–6]. Figure 13.11 shows cyclic voltammograms of VO2(B) in two different aqueous electrolytes, one containing Li ions, and the other not. Since the only appreciable reaction occurs when the Li ions are present in the water, it is obvious that the electrode reacts with Li, rather than hydrogen or oxygen.

If the lithium activity is too high in such an electrode, i.e., it has a potential less than 2.23 V versus pure lithium in water of pH 14, it will reduce water, forming H_2 and Li_2O.

13.17 Significance of Electrically Neutral Species

An important feature of the discussion of both nonaqueous and aqueous systems has been that the potentials and thermodynamics of electrically neutral species in the electrodes are important. Potentials and voltages are independent of the identity, or

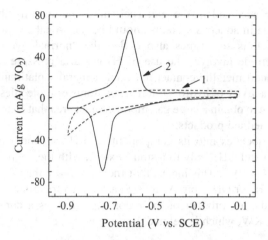

Fig. 13.11 Cyclic voltammograms of $VO_2(B)$ in two different aqueous electrolytes. Scan 1 was made in one without lithium ions, whereas lithium ions were present in the electrolyte of scan 2 [6]

even the character, of the electrolyte. They are directly related to the normal chemical thermodynamic properties of the electrode materials. Reference electrodes are typically elements or thermodynamically-related polyphase mixtures that are electrically neutral.

13.18   Reference Electrodes in Aqueous Electrochemical Systems

The examples discussed above indicate that the reference electrode situation is quite straightforward and is consistent with conventional thermodynamics in nonaqueous systems. However, this is quite different when dealing with aqueous systems, which are within the domain of traditional electrochemistry.

If one looks into the older electrochemical literature, he finds statements such as that reference electrodes were considered somewhat of a "black art" for many years, with information primarily passed on by word of mouth or in brief notes among workers in electrochemistry [7]. A significant step forward was the book entitled *Reference Electrodes* that was edited by Ives and Janz in 1961 [8]. Another source that is often cited is the review article by Butler in 1970 that dealt with reference electrodes in aprotic organic solvents [7].

This general approach to the reference electrode matter is quite different from that described earlier in this chapter. One major difference is that a property of the electrolyte, the hydrogen ion concentration, as expressed in the form of the pH, is generally considered important. This is different from the examples above, in which the electrolyte plays no role other than acting as an ion-pass and electronically-impervious filter.

It is generally accepted in the electrochemical community that the primary reference electrode in aqueous systems should be the so-called standard hydrogen electrode (SHE). It is sometimes also called the "normal hydrogen electrode" (NHE). This electrode involves the use of H_2 gas at a pressure of 1 atmosphere flowing over an inert metallic contact material (generally platinum) in an electrolyte in which the activity of hydrogen ions (not atoms or molecules) is unity. Pains are generally taken to obtain a large gas/metal contact area that is not blocked by the presence of intermediate products.

Pure water dissociates into its component ions H^+ (or more properly, H_3O+, $H_5O_2^+$, or $H_9O_4^+$) and OH^- only to a small extent, with the degree of dissociation equal to about 1.4×10^{-9}. This means that there are more than 7×108 molecules of water for each H^+ or OH^- ion. As in the case of electrically charged defect pair equilibria in solids, the product of their concentrations is a constant. The ionic product of water, KW, which is defined as:

$$K_W = [H^+][OH^-] \tag{13.22}$$

has been found to be approximately 10^{-14}.

In solutions of acids or bases the relative concentrations of these two ionic species can vary over many orders of magnitude, mostly much less than unity. A logarithmic function, pH, was introduced as a measure of the concentration of one of them, the H^+ ions. Because of the ionic product equilibrium, the value of the other follows. The definition of pH is:

$$pH = -\log[H^+] \tag{13.23}$$

Thus in neutral water, the concentration of H^+ ions is equal to that of the OH^- ions, and both are 10^{-7} per cm^3, so that the value of the pH is 7.

The assumption is generally made that the activity of H^+ ions is the same as their concentration, $[H^+]$. Thus the value of the pH at the SHE reference electrode where the H^+ ion activity is unity must be 0. In experiments it is often not convenient to actually have an SHE, and the associated pH 0, in an experiment. A number of other types of electrodes are thus generally employed as secondary references. Some of these are listed in Table 13.4.

Table 13.4 Examples of reference electrodes used in aqueous systems

Electrode	Voltage vs. SHE at pH $= 0$/V
Hg/HgO—0.1 M NaOH	0.926
Hg/Hg$_2$Cl$_2$—0.5 M H$_2$SO$_4$	0.68
Hg/Hg$_2$SO$_4$—sat. K$_2$SO$_4$	0.64
Hg/Hg$_2$Cl$_2$—0.1 M KCl	0.3337
Hg/Hg$_2$Cl$_2$—1 M KCl	0.2801
Hg/Hg$_2$Cl$_2$	0.2681
Calomel—sat. KCl	0.2412
Calomel—sat. NaCl	0.2360
Ag/AgCl	0.2223

13.19 Historical Classification of Different Types of Electrodes in Aqueous Systems

In the electrochemical literature one often finds that electrodes used in aqueous systems have been historically classified into three main types, electrodes of the first kind, electrodes of the second kind, and redox, or oxidation-reduction, electrodes.

13.19.1 Electrodes of the First Kind

Some of the common electrodes are sometimes called *cationic electrodes*, although there are also anionic examples. In addition to an inert electrical lead, they commonly consist of a single metal phase that is in contact with an electrolyte containing its cations. Common examples include metallic Ag, Bi, Cd, Hg, or Ni.

Another example is the *reversible hydrogen electrode* (RHE), in which bubbles of gaseous hydrogen at 1 atm pressure flow over a catalytic, but electrochemically inert, metallic surface that is in contact with the electrolyte. The general construction is the same as that of an SHE electrode, except that the pH of the electrolyte is not fixed at 0.

The potential of an electrode M of the first kind is generally given as:

$$E = \text{Constant} + \left(\frac{RT}{zF}\right)\ln[a(M^+)/a(M)] \qquad (13.24)$$

where z is the number of electrons per cation in the electrolyte. The constant is called the "standard electrode potential," E^0. If M is an element, its activity is defined as unity, or simply:

$$E = E^0 + \left(\frac{RT}{zF}\right)\ln a(M^+) \qquad (13.25)$$

Thus the electrode potential is a function of a property of the electrolyte, the activity, or concentration, of the M^+ ions.

In the case of the reversible hydrogen electrode, $a(H_2)$ is the pressure of the hydrogen gas. If this is 1 atm, this value is unity. Thus the reversible hydrogen electrode potential can be approximately stated as:

$$E = E^0 - 2.303\left(\frac{RT}{F}\right)(pH) \qquad (13.26)$$

If the pH is 0, this is equivalent to the SHE, so that the standard electrode potential E^0 is equal to zero. Thus the difference between the RHE and the SHE is

$$E_{RHE} = E_{SHE} = -2.303 \left({}^{RT}\!/_F \right) (pH) \qquad (13.27)$$

The value of the first term on the right-hand side is 0.059 V at 298 K. This difference (in volts) is then:

$$E_{RHE} - E_{SHE} = -0.059(pH) \qquad (13.28)$$

There are also analogous "anionic electrodes" of the first kind that contain an elemental gaseous species, such as Cl_2, that enters the electrolyte as anions. In that case, n will have a negative value.

13.19.2   Electrodes of the Second Kind

Electrodes of the second kind have two solid phases in contact with the (liquid) electrolyte, as well as an inert electrical lead. One of the phases is typically a metal, and the other is a *sparingly soluble salt*, or compound, of that metal. Examples of this type are (Ag, AgCl), (Hg, Hg_2Cl_2) (commonly called the *calomel electrode*), (Hg, $Hg_2 SO_4$), and (Hg/HgO).

These are sometimes called *anionic electrodes*, and generally also are in contact with, or contain, a solution that has a salt of the same anion as that in the second solid phase. As an example, the *standard calomel electrode* (SCE) generally contains solid Hg and solid Hg_2Cl_2 in contact with a saturated solution of KCl. Under these conditions the potential of this electrode is 0.242 V positive of the primary reference SHE potential.

The complication is that experiments are often not performed under the same conditions as those required for the reference electrode. The main issue is the electrolyte. If the experimental and the reference electrode electrolytes are different, they can be connected by use of an additional intermediate electrolyte. Traditionally, this involved the use of a *salt bridge* containing a liquid electrolyte, and care was taken that it did not introduce a significant liquid junction potential. Many modern electrodes use solid electrolytes, such as special ionically-conducting glasses, for this purpose.

If a reference electrode of this type is chemically isolated from the electrolyte, its constitution cannot change, and it will have an electric potential that is independent of the composition of the electrolyte being used in the experiment.

An electrode of the second kind containing solid Hg and solid HgO is often used in alkaline electrolytes, where it is in direct contact with the experimental electrolyte. Under these conditions it behaves differently from the electrodes that contain chloride and sulfate species. In this case the electrode potential is given by:

$$E = E^0 - \left({}^{RT}\!/_F \right) \ln a(OH^-) \qquad (13.29)$$

or

$$E = E^0 + 2.303 \left(\frac{RT}{F} \right) \log[H^+] \tag{13.30}$$

which is also

$$E = E^0 + 2.303 \left(\frac{RT}{F} \right) (pH) \tag{13.31}$$

This means that the potential varies with the pH of the electrolyte in the same way as does the reversible hydrogen electrode (RHE), although they have different values of E^0.

Thus there is a difference between the electrolyte pH dependence of the potentials of these two types of electrodes and those discussed above with isolated chloride or sulfate species. This can be understood by use of *Potential-pH plots*, often called *Pourbaix diagrams*, due to their development by *M.* Pourbaix [9]. These figures are very useful *thinking tools*, as they not only show how the potentials of various reactions vary with the pH of the electrolyte but also indicate domains of stability of the different phases present. The general form of such a diagram is shown in Fig. 13.12.

General Form of Pourbaix Diagram

Fig. 13.12 General form of an E vs. pH, or Pourbaix, diagram. The influence of the pH on the potentials at which H_2 and O_2 have unit activity is shown. When the pH is zero, the potentials of the SHE and RHE electrodes are equal

It can be seen that the potential of the SHE, with the requirement that the activity of the H^+ ions is unity, so that it must be physically isolated, is independent of the pH of the electrolyte, whereas the potential of the RHE varies with the pH. As shown above, the potential of the Hg/HgO electrode is pH-dependent as well, whereas the potentials of the Ag/AgCl, Hg,Hg_2Cl_2, and Hg,H_2SO_4 electrodes are not.

The Gibbs Phase Rule can be used to help understand these things, as well as the difference between the treatment of reference electrodes in nonaqueous and aqueous electrochemical systems.

13.20 The Gibbs Phase Rule

The Gibbs Phase Rule [10, 11] plays an important role in the consideration of phase equilibria, and was discussed in Chap. 4. It will be briefly reviewed here, and its application to reference electrodes discussed.

The Gibbs phase rule can be written as:

$$F = C - P + 2 \tag{13.32}$$

where C is the number of components (e.g., elements or electrically neutral stable entities), P is the number of phases present. F is the number of degrees of freedom, or the number of thermodynamic parameters that must be specified in order to define the system and all of its associated electrical and chemical properties.

The eligible thermodynamic parameters are the temperature, the total pressure, and the chemical composition of each phase present. These are all intensive variables, so their values are independent of the amount of any of the materials present.

In binary systems $C = 2$, so that if the temperature and the overall pressure are held constant, and there are two phases present so that $P = 2$, the value of F is zero. This means that all of the intensive variables then have fixed values. These are thus *zero-degree-of-freedom (ZDF) conditions,* and electrodes will have a fixed potential, regardless of the state of charge, and the amounts of the various phases present.

On the other hand, if there is only one phase present in a binary system, $C = 2$, $P = 1$, and thus $F = 1$. This means that the properties are dependent upon the composition within that phase. An example of this is the variation of the potential of an insertion reaction electrode as its composition changes during charge or discharge in a battery.

Similar considerations apply to electrodes containing three components. In this case a useful thinking tool is the *isothermal Gibbs triangle*, or its approximation, the *ternary phase stability diagram*. The electrical potential is independent of the composition within sub-triangles where three phases are in equilibrium. On the other hand, it is composition-dependent in 2-phase and 1-phase regions.

These conclusions have been thoroughly demonstrated experimentally in the case of binary alloy systems [12], and also ternary systems involving the reaction of lithium with binary metal oxides [13–15].

13.21 Application of the Gibbs Phase Rule to Reference Electrodes

13.21.1 Nonaqueous Systems

Application of these principles in nonaqueous systems is straightforward. If the temperature and total pressure are kept constant and there is only one component, e.g., a pure metal, the electrical potential must have a fixed value. On the other hand, if a binary phase, such as a metal oxide, is used as a reference, there must also be another phase in equilibrium with it, so that both C and P equal 2, in order to have a fixed potential. This second phase is often the metal component of the oxide, but it does not have to be. Instead, it could be a gas such as oxygen or air. The important thing is that this second phase should not introduce an additional component, for it would then be a ternary, instead of binary, system.

As mentioned above, if there are three components, i.e., a ternary system, there must be three phases in equilibrium with each other in order for $F = 0$.

13.21.2 Aqueous Systems

Aqueous systems introduce an additional feature. As it is expected that water will equilibrate with the electrode at their interface, the presence of water introduces an additional phase. In addition, it has two components, hydrogen and oxygen. These all have to be included in the consideration of the Gibbs Phase Rule.

In the case of a hydrogen gas electrode in contact with water, there are two components, hydrogen and oxygen, and two phases, water and hydrogen gas. Assuming constant temperature and pressure, $F = 0$, and the system is thermodynamically fixed.

However, experiments show that the electrical potential of such an electrode depends upon the value of the pH. When the pH is zero, the electrode is equivalent to the standard hydrogen electrode, the SHE. However, at other values of electrolyte pH its potential varies, as it is then a reversible hydrogen electrode, an RHE. The electrical potential difference between these two situations was shown above to be:

$$E_{RHE} - E_{SHE} = -0.059(\text{pH}) \qquad (13.33)$$

volts.

This hardly looks like a thermodynamically fixed situation. However, it must be recognized that normal chemical thermodynamics deals with the equilibria of electrically neutral species, and the pH is a measure of the concentration of a charged species, H^+ (or H_3O^+ or $H_5O_2^+$).

The chemical potential of a neutral species M, $\mu(M)$, can, in principle, be decomposed into two parts, the chemical potential of its ions, and the chemical potential of its electrons. This can be written as:

$$\mu(M) = \mu(M^+) + \mu(e^-) \qquad (13.34)$$

Thus if the value of $\mu(M)$ is held constant, as is the case if the system is thermodynamically fixed, the values of the chemical potentials of the ions and the electrons can both vary, but their values will depend upon each other.

In the case of hydrogen:

$$\mu(H_2) = 2\mu(H^+) + 2\mu(e^-) \qquad (13.35)$$

which can be rearranged to give

$$\mu(e^-) = \tfrac{1}{2}\mu(H_2) - \mu(H^+) \qquad (13.36)$$

The chemical potential of the electrons, $\mu(e^-)$, is related to the electrically measured quantity E by:

$$\mu(e^-) = zFE \qquad (13.37)$$

Electrons carry a negative charge, so $z = -1$ in this case. Actually, as mentioned already, one always measures differences in E and thus of $\mu(e^-)$, between electrodes, for absolute values of electrical potentials cannot be measured. The activities of individual ionic species also cannot be measured experimentally [16].

The chemical potential of the hydrogen ions can be written in terms of their concentration as:

$$\mu(H^+) = \mu(H^+)^0 + RT\ln a(H^+) \qquad (13.38)$$

where $\mu(H^+)^\circ$ is a constant. Substituting further,

$$\mu(H^+) = \mu(H)^0 + 2.303RT\log[H^+] \qquad (13.39)$$

or

$$\mu(H^+) = \mu(H^+)^0 - 2.303RT\,(pH) \qquad (13.40)$$

This can then be put back into the equation for the chemical potential of the electrons, giving:

$$\mu(e^-) = 1/2\mu(H_2) - \mu(H^+)^0 + 2.303(RT)\,(pH) \qquad (13.41)$$

so that the electrical quantity E is related to the pH by

$$E = \text{constant} - 2.303\left(\tfrac{RT}{F}\right)(pH) \qquad (13.42)$$

Fig. 13.13 General E vs. pH diagram showing the difference between ZDF and non-ZDF electrodes

where the value of the constant depends upon the identity of the neutral chemical species.

The result is that the potentials of all neutral species with zero degrees of freedom will lie along parallel lines with a slope of −0.059 V per pH unit in a plot of potential vs. pH, i.e., in *Pourbaix diagrams*. Their vertical locations will be determined by their potentials relative to the reversible hydrogen electrode.

Thus there is a clear differentiation between reference electrodes with zero degrees of freedom (ZDF electrodes) and those where this is not true that is readily seen in their dependence upon the pH of an aqueous electrolyte. This is indicated schematically in Fig. 13.13.

The result of this difference is that if one wants to compare electrodes in aqueous systems it is important to know whether they are ZDF electrodes or not, and if one of them is not, additional information, generally the pH of the electrolyte, is needed in order to specify the thermodynamic state.

13.22   Systems Used to Measure the pH of Aqueous Electrolytes

The difference in the potentials of ZDF electrodes and non-ZDF electrodes can be utilized to evaluate the pH of an electrolyte. An electrode to be used for this purpose will typically have a sealed inner compartment with a configuration that provides a

Fig. 13.14 Schematic drawing of the construction of a system that can be used to measure the pH of a liquid electrolyte. A chemically isolated calomel electrode is in contact with the electrolyte through an ionically-conducting glass membrane. A ZDF electrode (Hg/HgO) is in direct contact with the electrolyte

fixed potential relative to the SHE. This is surrounded by a solid electrolyte, generally a glass with a relatively high ionic conductivity, whose exterior is exposed to the electrolyte whose pH is to be evaluated. A second, ZDF, electrode is also present in the electrolyte, and the voltage between the two is measured.

The construction of such a pH-measuring system is shown schematically in Fig. 13.14.

13.23 Electrodes with Mixed-Conducting Matrices

As a final example, consider the use of a mixed-conducting matrix electrode containing a zero-degree-of-freedom (ZDF) reactant. Electrodes of this general class were first discussed some time ago [17, 18]. The microstructure contains a phase that has a high chemical diffusion coefficient for the atoms of the electroactive species and is also an electronic conductor. Although it is not necessary, such a phase will generally have a relatively small compositional width, so that it does not have an appreciable electrochemical capacity. In addition, the microstructure contains phases that can undergo a reconstitution chemical reaction. If the number of such phases is equal to the number of components within them, this reaction will have zero degrees of freedom, and thus a composition-independent potential.

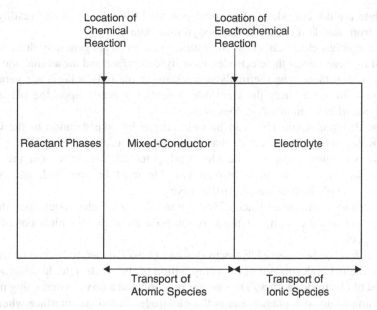

Fig. 13.15 Schematic one-dimensional model of a mixed-conductor matrix electrode. The potential is determined by the electrically neutral chemical reaction in the interior, whereas the electrochemical reaction takes place at the interface between the mixed-conductor and the electrolyte

The result is that this type of electrode has a potential that is determined by the internal ZDF chemical reaction, even though the electrochemical reaction takes place elsewhere, at the electrode/electrolyte interface on the outside of the mixed-conductor material. This is illustrated schematically in Fig.13.15.

Whereas the initial example of this principle involved a Li-Si constant potential reaction inside a Li-Sn alloy mixed conductor at elevated temperatures, it has been demonstrated [19, 20] that this concept also be used at ambient temperature.

Electrodes of this type might be useful as secondary references in cases where one or more of the reactive phases is not stable in contact with the electrolyte. So far as the electrical potential is concerned, the identity of the electrolyte and the details of the interfacial reaction are not important.

13.24 Closing Comments on Reference Electrodes

It is quite evident that the approach to reference electrodes is quite different in the nonaqueous and aqueous electrochemical communities. In the first case neutral chemical species are commonly used as references, and the details of the electrode/ electrolyte reaction and the identity and concentrations of the species in the

electrolyte are not considered. Electrical potential differences can be readily calculated from standard chemical thermodynamic data.

In the aqueous electrochemical community potentials are generally discussed in
terms of the reactions at the electrode/electrolyte interface and the atomic and ionic
species present there. The thermodynamic state of the bulk solid is not generally
considered. In some cases the electrode potential depends upon the pH of the
electrolyte, whereas in others it does not.

These different approaches can be rationalized by consideration of the Gibbs
Phase Rule. When the electrode has zero degrees of freedom (ZDF), all the
intensive variables, including the electrical potential, are fixed. On the other
hand, if this is not true an additional variable must be specified, and this is
commonly the pH in aqueous electrochemistry.

The characteristic dependence of the potential of ZDF electrodes upon the pH
has been explained in terms of the two components of the chemical potential of
neutral species.

The difference between ZDF electrodes and non-ZDF electrodes can be used to
measure the pH, independent of the composition of the electrolyte. In addition, the
potential of electrodes with a ZDF internal reaction and a mixed-conducting matrix
has nothing to do with phenomena at the electrode/electrolyte interface where the
electrochemical reaction takes place.

13.25 Potentials of Chemical Reactions

13.25.1 Introduction

It has been known for some time that ions can be inserted or removed from insertion
reaction materials by chemical, as well as electrochemical, means. This is one type
of soft chemistry, or *Chemie douce*, as it was initially called in France. Much of the
early attention to this possibility was focused upon materials based upon graphite,
and reviews of this work can be found in a number of places, e.g., in [21].

An important step in the development of the use of chemical methods to either
modify or synthesize advanced battery electrode materials was the work of
Armand, who used naphthalene complexes in a polar solvent to insert either sodium
or potassium, and n-butyl lithium dissolved in hexane to introduce lithium, into
insertion reaction materials [22]. He inserted these alkali metals into layer structures consisting of transition metal salts, such as CrO_3, between graphene planes.
The presence of these very covalent species gives these graphite-related materials a
very positive potential, so that they are interesting as potential positive electrode
reactants.

This is actually quite different from much of the other work on graphite
materials in lithium cells, in which the potential is much lower, so that they are

interesting for use as negative electrode materials. Nevertheless, the principles are the same.

The use of n-butyl lithium, which is commercially available as a solution in hexane, to insert lithium into a material M, forming Li_xM and octane, C_8H_{18}, can be simply written as

$$xLiC_4H_9 + M = Li_xM + \frac{x}{2}C_8H_{18} \tag{13.43}$$

In the sodium-naphthalene case the analogous reaction can be written as

$$xNaC_{10}H_8 + M = Na_xM + xC_{10}H_8 \tag{13.44}$$

As discussed earlier in this text, the standard Gibbs free energy change in a reaction involving electrically neutral species can be readily converted to an electrical potential difference, or voltage. The Gibbs free energy of formation of n-butyl lithium is about 96.5 kJ/mol, so that the potential that is attained upon its use to add lithium to electrode materials is about 1.0 V vs. elemental lithium. If the potential is greater than that value, it will tend to decompose, providing lithium to react with the material M. That is, Eq. (13.43) will tend to go to the right.

Thus it is possible to use such materials as reagents to chemically mimic electrochemical behavior, and thus screen or scan materials that might be considered as potential positive electrode reactants in lithium, or other alkali metal, cells. The amount of alkali metal uptake can be determined by assaying the resultant supernatant solution [23], and a rough indication of the kinetics of their reaction can be obtained by the observation of the temperature rise during the reaction [24]. In a number of cases an indication that a reaction has taken place is provided by a change in the color of the solution.

13.25.2 Relation Between Chemical Redox Equilibria and the Potential and Composition of Insertion Reaction Materials

This situation can be represented schematically, as shown in Fig. 13.16. If the potential of the solid is higher than the redox equilibrium potential, there will be a tendency for lithium to enter it from the adjacent liquid. On the other hand, if the potential of the solid is lower than the redox potential, there will be a tendency for the deletion of lithium, resulting in the potential increasing.

If two different reactants are used that have different redox potentials, the amount of lithium present in the solid can be changed between that characteristic of one of the redox potentials to that corresponding to the other. This is illustrated schematically in Fig. 13.17.

Fig. 13.16 Illustration of the relationship between the potential and the amount of lithium in Li_xM

Fig. 13.17 Illustration of the effect of using two different reagents, one at a high potential that causes a reduction in the amount of lithium in the solid, and the other, at a lower potential that increases the lithium content

13.25.3   Other Examples

An example of a chemical reaction that can be used to delete lithium is the reaction of a material containing lithium with iodine to form *LiI*, which was discussed in Chap. 2. The formation and decomposition of *LiI* is potentially reversible. Its standard Gibbs free energy of formation is −269.67 kJ/mol at 25 °C, which converts to 2.8 V. This provides a good approximation for its equilibrium potential in solutions. However, as in all of the cases discussed here, the actual potential may vary somewhat, depending upon the solvent, reagent concentration, and the amounts and identities of other species present.

This means that if iodine is available and a material Li_xM is present that has a potential lower than about 2.8 V there will be a tendency for iodine to extract lithium from it, forming LiI, and raising its potential. This can be represented by the equation

$$\frac{x}{2}I_2 + Li_yM = xLiI + Li_{(y-x)}M \qquad (13.45)$$

As an example of the use of this method, solutions of iodine in acetonitrile, CH_3CN, were employed by Murphy et al. [25] to delete lithium from Li_xVS_2 and raise its potential. They also used n-butyl lithium to add lithium and reduce its potential [26].

In addition to n-butyl lithium and iodine, there are other oxidation or reducing agents that can be used. Bromine, in a solution of chloroform, $CHCl_3$, has been used to oxidize, and therefore reduce the lithium content of, a number of materials. The standard Gibbs free energy of formation of lithium bromide is -341.6 kJ/mol, so its equilibrium potential is quite high, about 3.54 V. Early examples of the use of bromine were the deletion of lithium from $LiVO_2$ [27], and from the more positive electrode material $LiCoO_2$ [28].

Other, more highly oxidizing, reagents were discussed by [29] and [30]. One example is the hexafluorophosphate salt NO_2PF_6, which can be dissolved in acetonitrile and has a potential about 4.45 V above that of lithium. Its reaction with a lithium transition metal dioxide can be written as

$$LiMO_2 + xNO_2PF_6 = Li_{1-x}MO_2 + xNO_2 + xLiPF_6 \qquad (13.46)$$

A number of chemical reagents that have been used to modify mixed-conducting electrode materials in lithium systems are included in Table 13.5. Some of these chemical reagents are reversible, whereas others are not. More information about organolithium materials can be found in [31].

Table 13.5 Examples of lithium reaction materials and their approximate potentials vs. elemental lithium

Reagent	Solvent	E vs. Li Volts	Color, higher E	Color, lower E
MoF_6	Acetonitrile	4.75	None	None
NO_2PF_6	Acetonitrile	4.45	None	None
Bromine	Acetonitrile	3.54	Brown	None
DDQ	Acetonitrile	3.5	None	Red
Iodine	Acetonitrile	2.8	Purple	None
Benzophenone	Tetrahydrofuran	1.5	None	Blue
n-butyl lithium	Hexane	1.0	None	None
Benzophenone	Tetrahydrofuran	0.8	Blue	Purple
Naphthalene	Tetrahydrofuran	0.5	None	Green

Note: DDQ is 2,3-dichloro-4,5-dicyanobenzoquinone

13.25.4 Summary

There are a number of chemical equilibria that can act to influence the amount of inserted material; i.e., lithium, present in a mixed conductor in a manner analogous to the application of an electrochemical potential by the use of an electrochemical cell. Equilibria at low potentials are typically used to add lithium, and those with higher potentials are more commonly used to delete lithium from potential electrode materials.

13.26 Potential and Composition Distributions Within Components of Electrochemical Cells

13.26.1 Introduction

The electrostatic, chemical and electrochemical potentials inside condensed phases depend upon the nature and concentrations of the species that are present. These, in turn, vary with the local values of the relevant thermodynamic potentials, which are typically not uniform inside electrochemical cells.

A number of examples will be discussed in this chapter, including ionic conductors and mixed conductors between different types of non-blocking and selectively-blocking electrodes under the imposition of either electrical or chemical potential differences.

Under charge transport conditions the *transference numbers* of individual species vary with position if a gradient in thermodynamic potentials is present. This can be readily understood by use of *Defect Equilibrium Diagrams* as *thinking tools*.

These parameters can be experimentally evaluated by the use of proper sensors. One type measures the local value of the Fermi level of the electrons, and the other can be used to evaluate the local chemical potential or activity of neutral chemical species.

This is an important topic for several reasons. Local potentials, and their gradients, determine both the potentials and the kinetic behavior of electrodes in batteries. They also play critical roles in the properties of fuel cells.

13.26.2 Relevant Energy Quantities

The energy of species inside solids is the sum of their chemical and electrical energies. The chemical energy is expressed as the chemical potential, and for species i,

$$E_{\text{chem}} = \mu_i \tag{13.47}$$

The electrical energy is the product of the charge and the local value of the inner potential.

$$E_{\text{elect}} = z_i q \phi \tag{13.48}$$

The electrochemical potential η_i is the sum of the chemical and electrical energies

$$\eta_i = \mu_i + z_i q \phi \tag{13.49}$$

13.26.3   What Is Different About the Interior of Solids?

Chemical potentials outside of solids always are referenced to electrically neutral chemical species. But chemical species inside solids are typically electrically charged ions. In order to achieve internal charge balance, their charges must be balanced by the presence of either other charged ions or excess electrons or holes (a deficiency of electrons).

For equilibrium between an electrically neutral species M on the outside and the corresponding combination of an ionic species M^+ and an electron on the inside:

$$\mu_M = \mu_{M^+} + \mu_{e^-} \tag{13.50}$$

Each of these species has a corresponding electrochemical potential, the combination of chemical and electrostatic potentials, or potential energies

$$\eta_{M^+} = \mu_{M^+} + z_{M^+} q \phi \tag{13.51}$$

$$\eta_{e^-} = \mu_{e^-} + z_{e^-} q \phi \tag{13.52}$$

Here z is the charge number, q the electronic charge, and ϕ the local inner electrical potential. Since $Z_{M^+} = 1$ and $Z_{e^-} = -1$, if we add these equations the terms containing the inner potential cancel each other, giving

$$\eta_{M^+} + \eta_{e^-} = \mu_{M^+} + \mu_{e^-} \tag{13.53}$$

This can be simply rearranged to become

$$\mu_M = \eta_{M^+} + \mu_{e^-} \tag{13.54}$$

The external chemical potential of a neutral species M is equal to the sum of the electrochemical potentials of its two related internal species, M^+ and e^-.

The sum of the gradients of the electrochemical potentials of the charged species inside a solid can be observed as an externally measurable gradient in the chemical potential of the neutral species. In the case of a one-dimensional physical system with a distance parameter x this can be written as

$$\frac{d\mu_M}{dx} = \frac{d\eta_{M^+}}{dx} + \frac{d\eta_{e^-}}{dx} \tag{13.55}$$

Gradients in their respective electrochemical potentials constitute the driving forces for the transport of species.

13.26.4 Relations Between Inside and Outside Quantities

Information about potentials inside solids is typically obtained by use of external measurements. In the case of electronic species, this involves equilibration of the internal Fermi level with the Fermi level of an external metal probe. In the case of chemical species it is necessary to use an electrochemical cell and to balance chemical force with an equivalent electrical force

$$\Delta\mu_i = -z_i qE \tag{13.56}$$

where $\Delta\mu_i$ is the difference in chemical potential between neutral chemical species, z_i is the charge number of the ionic species under consideration, q is the elementary charge, and E is the voltage across the electrochemical cell

13.26.5 Basic Flux Relations Inside Phases

The particle flux density of any species i, J_i, is the number of particles of that type that cross a transverse area of 1 cm^2 per second.

This can be expressed in terms of the concentration $[i]$ (particles/cm^3) and the macroscopic drift velocity v_i (cm/s).

$$J_i = [i]v_i \tag{13.57}$$

The general mobility of species i, B_i, is defined as the ratio of the drift velocity and the negative gradient in the electrochemical potential, which is the force causing that drift.

$$B_i = -v_i \left/ \frac{d\eta_i}{dx} \right. \tag{13.58}$$

Note that the general mobility B_i is different from the electrical mobility b_i, which is defined as the drift velocity of species i per unit internal electrical field

$$b_i = -v_i \left/ \frac{d\phi}{dx} \right. \tag{13.59}$$

Thus the particle flux density of any species i can be written as

$$J_i = -[i]B_i \frac{d\eta_i}{dx} \tag{13.60}$$

Introducing the general definition of the electrochemical potential

$$J_i = -[i]B_i \left[\frac{d\mu_i}{dx} + z_i q \frac{d\phi}{dx} \right] \tag{13.61}$$

13.26.6 Two Simple Limiting Cases

To understand these matters, two types of materials as simple limiting cases can be considered,

1. A metal, in which there is no internal electrical field, and therefore

$$J_i = -[i]B_i \left[\frac{d\mu_i}{dx} \right] \tag{13.62}$$

2. A chemically homogeneous material in which $d\mu_i/dx$ is zero, so that

$$J_i = -[i]B_i \left[\frac{d\phi}{dx} \right] \tag{13.63}$$

13.26.7 Three Configurations

As examples, three simple configurations will be explored.

1. A solid electrolyte in which the transport of ionic species is blocked by the electrodes.
2. A mixed conductor in which the transport of electronic species is blocked by the electrodes.
3. A composite structure with a mixed conductor in series with a solid electrolyte.

13.26.8 Variation of the Composition with Potential

As mentioned above, ionic and electronic species are present inside a solid, and their respective concentrations vary with the values of the relevant electrochemical potentials.

The chemical compositions of solids also depend upon the chemical potentials of the species present.

These features can be readily understood for simple cases, and as an example, the concentrations of both ionic and electronic defects will be calculated here as a function of the overall composition and the electrical potential for a simple binary phase MX. This is an approach that was pioneered by workers in Philips Laboratories [32–34]. It is useful to express the results in a *Defect Equilibrium Diagram* (*DED*). It will be seen that such a graphical presentation of these matters can act as a very useful *thinking tool*.

13.26.9   Calculation of the Concentrations of the Relevant Defects in a Binary Solid MX That Is Predominantly an Ionic Conductor

In a binary solid, four defects have to be considered: two electronic defects, electrons and holes, as well as two ionic defects, which might be interstitials and vacancies of one of the components.

Since there are four unknowns, there must be four relevant equations.

Two of these are mass action relations, one for the formation of electron–hole pairs

$$K_e = [e^-]\,[h^o] \tag{13.64}$$

in which the brackets indicate concentrations (number of particles/cm^3), and one for the formation of ionic defect pairs

$$K_i = [D_{M^o}]\,[D_{X'}] \tag{13.65}$$

The notation that is used here is a modification of that generally referred to as *Kröger-Vink notation*. Instead of the common practice of using a dot to indicate a positive relative charge, a degree sign is used in this text, for that is a symbol that is readily available in typewriters and computers. In addition the symbol D is used here as a general symbol for an ionic defect. D_{M^o} is an M-rich ionic defect, and $D_{X'}$ is a defect whose presence makes the material more X-rich.

In addition to Eqs. (13.64) and (13.65), an expression is needed that relates to the overall chemical composition, and thus to the concentration of at least one of the ionic defects, in the phase *MX*. There are various ways in which this might be done. One is to assume that the *MX* is in equilibrium with its chemical environment, so that the chemical potentials of the species within it are the same as those in the environment. Consider the case in which it is assumed that the material is in equilibrium with a surrounding phase containing the species *X*.

The equilibrium between *X* species in the surrounding phase (assume that it is a gas and contains diatomic *X2* molecules) and singly charged *X* or *M* defect species inside the *MX* can be written as

$$X_{(g)} = 1/2\, X_{2(g)} = D_{X'} + h^o \qquad (13.66)$$

in which the X-rich defect species DX' could be either an interstitial X ion or a vacancy on the M lattice. Either one of these has a negative effective charge, and as mentioned above, it makes no difference which one is actually present in the material in this calculation. In either case, a positively charged electron hole would also have to be present in order to achieve charge balance.

An additional equation can be the law of mass action relation that corresponds to the incorporation of one of the chemical species, and its charge-balancing electronic species, into the nominally MX phase. An example is the X-incorporation relation

$$K_X = \frac{[D_{X'}]\,[h^o]}{(aX_2)^{1/2}} \qquad (13.67)$$

It would have been possible to use a relation that involves M-rich defects instead. In that case, however, an electron would have to be present in order to maintain charge balance.

In addition to these law-of-mass-action type relations, there must also be an expression that reflects the requirement for overall electrostatic charge balance. This is sometimes called the *electroneutrality condition*. The number of negative charges introduced by the presence of the defects carrying negative effective charges must be balanced by the positive charge due to the presence of species carrying positive charges. This can be written as

$$\left[e' \right] + [D_{X'}] = [h^o] + [D_{M^o}] \qquad (13.68)$$

Simultaneous solution of these four equations can be used to obtain expressions to use to evaluate the four defect concentrations.

Because of the composition dependence in the incorporation Eq. (13.20), the concentrations of all of the defects depend upon the composition of the phase, and thus upon the electrical potential.

To simplify matters it is useful to introduce a composition parameter F

$$F = K_X[a(X_2)]^{1/2} \qquad (13.69)$$

By substitution into the electroneutrality condition, the concentrations of the various defects can then be calculated in terms of the values of the equilibrium constants and the composition parameter F.

$$\left[e' \right] = K_e \left[\frac{K_i + F}{F(K_e + F)} \right]^{1/2} \qquad (13.70)$$

$$[h^o] = \left[\frac{F(K_e + F)}{K_i + F}\right]^{1/2} \tag{13.71}$$

$$[D_{X'}] = \left[\frac{F(K_i + F)}{K_e + F}\right]^{1/2} \tag{13.72}$$

$$[D_{M^o}] = K_i \left[\frac{K_e + F}{F(K_i + F)}\right]^{1/2} \tag{13.73}$$

These relations specify the concentrations of the four pertinent defects as functions of the overall composition of the MX phase, as expressed in terms of the value of the composition parameter F. They will each vary with temperature, as the values of the constants are temperature-dependent.

13.27 Defect Equilibrium Diagrams

The form of these relations is illustrated in the *Defect Equilibrium Diagram* shown in Fig. 13.18. The concentrations of the four types of defects are plotted on a logarithmic scale against the logarithm of the parameter F. It has been assumed in this example that the material MX is primarily an ionic conductor. This will be the case if the Gibbs free energy necessary to form the ionic defect pair is less than that necessary to form the electronic defect pair. Thus the value of K_i is significantly greater than the value of K_e. The constants used in this illustration are: $K_e = 10^{10}$, $K_i = 10^{40}$, and $K_X = 10^{20}$. It can be seen that there are three general regions of behavior, labeled as Region I at low values of F when the material will be relatively M-rich, Region II at intermediate values of F, and Region III at high values of F, when the material will be relatively X-rich.

13.27.1 Approximations Relevant in Specific Ranges of Composition or Activity

It is often useful to work with approximations to the general relations that are applicable over these three different ranges of composition or activity. As mentioned above, the important criterion for the determination of useful approximations is the value of the composition parameter F. In Region *I* the X activity is very small and the M activity correspondingly large, and F is very small, less than both K_e and K_i. In the central Region II, the value of F is between K_i and K_e. Likewise, when the value of the X activity is large and the M activity small, F will be larger than both K_e and K_i.

Fig. 13.18 Defect Equilibrium Diagram showing concentrations of defect species inside solid MX as functions of the composition parameter F

At very low values of X activity F will be smaller than both K_e and K_i, giving the following approximations for the defect concentrations.

$$\left[e'\right] = K_e \left(\frac{K_i}{FK_e}\right)^{1/2} \tag{13.74}$$

$$\left[h^o\right] = \left(\frac{FK_e}{K_i}\right)^{1/2} \tag{13.75}$$

$$\left[D_{X'}\right] = \left(\frac{FK_i}{K_e}\right)^{1/2} \tag{13.76}$$

$$\left[D_{M^o}\right] = K_i \left(\frac{K_e}{FK_i}\right)^{1/2} \tag{13.77}$$

Likewise, at intermediate values of X activity $K_i > F > K_e$, and the defect concentrations can be approximated by:

$$\left[e'\right] = \frac{K_e K_i^{\frac{1}{2}}}{F} \tag{13.78}$$

$$\left[h^o\right] = F K_i^{-\frac{1}{2}} \tag{13.79}$$

$$\left[D_{X'}\right] = K_i^{\frac{1}{2}} \tag{13.80}$$

$$\left[D^o\right] = K_i^{\frac{1}{2}} \tag{13.81}$$

When the X activity is very large, F becomes much greater than both K_e and K_i. The defect concentrations can be approximated by:

$$\left[e'\right] = K_e F^{-\frac{1}{2}} \tag{13.82}$$

$$\left[h^o\right] = F^{\frac{1}{2}} \tag{13.83}$$

$$\left[D_{X'}\right] = F^{\frac{1}{2}} \tag{13.84}$$

$$\left[D_{M^o}\right] = K_i F^{-\frac{1}{2}} \tag{13.85}$$

It can be seen from observation of the example Defect Equilibrium Diagram shown in Fig. 13.18 that the transition between Regions I and II occurs when $F = K_e$, and the transition between Regions II and III is where $F = K_i$. The slopes in Region II are ± 1, and in Regions I and III are $\pm 1/2$. Thus it is very easy to draw the general form of such a figure, even without knowing the values of the relevant constants.

Figure 13.18 shows that the defects that have the greatest concentrations, and which therefore play the dominant role in the electroneutrality relation and in determining the properties, are different in the three Regions. The species with the highest concentrations in the central Region II are both ionic defects. This indicates that the material may be primarily an ionic conductor in this range of composition. This is not true in the other Regions, where the composition has more extreme values, either relatively M-rich in Region I, or relatively X-rich in Region III. In those cases, the dominant defect concentrations include one ionic defect and one electronic defect, and it is likely that there is important, if not dominant, electronic conductivity.

Since the mobilities of the electronic defects are generally much higher than those of ionic defects, this material will be an n-type conductor at higher

Fig. 13.19 Schematic Defect Equilibrium Diagram Showing Central Region of Solid Electrolyte Behavior

values of F, and a p-type conductor at lower values of F than the transitions between the respective regions in Fig. 13.1. Therefore it is obvious that if it is desired that a material act as a solid electrolyte in an electrochemical cell it is important that the potentials of the two electrodes both be well within the central region of the *Defect Equilibrium Diagram* for the material in question. This is shown schematically in Fig. 13.19.

It will be shown in a later chapter that if this is not true, the measured voltage across the cell will be different from (less than) that expected from the difference in the chemical potentials at the electrodes.

13.27.2 *Situation in Which an Electrical Potential Difference Is Applied Across a Solid Electrolyte Using Electrodes That Block the Entry and Exit of Ionic Species*

As discussed in a later chapter, electrical measurement methods are often used in order to determine the ionic conduction properties of materials that are being considered as solid electrolytes. There are two general strategies. One is to utilize electrodes that are essentially transparent to the ionic species, so that the overall impedance is dominated by what happens within the solid being measured. This often involves *DC measurements*. The other strategy is to use electrodes that are deliberately blocking to the ionic species, and to measure the system response to the application of alternating potential differences. Such AC measurement methods are often called *impedance spectroscopy*. The blocking-electrode case will be discussed here.

The flux of any species i is proportional to the gradient of its electrochemical potential.

$$J_i = -[i]B_i \frac{d\eta_i}{dx} \tag{13.86}$$

If one, or both, of the electrodes block the passage of the ionic species, there can be no ionic flux in the solid electrolyte material being investigated. Therefore, the gradient in the electrochemical potential of the ions inside the material must be zero.

$$\frac{d\eta_i}{dx} = \frac{d\mu_i}{dx} + z_i q \frac{d\phi}{dx} = 0 \tag{13.87}$$

And if the potentials of both electrodes fall within the central region of the Defect Equilibrium Diagram for the electrolyte, there is no gradient in the concentrations of the ionic species, so that

$$\frac{d\mu_i}{dx} = 0 \tag{13.88}$$

And thus

$$\frac{d\phi}{dx} = 0 \tag{13.89}$$

So there is no internal electrical field inside the solid, despite the imposition of an external electrical potential difference.

There will be gradients in the chemical potentials, and thus of the concentrations, of the electrons and holes, however. The result is that there is an electronic current across the cell, but it is due to the composition gradients of the holes and electrons in the interior of the electrolyte, not the presence of an electrical field.

Experimental observation of the magnitude and the voltage dependence of this current provide information about the separate contributions of the holes and electrons. This is known as the Hebb-Wagner experiment, and is discussed elsewhere in this text.

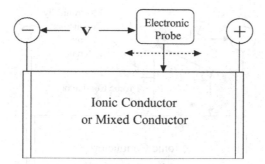

Fig. 13.20 Use of an electronic probe (e.g., a metal wire) to measure the variation of the electrochemical potential of the electrons with position

13.27.3   The Use of External Sensors to Evaluate Internal Quantities in Solids

Electronic probes can be used to evaluate the electrochemical potential (Fermi level) of the electrons at specific locations. If there is no passage of current between the probe and the solid, their Fermi levels must be equal. Such measurements can be made as a function of position, and referenced to one of the electrodes of the cell, in order to provide information about the spatial variation of the electronic Fermi level along the material being investigated. This is shown schematically in Fig. 13.20.

But information about the potential of the ionic species within the solid requires a different approach. As mentioned earlier, it is not possible to independently measure the properties of ions. Chemical potentials and forces within solids always relate to neutral species or combinations of species. The way to acquire this information is to use a probe that employs a suitable ionic conductor as electrolyte and an electronically-conducting chemical reference electrode. By measurement of the voltage across this ionically-conducting probe the difference in the chemical potential of the reference and the material being investigated can be obtained. If this is done as a function of position, information can be obtained about the chemical potential of the neutral chemical species present. This is shown schematically in Fig. 13.21.

13.27.4   Another Case, A Mixed Conductor in Which the Transport of Electronic Species Is Blocked

Instead of the electrodes acting to block the transport of ionic species, it is possible to block the passage of electrons into and out of a mixed conductor or ionic conductor. This can be accomplished by putting an ionic conductor in the system,

Fig. 13.21 Use of an ionically conducting probe and reference electrode in order to obtain information about the variation of the chemical potential of the neutral chemical species present with position

so that if current is passed between the electrodes there will be an ionic flux, but no electronic flux.

Even though there may be current flow through the system, external measurement of the Fermi Level with an electronic probe will show that there is no internal electrical field.

Instead of the relation

$$\frac{d\mu_M}{dx} = \frac{d\eta_{M^+}}{dx} + \frac{d\eta_{e^-}}{dx}$$ (13.90)

if there is no gradient in the electrochemical potential of the electrons, this simplifies to

$$\frac{d\mu_M}{dx} = \frac{d\eta_{M^+}}{dx}$$ (13.91)

An experimental example is shown in Fig. 13.22 [35]. In that case, an ionic probe was used to evaluate the electrochemical potential of silver ions within the mixed conductor Ag_2S, which was placed between two slabs of AgI, which is a pure silver ionic conductor. Thus current flow resulted in the gradient within the Ag_2S when silver ions were transported from one silver metal electrode to the other.

Fig. 13.22 Variation of the chemical potential of silver with position within Ag$_2$S, a mixed conductor, during current passage, evaluated by the use of an ionic probe [4]

13.27.5   Further Comments on Composite Electrochemical Cells Containing a Mixed Conductor in Series with a Solid Electrolyte

One of the concepts that has been proposed for use in fuel cells involves the use of a *monolithic structure* that is a composite of an ionic conductor and a mixed conductor that both have the same crystal structure. The mixed conductor can then act as an electrode. The potential advantage of this approach is that it would minimize the generation of stresses due to local differences in thermal expansion.

Assuming that these two components have the same crystal structure and are of approximately the same composition, there are two general ways to do this.

One is by the use of localized doping of an ionic conductor to produce mixed conductor regions at the surface with increased electronic conductivity. These doped regions can then act as mixed-conducting electrodes in series with the adjacent solid electrolyte region.

An alternative might be possible in some cases that would not require doping. This can be understood by consideration of the *Defect Equilibrium Diagrams* discussed earlier. It can be seen that if the local electrical potential is sufficiently negative, there is the tendency for the presence of excess electrons, making the material an n-type mixed conductor. Alternatively, it may be possible to induce the local presence of holes at very positive potentials to make the material a p-type electronic conductor. The first of these is shown schematically in Fig. 13.23.

When this is the case, the measured voltage is reduced below that which would be the case if the chemical potential difference were placed upon a purely ionic conductor. This is illustrated in Fig. 13.24. Unfortunately, this means that the use of this monolithic concept in a solid electrolyte fuel cell necessarily means that the output voltage is reduced.

The spatial distribution of the chemical potential of the neutral chemical species *M* is shown in Fig. 13.25.

Fig. 13.23 Schematic illustration of the case of an electrochemical cell in which an imposed chemical (or electrical) potential difference is such that the potential of the negative electrode is in the region of the DED in which the concentration of electrons is large enough to cause local mixed-conduction

Fig. 13.24 Schematic illustration showing that the electrical voltage is less than that corresponding to the imposed chemical potential difference if a portion of the material is mixed-conducting

Fig. 13.25 Spatial distribution of the chemical potential of the neutral chemical species M across the cell illustrated in Figs 13.23 and 13.24

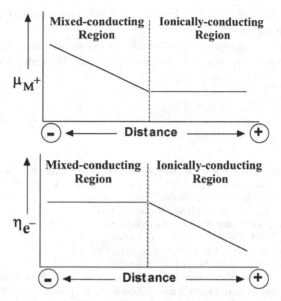

Fig. 13.26 Spatial distribution of the chemical potential of the M+ ions and the electrochemical potential of the electrons (Fermi level) is illustrated for the cell illustrated in Figs. 13.23 and 13.24

The corresponding distributions of the chemical potentials of the M^+ ions and the electrochemical potential of the electrons are illustrated in Fig. 13.26 for this situation.

It is seen that, although the chemical potential of the neutral chemical species, an externally measurable quantity, is a linear function of distance across the cell, the positional variations of the internal potentials of the two types of species are quite different.

13.28 Transference Numbers of Particular Species

The several situations that have been discussed here clearly indicate that the charge transport properties of a given material can vary significantly, depending upon the potentials imposed by its electrodes, as well as whether the electrodes—and other phases present—limit the passage of either ionic or electronic species.

A term used in this connection is the *Hittorf transference number* of a particular species. It indicates the fraction of the total charge current that is transported by that species.

Consideration of the Defect Equilibrium Diagram shows that the concentrations of the various charged species present vary with the potentials applied by the electrodes, and the relation between chemical potentials and electrical potentials. Variation of these concentrations means that there is a variation in the transference numbers of the different species. Transference numbers can thus also vary with position within a solid.

But in addition, it is obvious that transference numbers also depend upon the experimental conditions—e.g., the properties of the electrodes used in a given experiment.

References

1. Wen CJ, Boukamp BA, Weppner W, Huggins RA (1979) J Electrochem Soc 126:2258
2. Barin I (1995) Thermochemical Data of Pure Substances, 3rd edn. VCH, Weinheim, Published Online 24 Apr 2008. ISBN 9783527619829783527619825
3. Li W, Dahn JR, Wainwright DS (1994) Science 264:1115
4. Li W, McKinnon WR, Dahn JR (1994) J Electrochem Soc 141:2310
5. Li W, Dahn JR (1995) J Electrochem Soc 142:1742
6. Zhang M, Dahn JR (1996) J Electrochem Soc 143:2730
7. Butler JN (1970) Adv Electrochem Electrochem Eng 7:77
8. Ives DJG, Janz GJ (1961) Reference Electrodes. Academic Press, New York
9. Pourbaix M (1966) Atlas of Electrochemical Equilibria. Pergamon Press, Oxford
10. Gibbs JW (1875) Scientific Papers. Transactions of the Connecticut Academy of Science, 108
11. Gibbs JW (1961) The Scientific Papers of J. W. Gibbs, vol 1. Dover Publications, New York
12. Weppner W, Huggins RA (1978) J Electrochem Soc 125:7
13. Godshall NA, Raistrick ID, Huggins RA (1980) Mater Res Bull 15:561
14. Godshall NA, Raistrick ID, Huggins RA (1981) Proc. 16th IECEC, 769
15. Godshall NA, Raistrick ID, Huggins RA (1984) J Electrochem Soc 131:543
16. Koryta J, Dvorak J, Bohackova V (1966) Electrochemistry. Methuen, London
17. Boukamp BA, Lesh GC, Huggins RA (1981) J Electrochem Soc 128:725
18. Boukamp BA, Lesh GC, Huggins RA (1981) In Proceedings of the Symposium on Lithium Batteries, H. V. Venkatasetty, Ed., Electrochem. Soc. p. 467
19. Anani A, Crouch-Baker S, Huggins RA (1987) In Proceedings of the Symposium on Lithium Batteries. A. N. Dey, Ed., Electrochem. Soc. p. 382.
20. Anani A, Crouch-Baker S, Huggins RA (1988) J Electrochem Soc 135:2103
21. Ebert LB (1976) Intercalation Compounds of Graphite. In Annual Review of Materials Science, ed. by R.A. Huggins, Annual Reviews, Inc. p. 181
22. Armand M (1973) New Electrode Material. In Fast Ion Transport in Solids, ed. by W. van Gool, North-Holland. p. 665
23. Dines MB (1975) Mater Res Bull 10:287
24. Whittingham MS, Dines MB (1977) J Electrochem Soc 124:1387
25. Murphy DW, Cros C, DiSalvo FJ, Waszczak JV (1977) Inorg Chem 16:3027
26. Murphy DW, Carides JN, DiSalvo FJ, Cros C, Waszczak JV (1977) Mater Res Bull 12:825
27. Vidyasagar K, Gopalakrishnan J (1982) J Solid State Chem 42:217
28. Mendiboure A, Delmas C, Hagenmuller P (1984) Mater Res Bull 19:1383
29. Anderson GM, Iqbal J, Sharp DWA, Winfield JM, Cameron JH, McLeod AG (1984) J Fluor Chem 24:303
30. Wizansky AR, Rauch PE, Disalvo FJ (1989) J Solid State Chem 81:203
31. Wakefield BJ (1974) The Chemistry of Organolithium Compounds. Pergamon Press, New York
32. Kröger FA, Vink HJ, van den Boomgard J (1954) Z Phys Chem 203:1
33. Brouwer G (1954) Philips Res Rep 9:366
34. Kröger FA (1974) The Chemistry of Imperfect Crystals, 2nd edn. North Holland, Amsterdam
35. Schmalzried H, Ullrich M, Wysk H (1992) Solid State Ion 51:91

Chapter 14
Insertion Reaction Electrodes

14.1 Introduction

The topic of *insertion reaction electrodes* did not even appear in discussions of batteries and related phenomena just a few years ago, but is a major feature of some of the most important modern battery systems today. Instead of reactions occurring on the surface of solid electrodes, as in traditional electrochemical systems, what happens *inside* the electrodes is now recognized to be of critical importance.

A few years after the surprise discovery that ions can move surprisingly fast inside certain solids, enabling their use as solid electrolytes, it was recognized that some ions can move rapidly into and out of some other (electrically conducting) materials. The first use of insertion reaction materials was for non-blocking electrodes to assist the investigation of the ionic conductivity of the (then) newly discovered ambient temperature solid electrolyte, sodium beta alumina [1–3]. Their very important use as charge-storing electrodes began to appear shortly thereafter.

This phenomenon is a key feature of the electrodes in many of the most important battery systems today, such as the lithium-ion cells. Specific examples will be discussed in later chapters.

Many examples are now known in which a mobile guest species can be *inserted into*, or *removed (extracted) from*, a *stable host crystal structure*. This phenomenon is an example of both *soft chemistry* and *selective equilibrium*, in which the mobile species can readily come to equilibrium, but this may not be true of the host, or of the overall composition. The mobile species can be atoms, ions or molecules, and their concentration is typically determined by equilibrium with the thermodynamic conditions imposed on the surface of the solid phase.

In the simplest cases, there is little, if any, change in the structure of the host. There may be modest changes in the volume, related to changes in bond distances, and possibly directions, but the general character of the host is preserved. In many

© Springer International Publishing Switzerland 2016   261
R.A. Huggins, *Energy Storage*, DOI 10.1007/978-3-319-21239-5_14

cases the *insertion* of guest species is reversible, and they can also be *extracted*, or *deleted*, returning the host material to its prior structure.

The terms "*intercalation*" and "*de-intercalation*" are often used for reactions involving the insertion and extraction of guest species for the specific case of host materials that have *layer-type crystal structures*. On the other hand, "*insertion*" and "*extraction*" are more general terms. Reactions of this type are most likely to occur when the host has an open-framework or a layered type of crystal structure, so that there is space available for the presence of additional small ionic species. Since such reactions involve a change in the chemical composition of the host material, they can also be called *solid solution reactions*.

Insertion reactions are generally *topotactic*, with the guest species moving into, and residing in, specific sites within the host lattice structure. These sites can often be thought of as *interstitial sites* in the host crystal lattice that are otherwise empty. The occurrence of a *topotactic* reaction implies some three-dimensional correspondence between the crystal structures of the parent and the product. On the other hand, the term *epitaxy* relates to a correspondence that is only two-dimensional, such as on a surface.

It has been known for a long time that large quantities of hydrogen can be inserted into, and extracted from, palladium and some of its alloys. Palladium-silver alloys are commonly used as hydrogen-pass filters, i.e., filters that let only hydrogen pass through. Several types of materials with layer structures, including graphite and some clays, are also often used to remove contaminants from water by absorbing them between the layers in their crystal structures.

The most common examples of interest in connection with electrochemical phenomena involve the insertion or extraction of relatively small guest cationic species, such as H^+, Li^+, and Na^+. However, it will be shown later that there are also some materials in which anionic species can be inserted into a host structure.

It should be remembered that electrostatic energy considerations dictate that only neutral species, or neutral combinations of species, can be added to, or deleted from, solids. Thus the addition of cations requires the concurrent addition of electrons, and the extraction of cations is accompanied by either the insertion of holes or the deletion of electrons. Thus this phenomenon almost always involves materials that have at least some modicum of electronic conductivity.

The term "*soft chemistry*," or *chimie douce* in French, as much of the early work took place in France [4, 5], is sometimes used to describe reactions or chemical changes that involve only the relatively mobile components of the crystal structure, leaving the balance of the structure relatively unchanged.

Such reactions are often highly reversible, but in some cases, the insertion or extraction of mobile atomic or ionic species causes irreversible changes in the structure of the host material, and the reversal of this process does not return the host to its prior structure. In extreme cases, the structure may be so distorted that it becomes *amorphous*. These matters will be discussed below.

Insertion reactions are much more prevalent at lower temperatures than at high temperatures. The mobility of the component species in the host structure generally increases rapidly with temperature. This allows much more significant changes in the overall structure to occur, leading to *reconstitution reactions*, with substantial structural changes, rather than only the motion of the more mobile species, at elevated temperatures. Reconstitution reactions typically can be thought of as involving bond breakage, atomic reorganization, and the formation of new bonds.

14.2   Examples of the Insertion of Guest Species into Layer Structures

A number of materials have crystal structures that can be characterized as being composed of rather stiff *covalently bonded slabs* containing several layers of atoms. These slabs are held together by relatively strong, e.g., covalent, bonds. But adjacent slabs are bound to each other by relatively weak *van der Waals* forces. The space between the tightly-bound slabs is called the *gallery space*, and additional species can reside there. Depending upon the identity, size and charge of any inserted species present, the inter-slab dimensions can be varied.

Materials with the CdI_2 structure represent a simple example. They have a basic stoichiometry MX_2, and can be viewed as consisting of close-packed layers of negatively charged X ions held together by strong covalent bonding to positive M cations. In this case, the cations are octahedrally coordinated by six X neighbors, and the stacking of the X layers is hexagonal, with alternate layers directly above and below each other. This is generally described as ABABAB stacking.

This structure can be depicted as shown schematically in Fig. 14.1. Examples of materials with this type of crystal structure are CdI_2, $Mg(OH)_2$, $Fe(OH)_2$, $Ni(OH)_2$,

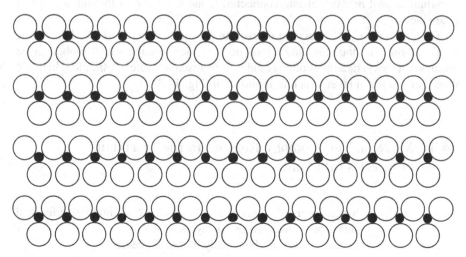

Fig. 14.1 Simple schematic model of a layer-type crystal structure with hexagonal ABABAB stacking. The *empty areas* between the covalently bonded slabs are called galleries

Fig. 14.2 Another type of model of a layer-type crystal structure. The example is TiS_2

and TiS_2. Another, simpler, way to depict these structures is illustrated in Fig. 14.2 for the case of TiS_2.

14.3   Floating and Pillared Layer Structures

In many cases, the mobile species move into and through sites in the previously empty gallery space between slabs of host material that are held together only by relatively weak van der Waals forces. The slabs can then be described as floating, and the presence of guest species often results in a significant increase in the inter-slab spacing.

However, in other cases the slabs are already rigidly connected by *pillars*, which partially fill the galleries through which the mobile species move. The pillar species are typically immobile, and thus are different from the mobile guest species. Because of the presence of the static pillars, the mobile species move through a two-dimensional network of interconnected tunnels, instead of through a sheet of available sites.

The presence of pillars acts to determine the spacing between the slabs of the host material, and thus the dimensions of the space through which mobile guest species can move. Examples of this kind will be discussed later. A simple schematic model of a pillared layer structure is shown in Fig. 14.3.

14.4   More on Terminology Related to the Insertion
of Species into Solids

Sheets	Single layers of atoms or ions. In the case of graphite, individual sheets are called *graphene* layers.
Stacks	Parallel sheets of chemically identical species.
Slabs or Blocks	Multilayer structures tightly bound together, but separated from other structural features.

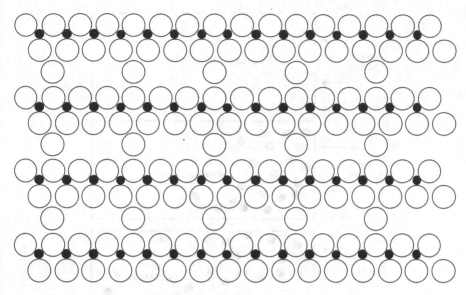

Fig. 14.3 Schematic model of pillared layer structure

	Example: covalently bonded MX_2 slabs such those shown above in the CdI_2 structure.
Galleries	The spaces between slabs in which the bonding is relatively weak, and in which guest species typically reside.
Pillars	Immobile species within the galleries that serve to support the adjacent slabs and to hold them together.
Tunnels	Connected interstitial space within the host structure in which the guest species can move and reside. Tunnels can be empty, partly occupied, or fully occupied by guest species.
Cavities	Empty space larger than the size of a single atom vacancy.
Windows	Locations within the host structure through which the guest species have to move in order to go from one site to another. Windows are typically defined by structural units of the host structure.

14.5 Types of Inserted Guest Species Configurations

There are several types of insertion reactions. In one case the mobile guest species randomly occupy sites within all of the galleries, gradually filling them all up as the guest population increases. When this is the case the variation of the electric potential with composition indicates a single-phase solid solution reaction, and there can be transient composition gradients within the gallery space.

Fig. 14.4 Random diffusion of guest species into gallery space

If, however, the presence of the guest species causes a modification of the host structure, the insertion process can occur by the motion of an interface that separates the region into which the guest species have moved from the area in which there are no, or fewer, guest species. Thermodynamically, this has the characteristics of a polyphase reconstitution reaction, and occurs at a constant potential.

Alternatively, there can be two or more types of sites in the gallery space, with different energies, and the guest species can occupy an ordered array of sites, rather than all of them. When this is the case, changes in the overall concentration of mobile species requires the translation of the interface separating the occupied regions from those that are not occupied, again characteristic of a constant-potential reconstitution reaction. These moving interfaces can remain planar, or they can develop geometrical roughness. Several possibilities are illustrated schematically in Figs. 14.4, 14.5, and 14.6.

14.6   Sequential Insertion Reactions

If there are several different types of sites with different energies, insertion generally occurs on one type of site first, followed by the occupation of the other type of site. Figure 14.7 shows the potential as a function of composition during the

Fig. 14.5 Motion of two-phase interface when the guest species is not ordered upon possible sites

Fig. 14.6 Motion of two-phase interface when the guest species is ordered upon possible sites in the gallery space

Fig. 14.7 Coulometric titration curve related to the insertion of lithium into NiPS$_3$. There is random filling of the first two types of sites. A reconstitution reaction occurs above about 1.25 Li [6]

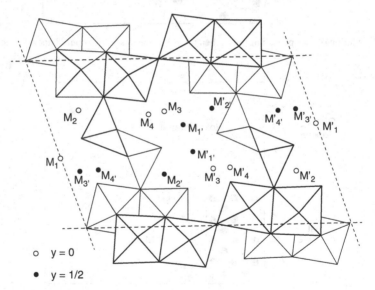

Fig. 14.8 (010) projection of the K$_x$V$_2$O$_5$ structure, showing the different types of sites for the guest species. After [6]

insertion of lithium into NiPS$_3$, in which there are two types of sites available. They are occupied in sequence, with random occupation in both cases.

Another example in which there are also different types of sites available for the insertion of Li ions involves the host K$_x$V$_2$O$_5$ structure. The host crystal structure illustrating the several different types of sites for guest ions is shown schematically in Fig. 14.8 [6].

Fig. 14.9 Coulometric titration curve for the insertion of Li into $K_{0.27}V_2O_5$. After [6]

The experimentally measured coulometric titration curve for the insertion of Li ions into a member of this group of materials is shown in Fig. 14.9 [7]. It shows that the reaction involves three sequential steps. Up to about 0.4 Li can be incorporated into the first set of sites randomly. This is followed by the insertion of another 0.4 Li into another set of sites in an ordered arrangement. This means that there are two different lithium arrangements, with a moving interface between them. Thus there are two phases present, so this corresponds to a reconstitution reaction. This is then followed by another reconstitution reaction, the insertion of about one additional Li into another ordered structure.

A different type of ordered reaction involves selective occupation of particular galleries, and not others, in a material with a layered crystal structure. This phenomenon is described as "staging." If alternate galleries are occupied and intervening ones are not, the material is described as having a "second-stage" structure. If every third gallery is occupied, the structure is "third-stage," and so forth. A simple model depicting staging is shown in Fig. 14.10.

14.7   Co-insertion of Solvent Species

In some cases it is found that species from the electrolyte can also move into the gallery space. This tends to be the case when the electrolyte solvent molecules are relatively small, so that they can enter without causing a major disruption of the host structure. This is found to occur in some organic solvent systems, and also some aqueous electrolyte systems where the electroactive ion is surrounded by a water hydration sheath. This is a matter of major concern in the case of negative electrodes in lithium systems, and will be discussed at much greater length in a later chapter.

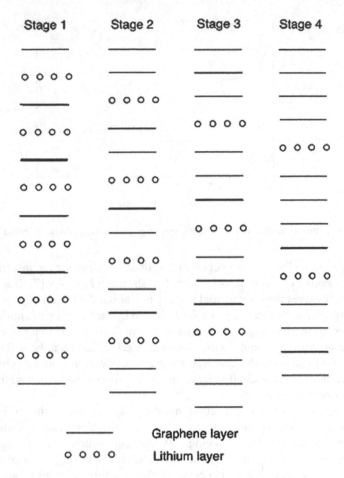

Stage 1   Stage 2   Stage 3   Stage 4

──────── Graphene layer

o o o o   Lithium layer

Fig. 14.10 Simple model depicting staging when potassium is inserted in the galleries of graphite

14.8   Insertion into Materials with Parallel Linear Tunnels

The existence of staging indicates that, at least in some materials, the presence of inserted species in one part of the structure is "seen" in other parts of the structure. An interesting example of this involves the presence of mobile guest species in the material Hollandite that has a crystal structure with parallel linear tunnels, rather than slabs.

A drawing of this structure is shown in Fig. 14.11. At low temperatures the interstitial ions within the tunnels are in an ordered arrangement upon the available sites. In addition, there is coordination between the arrangement in one tunnel with that of other nearby tunnels. Thus there is three-dimensional ordering of the guest species.

Fig. 14.11 Hollandite structure. Viewed along the c-axis

Fig. 14.12 Influence of temperature upon various types of order in a structure with parallel tunnels

As the temperature is raised somewhat, increased thermal energy causes the ordered interaction between the mobile ion distributions in nearby tunnels to relax, although the ordering within tunnels is maintained.

At even higher temperatures the in-tunnel ordering breaks down, so that the species are distributed randomly inside the tunnels, as well. The influence of temperature is illustrated schematically in Fig. 14.12.

14.9   Changes in the Host Structure Induced by Guest Insertion or Extraction

It was mentioned earlier that the insertion or extraction of mobile guest species can cause changes in the host structure. There are several types of such structural changes that can occur. They will be briefly discussed in the next sections.

14.9.1   Conversion of the Host Structure from Crystalline to Amorphous

There are a number of examples in which an initially crystalline material becomes amorphous as the result of the insertion of guest species, and the corresponding mechanical strains in the lattice. This often occurs gradually as the insertion/ extraction reaction is repeated, e.g., upon electrochemical cycling. One example of this, the $Li_x V_6 O_{13}$ binary system, is shown in Figs. 14.13 and 14.14 [8]. In this case, the shape of the potential curve during the first insertion of lithium into crystalline $V_6 O_{13}$ shows that a sequence of reconstitution reactions take place that give rise to a series of different phases, and a discharge curve with well-defined features.

After a number of cycles, however, the discharge curve changes, with a simple monotonous decrease in potential, indicative of a single-phase insertion reaction. X-ray diffraction experiments confirmed that the structure of the material had become amorphous.

Another example of changes resulting from an insertion reaction is shown in Fig. 14.15. In this case, lithium was inserted into a material that was initially

Fig. 14.13 Discharge curve observed during the initial insertion of lithium into a material that was initially $V_6 O_{13}$. After [8]

Fig. 14.14 Discharge curve observed during the 20th insertion of lithium into a material that was initially V_6O_{13}. After [8]

Fig. 14.15 The variation of the potential as lithium is added to V_2O_5. When the composition reached $Li_3V_2O_5$ an amorphous phase was formed. After [9]

V_2O_5 [9]. The result is similar to the V_6O_{13} case, with clear evidence of the formation of a series of different phases as lithium was added. It was found that the insertion reaction was reversible, forming the ε and δ structures, if only up to about 1 Li was inserted into α-V_2O_5. The addition of more lithium resulted in the formation of different structural modifications, called the γ and ω structures, which have nominal compositions of $Li_2V_2O_5$ and $Li_3V_2O_5$, respectively. These two reactions are not reversible, however.

When lithium was extracted from the ω phase, its charge–discharge curve became very different, exhibiting the characteristics of a single phase with a wide range of solid solution. The amount of lithium that could be extracted from this phase was quite large, down to a composition of about $Li_{0.4}V_2O_5$. Upon the reinsertion of lithium, the discharge curve maintained the same general form, indicating that a reversible amorphous structure had been formed during the first insertion process.

14.9.2   Dependence of the Product upon the Potential

It has been found that displacement reactions can occur in a number of materials containing silicon when they are reacted with lithium to a low potential (high lithium activity). An irreversible reaction occurs that results in the formation of fine particles of amorphous silicon in an inert matrix of a residual phase that is related to the precursor material [10, 11]. Upon cycling, the amorphous Li-Si structure shows both good capacity and high reversibility.

However, it has also been shown [12] that if further lithium is inserted, going to a potential below 50 mV, a crystalline Li-Si phase forms instead of the amorphous one.

Because of the light weight of silicon, the large amount of lithium that can react with it, and the attractive potential range, silicon or its alloys may play an important role as a negative electrode reactant in lithium batteries in the future.

14.9.3   Changes upon the Initial Extraction of the Mobile Species

Similar phenomena can also occur during the initial extraction of a mobile species that is already present in a solid. This is shown in Fig. 14.16 for the case of a material with an initial composition of about $Li_{0.6}V_2O_4$ [13]. It can be seen that the

Fig. 14.16 Initial charging and discharge curves of a material with a composition of $Li_{0.6}V_2O_4$. After [13]

reaction starts between 3.0 and 3.5 V vs. pure Li, as is generally found for materials that have come into equilibrium with air. The reason for this will be discussed later.

The initial lithium could be essentially completely deleted from the structure, causing the potential to rise to over 4 V vs. pure lithium. When lithium was subsequently reinserted, the discharge curve had a quite different shape, indicating the presence of a reconstitution reaction resulting in the formation of an intermediate phase.

14.10   The Variation of the Potential with Composition in Insertion Reaction Electrodes

14.10.1   Introduction

The externally measured electrical potential of an electrode is determined by the electrochemical potential of the electrons within it, η_{e^-}. This is often called the *Fermi level*, E_F. Since potentials do not have absolute values, they are always measured as differences. The voltage of an electrochemical cell is the electrically measured difference between the Fermi levels of the two electrodes:

$$\Delta E = \Delta \eta_{e^-} \qquad (14.1)$$

As has been demonstrated many times in this text already, the measured potential of an electrode often varies with its composition, e.g., as guests species are added to, or deleted from, a host material. The relevant compositional parameter is the chemical potential of the electrically neutral electroactive species. If this species exists as a cation M^+ within the electrode, the important parameter is the chemical potential of neutral M, μ_M. This is related to the electrochemical potentials of the ions and the electrons by

$$\mu_M = \eta_{M^+} + \eta_{e^-} \qquad (14.2)$$

Under open circuit conditions there is no flux of ions through the cell. Since the driving force for the ionic flux through the electrolyte is the gradient in the electrochemical potential of the ions, for open circuit

$$\frac{d\eta_{M^+}}{dx} = \Delta \eta_{M^+} = 0 \qquad (14.3)$$

Therefore, the measured voltage across the cell is simply related to the difference in the chemical potential of the neutral electroactive species in the two electrodes by

$$\Delta E = \Delta \eta_{e^-} = \frac{-1}{z_{M^+}q} \Delta \mu_M \qquad (14.4)$$

The common convention is to express both the difference in the electrical potential (the voltage) and the difference in chemical potential as the values in the right-hand (positive) electrode less those in the left-hand electrode.

A general approach that is often used is to understand the potentials of electrons is based upon the *electron energy band model*. The critical features are the variation of the density of available states with the energy of the electrons, and the filling of those states up to a maximum value that is determined by the chemical composition. The energy at this maximum value is the *Fermi level*.

In the case of metals the variation of the potential of the available states is a continuous function of the composition, and the *free electron theory* can be used to express this relationship.

In nonmetals, semiconductors and insulators, the density of states is not a continuous function of the chemical composition. Instead, there are potential ranges in which there are no available states that can be occupied by electrons. In the case of the simple semiconductors such as silicon or gallium arsenide, one speaks of a *valence band*, in which the states are generally fully occupied, an *energy gap* within which there are no available states, and a *conduction band* with normally empty states. The concentrations of electrons in these energy bands varies with the temperature due to *thermal excitation*, and can also be modified by the presence of aliovalent species, generally called *dopants*. *Optical excitation* has an effect similar to that of *thermal excitation*.

In a number of materials, particularly those in which the electronic conductivity is relatively low, it is convenient to think of the relation between the occupation of energy states and a change in the formal valence, or charge, upon particular species within the host structure. For example, the addition of an extra electron could result in a change of the formal charge of W^{6+} to W^{5+}, Ti^{4+} to Ti^{3+}, Mn^{4+} to Mn^{3+}, or Fe^{3+} to Fe^{2+} in a transition metal oxide. Such phenomena are called *redox* reactions.

These different cases will be discussed below, and it will be seen that there is a clear relationship between them.

14.10.2   The Variation of the Electrical Potential with Composition in Simple Metallic Solid Solutions

There are a number of metals in which insertion of mobile guest species can occur. As mentioned already, this can be described as a solid solution of the guest species in the host crystal structure. The important quantity controlling the potential is the variation of the chemical potential of the neutral guest species as a function of its concentration. This can be formally divided into the influence of the change in the electron concentration in the host material, and the effect due to a change in the concentration of the ionic guest species, M^+.

In the case of a random solid solution in a material with a high electronic conductivity the two major contributions are the contribution from the composition dependence of the Fermi level of the degenerate electron gas that is characteristic of such mixed conductors, and that due to the composition dependence of the enthalpy and configurational entropy of the guest ions in the host crystal lattice [14].

14.10.3 *Configurational Entropy of the Guest Ions*

If the guest ions are highly mobile and can readily move through the host crystal structure we may assume that all of the identical crystallographic sites are equally accessible. There will be a contribution to the total free energy due to the *configurational entropy* S_c which is related to the random distribution of the guest atoms over the available sites. This can be expressed as

$$S_c = -k\left(\ln\frac{x}{x_0 - x}\right) \tag{14.5}$$

where x is the concentration of guest ions and x_o is the concentration of identical available sites. k is Boltzmann's constant. The configurational entropy contribution to the potential is the product of the absolute temperature and the entropy. This is plotted in Fig. 14.17.

Fig. 14.17 Contribution to the potential due to the configurational entropy of a random distribution of guest ions upon the available identical positions in a host crystal structure. The values on the abscissa are the fractional site occupation, and those on the ordinate are mV [14]

This model assumes that there is no appreciable interaction between nearby guest species, and that there is only one type of site available for them to occupy.

14.10.4   The Concentration Dependence of the Chemical Potential of the Electrons in a Metallic Solid Solution

In a simple metal the electron concentration is typically sufficiently high that at normal temperatures the electrochemical potential of the electrons can be approximately by the energy of the Fermi level E_F.

In the free electron model this can be expressed as

$$E_F = \frac{h^2}{8m\pi^2}\left(\frac{3\pi^2 N_A}{V_m}\right)^{2/3} N^{2/3} \tag{14.6}$$

where m is the electron mass, N_A is Avogadro's number, V_m is the molar volume, and E_F is calculated from the bottom of the conduction band.

Thus the electronic contribution to the total chemical potential is proportional to the 2/3 power of the guest species concentration if the simple free electron model is valid. More generally, however, the electron mass is replaced by an effective mass m^*. This takes into account other effects, such as the influence of the crystal structure upon the conduction band.

Then the chemical potential of the electrons is directly related to the Fermi level, E_F, which can be written as

$$E_F = (\text{Constant})\left(\frac{x^{2/3}}{m^*}\right) \tag{14.7}$$

where x is the guest ion concentration and m^* is the effective mass of the electrons.

14.10.5   Sum of the Effect of These Two Components upon the Electrical Potential of a Metallic Solid Solution

Thus the composition dependence of the electrode potential in a metallic solid solution can be written as the sum of the influence of composition upon the configurational entropy of the guest ions, and the composition dependence of the Fermi level of the electrons. This can be simply expressed as

Fig. 14.18 Calculated influence of the value of the electronic effective mass upon the composition dependence of the potential in an insertion reaction in a simple metal [14]

$$E = (\text{Constant})\left(\frac{x^{2/3}}{m^*}\right) - \left(\frac{RT}{zF}\right)\ln\left(\frac{x}{1\text{-}x}\right) \qquad (14.8)$$

This relationship is illustrated in Fig. 14.18 for several values of the electron effective mass. It also shows the influence of the value of the electron effective mass upon the general slope of the curve. From Eq. (14.8) it can be seen that smaller effective masses make the first term larger, and this results in the potential being more composition dependent.

An example showing experimental data [14] that illustrate the general features of this model is shown in Fig. 14.19. The host material in this case was an oxide, a "tungsten bronze." In this family of oxides the electronic conductivity is very high, and the electron energy spectrum approximates that of a free electron metal.

It should be remembered that although the band diagrams commonly used in discussing semiconductors are plotted with greater energy values higher, the scale of the electrical potential is in the opposite direction. This is because the energy of a charged species is the product of its charge and the electrical potential, and the charge on electrons is negative.

Fig. 14.19 Variation of the electrical potential as a function of lithium concentration in $Li_xNa_{0.4}WO_3$. After [14]

14.10.6 The Composition Dependence of the Potential in the Case of Insertion Reactions That Involve a Two-Phase Reconstitution Reaction

The earlier discussion of the influence of the Gibbs Phase Rule upon the compositional variation of the potentials in electrodes pointed out that when there are two phases present in a two-component system, the potential will have a fixed, or constant, value, independent of the composition. This will also be the case for materials that act as pseudo-binaries, regardless of how many different species are actually present. A number of insertion reaction materials are of this type, with one relatively mobile species inside a relatively stable host structure. If, in the time span of interest, the host structure does not undergo any changes it can be considered to be a single component thermodynamically. This is the case in a number of materials in which the host structure is a transition metal oxide.

One example in which the potential is composition-independent involves the insertion and extraction of lithium in materials with the nominal composition $Li_4Ti_5O_{12}$, which has a defect spinel crystal structure [15].

The normal composition of spinel structure materials can be described as AB_2O_4, where the A cation species resides on tetrahedral sites, and the B cation species on octahedral sites within a close-packed face-centered cubic oxygen lattice. This composition can also be written as $A_3B_6O_{12}$.

If this material were to have the $Li_4Ti_5O_{12}$ stoichiometry, an extra lithium ion must be present, and a titanium ion is missing.

This can be accomplished replacing a missing titanium ion with an extra lithium ion on an octahedral B site in the structure. The other three lithium ions would

Fig. 14.20 Charge and discharge curve of $Li_4Ti_5O_{12}$ cell. After [15]

remain on their normal tetrahedral A sites. Thus the composition can be written as $Li_3[LiTi_5]O_{12}$, or alternatively, $Li[Li_{1/3}Ti_{5/3}]O_4$.

It has been found that an additional lithium ion can react with this material, and this can be written as

$$Li + Li\left[Li_{1/3}Ti_{5/3}\right]O_4 = Li_2\left[Li_{1/3}Ti_{5/3}\right]O_4 \qquad (14.9)$$

X-ray diffraction data have indicated that all the lithium ions now occupy octahedral sites, instead of tetrahedral sites. Since there are only as many octahedral sites available as oxide ions in this structure, they must now be all filled. This is likely why the capacity of this electrode material is limited to this composition.

Experiments were performed on samples of these materials that were prepared in air, and were white in color. As with all essentially all materials prepared in air, their potential was initially near 3 V versus lithium. In electrochemical experiments, when lithium was added by transfer from the negative electrode, lithium in carbon, the potential went rapidly down to 1.55 V, and remained there until the reaction was complete. Thus this insertion reaction has the characteristics of a moving-interface reconstitution reaction.

Upon deletion of the inserted lithium the potential retraced the discharge curve closely, with very little hysteresis. This is illustrated in Fig. 14.20 [15]. Because of the small volume change, negligible hysteresis and rapid kinetics this material acts as a very attractive electrode in lithium cells. The one disadvantage is that its potential is, unfortunately, about half way between the negative and positive electrode potentials in most lithium batteries.

As will be discussed later, hysteresis, which leads to a difference in the composition-dependence of the potential when charging and discharging, is often related to mechanical strain energy, i.e., dislocation generation and motion, as a

Fig. 14.21 Charge and discharge curves for the reaction of lithium with FePO$_4$. After [16]

consequence of volume changes that occur due to the insertion and extraction of guest ions.

Another example of an insertion-driven reconstitution reaction is the reaction of lithium with FePO$_4$, which also happens readily at ambient temperature. This also has a very flat reaction potential, as shown in Fig. 14.21 [16]. In this case the material is prepared (in air) as LiFePO$_4$, and the initial reaction within the cell involves charging, i.e., deleting lithium from its crystal structure. This lithium goes across the electrochemical cell and into the carbon material in the negative electrode. The reaction that occurs at the operating potential during the initial charge can be simply written as

$$LiFePO_4 = Li + FePO_4 \qquad (14.10)$$

Upon discharge of the cell, the reaction goes, of course, in the opposite direction.

This material is now one of the most important positive electrode reactants in lithium batteries, and will be discussed further in a later chapter.

14.11   Final Comments

This chapter is intended to be only a general introduction to the scope of insertion reactions in electrode materials. This is a very important topic, and will be addressed further in the discussions of specific materials in later chapters.

References

1. Whittingham MS, Huggins RA (1971) J Electrochem Soc 118:1
2. Whittingham MS, Huggins RA (1971) J Chem Phys 54:414
3. Whittingham MS, Huggins RA (1972) Solid State Chemistry, ed. by R.S. Roth and S.J. Schneider, Nat. Bur. of Stand. Special Publication 364, p. 139
4. J. Livage, Le Monde, October 26, 1977
5. J. Rouxel, Materials Science Forum (1994), p. 152
6. J. Goodenough, Annual Review of Matls Sci., Vol. 1, ed. by R.A. Huggins (1970), p. 101
7. Raistrick ID, Huggins RA (1983) Mat Res Bull 18:337
8. Macklin WJ, Neat RJ, Sandhu SS (1992) Electrochim Acta 37:1715
9. Delmas C, Cognac-Auradou H, Cocciantelli JM, Menetrier M, Doumerc JP (1994) Solid State Ionics 69:257
10. Netz A, Huggins RA, Weppner W (2003) J Power Sources 119–121:95
11. Netz A, Huggins RA (2004) Solid State Ionics 175:215
12. Obrovac MN, Christensen L (2004) Electrochem Solid State Lett 7:A93
13. Chirayil TA, Zavalij PY, Whittingham MS (1996) J Electrochem Soc 143:L193
14. Raistrick ID, Mark AJ, Huggins RA (1981) Solid State Ionics 5:351
15. Ohzuku T, Ueda A, Yamamoto N (1995) J Electrochem Soc 142:1431
16. Yamada A, Chung SC, Hinokuma K (2001) J Electrochem Soc 148:A224

Chapter 15
Electrode Reactions That Deviate from Complete Equilibrium

15.1 Introduction

The example that was discussed earlier, the reaction of lithium with iodine to form LiI, dealt with elements and thermodynamically stable phases. By knowing a simple parameter, the Gibbs free energy of formation of the reaction product, the cell voltage under equilibrium and near-equilibrium conditions can be calculated for this reaction. If the cell operates under a fixed pressure of iodine at the positive electrode and at a stable temperature, the Gibbs phase rule indicates that the number of the residual degrees of freedom F in both the negative and positive electrodes is zero. Thus the voltage is independent of the extent of the cell reaction in both cases.

This is a case in which the reaction involves species that are *absolutely stable*. The description of a phase as absolutely stable means that it is in the thermodynamic state, e.g., crystal structure, with the lowest possible value of the Gibbs free energy for its chemical composition.

15.2 Stable and Metastable Equilibrium

On the other hand, there could be several versions of a phase with different structures that might be stable in the sense that they have lower values of the Gibbs free energy than would be the case with minor changes. Such a situation, in which a phase is stable against small perturbations, is described by the term *metastable*. On the other hand, it may be less stable than the *absolutely stable* modification. This can be illustrated schematically by the use of a simple mechanical model, as is illustrated in Fig. 15.1.

This situation can also be described in terms of the changes in the potential energy of a simple block. If the block sits on its end, it is in a metastable state, and if it is tipped a small amount, its potential energy will be increased, but it will tend to

© Springer International Publishing Switzerland 2016
R.A. Huggins, *Energy Storage*, DOI 10.1007/978-3-319-21239-5_15

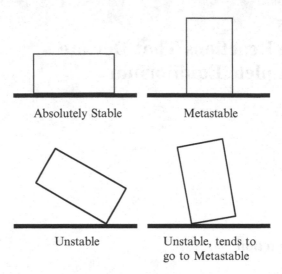

Fig. 15.1 Simple mechanical model illustrating metastable and absolutely stable states

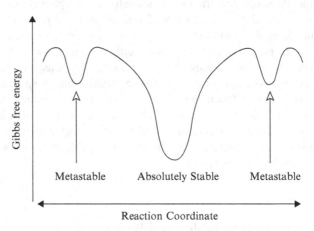

Fig. 15.2 Reaction coordinate representation of a system with metastable and absolutely stable states

revert back to its initial metastable condition. But a larger perturbation will get it over this potential energy hump so that it will tip over and land in the flat position, the absolutely stable state.

This situation can also be illustrated by the use of a reaction coordinate diagram of the type often used in discussions of chemical reaction kinetics, as shown in Fig. 15.2.

This discussion does not only apply to single phases, for it is possible to have a situation in which a material has a microstructure that consists of a metastable single phase, whereas the absolutely stable situation involves the presence of two, or perhaps more, phases. In order for the system to go from the metastable single-phase situation to the more stable *polyphase* structure it is necessary to *nucleate* the additional phase or phases as well as to change the composition of the initial metastable phase. This may be kinetically very difficult.

In the case of the Li/I system, where there is only one realistic structure for the reactant and product phases, only the absolutely stable situation has to be considered.

However, in other materials systems the situation is often different at lower temperatures from that at high temperatures, where absolutely stable phases are generally present. As will be discussed later, metastable phases and metastable crystal structures often play significant roles at ambient temperatures.

15.3 Selective Equilibrium

There is also the possibility that a material may attain equilibrium in some respects, but not in others. Some of the most interesting and important ambient temperature materials fall into this category.

A number of the reactants in ambient temperature battery systems have crystal structures that can be described as a composite consisting of a highly mobile ionic species within a relatively stable host structure.

Such structures are sometimes described as having two different *sub-lattices*, one of which has a high degree of mobility, and the other is highly stable, for its structural components are rigidly bound. The guest species with high mobilities are typically rather small and move about through interstitial tunnels in the surrounding rigid host structure. The species in the mobile sub-lattice can readily come to equilibrium with the thermodynamic forces upon them, whereas the more tightly bound parts of the host structure cannot.

The term *selective equilibrium* can be used for this situation. Under these conditions, the stable part of the crystal structure can be treated as a single component when considering the applicability of the Gibbs phase rule. An example that will be discussed later is the phase Li_xTiS_2. The structure of this material can be thought of as consisting of rigid planar slabs of covalently bonded TiS_2, with mobile lithium ions in the space between them. The lithium species readily attain equilibrium with the external environment at ambient temperatures, whereas the TiS_2 part of the structure is relatively inert so that it can be considered to be a single component. Thus at a fixed temperature and total pressure the number of residual degrees of freedom is 1. This means that the value of one additional thermodynamic parameter will determine all of the intensive variables. As an example, a change in the electrical potential causes a change in the equilibrium amount of lithium in the structure, the value of x in Li_xTiS_2, but has no influence upon the TiS_2 slabs.

15.4   Formation of Amorphous vs. Crystalline Structures

An amorphous structure can result when a phase is formed under conditions in which complete equilibrium and the expected crystalline structure cannot be attained. Although they may have some localized ordered arrangements, amorphous structures do not have regular long-range arrangements of their constituent atoms or ions. Amorphous structures are always less stable than the crystalline structure with the same composition. Thus they have less negative values of the Gibbs free energy of formation than their crystalline cousins.

If the phase LiM can be electrochemically synthesized by the reaction of lithium with species M, a type of reconstitution reaction, there will be a corresponding constant voltage two-phase plateau in the titration curve related to that reaction. The magnitude of the plateau voltage is determined by the Gibbs free energy of the product phase, as described earlier. Because of its less negative Gibbs free energy of formation, the potential of the plateau related to the formation of an amorphous LiM phase must always be lower than that of the corresponding crystalline version of LiM. This is illustrated schematically in Fig. 15.3.

This has interesting consequences for the case in which two intermediate phases can be formed. As an example, assume that lithium can react with material M to form two phases in sequence, LiM and Li_2M. The reaction for the formation of the first phase, LiM is

$$Li + M = LiM \tag{15.1}$$

and if the phase LiM has a very narrow range of composition, the equilibrium titration curve, a plot of potential E versus composition, will look like that shown in Fig. 14.3.

Fig. 15.3 Schematic drawing of the voltage of galvanic cell as a function of overall composition for a simple formation reaction $Li + M = LiM$ for two cases, one in which the LiM product is crystalline, and the other in which it is amorphous

Fig. 15.4 Schematic titration curve for a sequence of two reactions of Li with M, first forming LiM, and then forming Li_2M

The plateau voltage is given by

$$E = -\Delta G_f^{\,\circ}(\text{LiM})/F \tag{15.2}$$

If additional lithium can react with LiM to form the phase Li_2M there will be an additional voltage plateau, whose potential is determined by the reaction

$$\text{Li} + \text{LiM} = \text{Li}_2\text{M} \tag{15.3}$$

This is shown schematically in Fig. 15.4.

The voltage of the second plateau is lower than that of the first, and is given by

$$E = -\left[\Delta G_f^{\,\circ}(\text{Li}_2\text{M}) - \Delta G_f^{\,\circ}(\text{LiM})\right]/F \tag{15.4}$$

But what if the first phase, LiM, is amorphous, rather than crystalline? As mentioned above, this means that Gibbs free energy of formation of that phase is smaller and the voltage of the first plateau is reduced.

The total Gibbs free energy of the two reactions is determined, however, by the Gibbs free energy of formation of the final phase, Li_2M. This is not changed by the formation of the intermediate phase LiM. The total area under the curve is thus a constant. The interesting result is that if the voltage of the first plateau is reduced, the voltage of the second one must be correspondingly increased. This can be depicted as in Fig. 15.5.

Thus the lower stability of the intermediate phase reduces the magnitude of the step in the titration curve. Therefore the overall behavior approaches what it would be if the intermediate phase did not form at all, and there would only be one reaction, the direct formation of phase Li_2M.

Fig. 15.5 Change in the schematic titration curve if the first product, LiM, is amorphous. The voltage of the second plateau must be higher to compensate for the reduced voltage of the first plateau

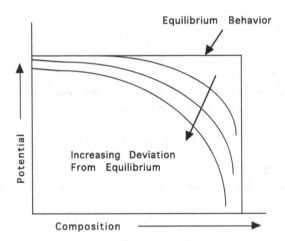

Fig. 15.6 Schematic representation of the influence of kinetic limitations upon both the potential and capacity of an electrode reaction

15.5   Deviations from Equilibrium for Kinetic Reasons

The observed potentials and capacities of electrodes are often displaced from those that would be expected from equilibrium thermodynamic considerations because of kinetic limitations. There may not be sufficient time to attain the compositional and/or structural changes that should, in principle, occur. This is more likely to occur at lower temperatures, and under higher current conditions.

The influence of increasing deviations from equilibrium conditions upon the behavior of a simple reconstitution reaction is shown schematically in Fig. 15.6. It is seen that both the potential and the apparent capacity can deviate significantly from equilibrium values.

The kinetics of electrode reactions, and methods that can be used to evaluate them, will be discussed in subsequent chapters.

Chapter 16
Primary, Non-rechargeable Batteries

16.1 Introduction

Except for the discussions of the lithium/iodine cell in Chap. 10, all of the discussion concerning batteries for energy storage has been oriented toward understanding the properties of individual cell components and systems. The emphasis has been upon those that are most interesting for use in rechargeable batteries.

There are, however, a number of types of batteries that are very common and important, even though they cannot be readily recharged. They are often called primary batteries, and are typically discarded when they become discharged. Several of these will be discussed in this chapter.

Because some primary cells have higher values of specific energy than current rechargeable systems, there is continual interest in finding methods to electrically recharge them, rather than having to refurbish them chemically by reprocessing one or more of their components.

16.2 The Common Zn/MnO₂ "Alkaline" Cell

16.2.1 Introduction

Elemental zinc is used as the negative electrode in a number of aqueous electrolyte batteries. The most prominent example is the very common Zn/MnO_2 primary "alkaline cell" that is used in a wide variety of small electronic devices. Elemental zinc is the negative electrode reactant, MnO_2 the positive reactant, and the electrolyte is a solution of KOH. These cells are available in great numbers in standard AA and AAA sizes.

© Springer International Publishing Switzerland 2016   291
R.A. Huggins, *Energy Storage*, DOI 10.1007/978-3-319-21239-5_16

As will be discussed later, the positive electrode reaction in this case involves the insertion of hydrogen into the MnO_2 crystal structure, forming H_xMnO_2. A discussion of Zn/MnO_2 technology can be found in [1].

The initial open circuit voltage of these cells is in the range 1.5–1.6 V. This is greater than the decomposition voltage of water, which can be calculated from its Gibbs free energy of formation, 237.1 kJ/mol, to be 1.23 V at ambient temperatures from

$$\Delta E = -\left(\frac{\Delta G_f^*(H_2O)}{2F}\right) \tag{16.1}$$

It will be shown here that this is possible because the zinc negative electrode is covered by a thin layer of ionically conducting ZnO, the thermodynamic result is that its potential is several hundred millivolts lower than the potential at which gaseous hydrogen is normally expected to evolve if an unoxidized metal electrode were to be in contact with water.

The open circuit voltage decreases as energy is extracted and the residual capacity becomes reduced. This reduction in cell voltage is due to the change of the potential of the MnO_2 positive electrode due to the insertion of protons from the electrolyte. This can be described as changing the value of x in the composition H_xMnO_2 from zero to about 1. The proton content can be increased until the value of x becomes 2. However, as will be shown later, the second proton reaction occurs at a cell voltage of about 1 V, which is too low to be of practical use.

16.2.2 Thermodynamic Relationships in the H-Zn-O System

The potential and stability of the zinc electrode can be understood by consideration of the thermodynamics of the ternary H-Zn-O system, and its representation in a ternary phase stability diagram.

In addition to the elements and water, the only other relevant phase in this system is ZnO, and the value of its Gibbs free energy of formation is −320.5 kJ/mol at 25 °C.

As discussed earlier, one can use the values of the Gibbs free energy of formation of the different phases to determine which tie lines are stable in a ternary phase stability diagram. In this case the only possibilities would be either a line between Zn and H_2O or a line between ZnO and hydrogen. Because the Gibbs free energy of formation of ZnO is more negative than that of water, the second of these possible tie lines must be the more stable. The simple result in this case is shown in Fig. 16.1. It shows that a sub-triangle is formed that has Zn, ZnO, and H_2 at its corners. Another has water, ZnO, and hydrogen at its corners. The potentials of all compositions in the first triangle with respect to oxygen can be calculated from the Gibbs free energy change related to the simple binary reaction along its edge,

Fig. 16.1 Ternary phase stability diagram for the H-Zn-O system. The numbers within the ternary sub-triangles are the potentials relative to pure oxygen

$$Zn + \tfrac{1}{2}O_2 = ZnO \qquad\qquad (16.2)$$

and the result is -1.66 V. The potential of all compositions in the second triangle is likewise related to the Gibbs free energy of formation of water, or -1.23 V relative to pure oxygen. That means that zinc has a potential that is 0.43 V more negative than the potential of pure hydrogen in aqueous electrolytes.

Because of the presence of the thin ionically conducting, but electronically insulating, layer of ZnO, water is not present at the electrochemical interface, the location of the transition between ionic conduction and electronic conduction, and hydrogen gas is not formed on the zinc. Thus the effective stability range of the electrolyte is extended, as discussed later.

16.2.3   Problems with the Zinc Electrode

Whereas its low potential is very attractive, there are two negative features of the use of zinc electrodes in aqueous systems. Both relate to its rechargeability in basic aqueous electrolytes.

One of these is that ZnO dissolves in KOH electrolytes, producing an appreciable concentration of zincate ions, $Zn(OH)_4{}^{2-}$, in which the Zn^{2+} cations are tetrahedrally surrounded by four OH⁻ groups. Nonuniform zincate composition gradients during recharging, as well as the ZnO on the surface, lead to the formation of *protrusions*, *filaments*, and *dendrites* during the re-deposition of zinc from the electrolyte at appreciable currents.

The other is that the zinc has a tendency to not redeposit upon the electrode at the same locations during charging of the cell as those from which it was removed during discharge. Gravitational de-mixing causes the concentration of zincate ions to increase at lower locations, leading to slight differences in the electrolyte conductivity. The result is that there is a gradual redistribution of the zinc, so that the lower portions of the electrode become somewhat thicker or denser as it is discharged and recharged. This effect is often called *shape-change*.

MnO_2 is polymorphic, which means that it can exist with a number of different crystal structures. It has been known for many years that they exhibit very different electrochemical behavior. The form found in mineral deposits has the rutile (beta) structure, and is called pyrolusite. It is relatively inactive as a positive electrode reactant in KOH electrolytes. It can be given various chemical treatments to make it more reactive, however. One of these produces a modification containing some additional cations that is called birnessite. Manganese dioxide can also be produced chemically, and then generally has the delta structure. The material that is currently much more widely used in batteries is produced electrolytically, and is called EMD. It has the gamma (ramsdellite) structure.

The reason for the differences in the electrochemical behavior of the several morphological forms of manganese dioxide presented a quandary for a number of years. It was known, however, that the gamma electrochemically active materials contain about 4 % water in their structures that can be removed by heating to elevated temperatures (100–400 °C), but the location and form of that water remained a mystery. This problem was solved by Ruetschi, who introduced a cation vacancy model for MnO_2 [2, 3].

The basic crystal structure of the various forms of MnO_2 contains Mn^{4+} ions in octahedral holes within hexagonally (almost) close-packed layers of oxide ions. That means that each Mn^{4+} ion has six oxygen neighbors, and these MnO_6 octahedra are arranged in the structure to share edges and corners. Differences in the edge- and corner-sharing arrangements result in the various polymorphic structures mentioned above.

If some of the Mn^{4+} ions are missing (cation vacancies are present), their missing positive charge has to be compensated by something else in the crystal structure. The Ruetschi model proposed that this charge balance is accomplished by the local presence of four protons. These protons would be bound to the neighboring oxide ions, forming a set of four OH^- ions. This local configuration is sometimes called a Ruetschi defect. There is very little volume change, as OH^- ions have essentially the same size as O_{2-} ions, and these species play the central role in determining the size of the crystal structure.

On the other hand, reduction of the MnO_2 occurs by the introduction of additional protons during discharge, as first proposed by Coleman [4], and does produce a volume change. The charge of these added mobile protons is balanced by a reduction in the charge of some of the manganese ions present from Mn^{4+} to Mn^{3+}. Mn^{3+} ions are larger than Mn^{4+} ions, and this change in volume during reduction has been observed experimentally.

The presence of protons (or OH^- ions) related to the manganese ion vacancies facilitates the transport of additional protons as the material is discharged. This is why these materials are very electrochemically reactive.

16.2.4   The Open Circuit Potential

The EMD is produced by oxidation of an aqueous solution of manganous sulfate at the positive electrode of an electrolytic cell. This means that the MnO$_2$ that is produced is in contact with water.

The phase relations, and the related ternary phase stability diagram, for the H-Mn-O system can be determined by use of available thermodynamic information [5, 6], as discussed in previous chapters. From this information it becomes obvious which neutral species reactions determine the potential ranges of the various phases present, and their values.

Following this approach, it is found that the lower end of the stability range of MnO$_2$ is at a potential that is 1.014 V vs. one atmosphere of H$_2$. The upper end is well above the potential at which oxygen evolves by the decomposition of water.

Under equilibrium conditions all oxides exist over a range of chemical composition, being more metal-rich at lower potentials, and more oxygen-rich at higher potentials. In the higher potential case, an increased oxygen content can result from either the presence of cation (Mn) vacancies or oxygen interstitials. In materials with the rutile, and related, structures that have close-packed oxygen lattices the excess energy involved in the formation of interstitial oxygens is much greater than that for the formation of cation vacancies. As a result, it is quite reasonable to assume that cation vacancies are present in the EMD MnO$_2$ that is formed at the positive electrode during electrolysis.

Due to the current that flows during the electrolytic process the potential of the MnO$_2$ that is formed is actually higher than the equilibrium potential for the decomposition of water. A number of other oxides with potentials above the stability range of water have been shown to oxidize water. Oxygen gas is evolved, and they become reduced by the insertion or protons. Therefore, it is quite reasonable to expect that EMD MnO$_2$ would have Mn vacancies, and that there would also be protons present, as discussed by Ruetschi [2, 3].

When such positive oxides oxidize water and absorb hydrogen as protons and electrons, their potentials decrease to the oxidation limit of water, 1.23 V vs. H$_2$ at 25 °C. This is the experimentally observed value of the open circuit potential of MnO$_2$ electrodes in Zn/MnO$_2$ cells.

This water oxidation phenomenon that results in the insertion of protons into MnO$_2$ is different from the insertion of protons by the absorption of water into the crystal structure of materials that initially contain oxygen vacancies, originally discussed by Stotz and Wagner [7]. It has been shown that both mechanisms can be present in some materials [8, 9].

16.2.5   Variation of the Potential During Discharge

As mentioned above, this electrode operates by the addition of protons into its crystal structure. This is a single-phase insertion reaction, and therefore the potential varies with the composition, as discussed earlier.

Fig. 16.2 Schematic discharge curve of Zn/MnO₂ cell

If all of the initially present Mn^{4+} ions are converted to Mn^{3+} ions, the overall composition can be expressed as $HMnO_2$, or $MnOOH$.

It is also possible to introduce further protons, so that the composition moves in the direction of $Mn(OH)_2$. In this case, however, there is a significant change in the crystal structure, so that the mechanism involves the translation of a two-phase interface between $MnOOH$ and $Mn(OH)_2$. This is analogous to the main reaction involved in the operation of the nickel electrode, as will be discussed later.

The sequence of these two types of reactions during discharge of a MnO_2 electrode is illustrated in Fig. 16.2.

The second, two-phase, reaction occurs at such a low cell voltage that the energy that is available is generally not used. Such cells are normally considered to only be useful down to about 1.2 V.

Although these Zn/MnO₂ cells are generally considered to be non-rechargeable, there have been some developments that make it possible to recharge them a modest number of times, and a small fraction of the alkaline cell market has been oriented in this direction. This involves modifications in the design and proprietary changes in the composition of the materials. Their rechargeability depends upon the depth to which they have been discharged, and there is a gradual reduction in the available capacity.

16.3   Ambient Temperature Li/FeS₂ Cells

Another type of consumer battery that is gradually becoming more popular in a number of consumer markets is the Li/FeS₂ cell. In this case the negative electrode is lithium metal. The electrolyte is a lithium salt dissolved in an organic solvent similar to that used in rechargeable lithium batteries.

The potential of the elemental lithium negative electrode is constant, but that of the positive electrode varies with the state of charge. At very low current drain, it shows two voltage plateaus, at about 1.7 V and 1.5 V versus lithium, as seen in Fig. 16.3. This indicates the formation of an intermediate phase, and therefore a sequence of two different reactions.

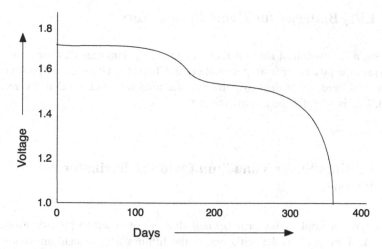

Fig. 16.3 Variation of the cell voltage with state of charge for Li/FeS₂ cells

Fig. 16.4 Variation of the capacity of typical Li/FeS₂ cells with the current drain

At moderate, or greater, currents, the plateau structure disappears, and the output voltage drops steadily from about 1.6 to 1.5 V with the state of charge. This means that the intermediate phase does not form under those conditions.

The voltage of this type of battery is about 0.1 V higher than that of the common alkaline cells, and there is less fade as the cell becomes discharged. The primary advantage of the Li/FeS₂ cells over the less expensive Zn/MnO₂ cells is their ability to handle higher currents. This is shown in Fig. 16.4. This property makes them especially useful for pulse applications, such as in cameras.

16.4   Li/I$_2$ Batteries for Heart Pacemakers

There was a discussion of the non-rechargeable Li/I$_2$ batteries that are commonly used to provide power for heart pacemakers in Chap. 10. There is no need to repeat that material here, other than to point out the unusual situation that the reaction product, LiI, is actually the electrolyte in this case.

16.5   Lithium/Silver Vanadium Oxide Defibrillator Batteries

Another type of implantable primary cell that is now used to provide power for medical devices, such as defibrillators, is the lithium/silver vanadium oxide cell. The attractive features of this chemistry were first recognized in 1979 [10, 11]. The person responsible for the commercial development of these batteries, Esther Takeuchi, received the National Medal of Technology and Innovation from President Obama in October, 2009.

The negative electrode in these cells is elemental lithium, and the electrolyte is the lithium salt LiBF$_4$ in an organic solvent, propylene carbonate. The positive electrode starts as AgV$_2$O$_{5.5}$, a member of the family of electronically conducting oxides called vanadium bronzes [12].

Lithium reacts with this positive electrode material by an insertion reaction that can be written as

$$xLi + AgV_2O_{5.5} = Li_xAgV_2O_{5.5} \qquad (16.3)$$

This reaction occurs over several steps, with corresponding values of x. This can be seen from the plot of the cell voltage as a function of the extent of this reaction shown in Fig. 16.5.

Charge balance is accomplished by a change in the effective charge of the cations originally in the vanadium bronze. In AgV$_2$O$_{5.5}$ all of the vanadium ions have an effective charge of 5+. Upon adding Li+ ions, the system moves into a constant-potential 2-phase regime, where both AgV$_2$O$_{5.5}$ and LiAgV$_2$O$_{5.5}$ are present. When the overall composition reaches the end of that voltage plateau, it moves into a variable-potential composition range in which only the phase LiAgV$_2$O$_{5.5}$ is present, and half of the vanadium ions have a charge of 5+, and the other half have a charge of 4+. The addition of 2 additional lithium ions causes the composition to move into another two-phase plateau in which the phase LiAgV$_2$O$_{5.5}$ is in equilibrium with a composition that is nominally Li$_3$AgV$_2$O$_{5.5}$. The effective charge of the vanadium ions in this latter phase is still 4+, but the nominal charge upon the silver ions has become zero. This means that there must be some particles of elemental silver present in the microstructure in addition to a phase of composition Li$_3$V$_2$O$_{5.5}$. Experiments have shown that the lower-lithium

Fig. 16.5 Schematic illustration of the variation of the cell voltage as a function of the amount of lithium reacted

reactions are reversible, but the last, which involves the precipitation of a new phase, is not.

The decrease of voltage as the cell is discharged allows the state of charge to be readily determined by voltage measurements. This is important when such power sources are used in implantable medical devices. These cells exhibit a very low rate of self-discharge, have a long shelf life, and store a large amount of energy per unit volume, 930 Wh/l. The latter feature is attractive for applications in which battery size is important.

16.6 Zn/Air Cells

Primary cells based upon the reaction of zinc with air have been available commercially for a number of years. This chemistry can produce a rather large value of specific energy, is relatively inexpensive, and presents no significant environmental problems. One of the major applications is as a power source for small hearing aids.

A cell with metallic zinc as the negative electrode and oxygen (or air) on the positive side is shown schematically in Fig. 16.6.

There must be a mechanism for the flow of electrons into and out of an external electronic circuit from both electrodes. This is accomplished on the negative side by contact with metallic zinc. On the positive side there is a porous metallic conductor

Fig. 16.6 Schematic representation of a Zn/O₂ cell

Fig. 16.7 Schematic representation of a Zn/O₂ cell that is partially discharged

in contact with both the oxygen reactant and the alkaline electrolyte. Although this metal plays no role in the overall cell reaction, the three-phase contact allows the electrochemical reaction that converts neutral atoms into ions and electrons.

The discharge reaction mechanism involves the transport of oxygen across the cell from the positive to the negative electrode, with the formation of ZnO on top of the Zn. A cell that is partially discharged is shown schematically in Fig. 16.7.

ZnO is an electronic conductor, so the electrochemical interface, where the electrical charge transport mechanism is converted from ions to electrons, is at the interface between the ZnO and the electrolyte. It is the electric potential at that interface that determines the externally measurable electrical potential of the negative electrode.

The reaction that determines the potential is generally assumed to be the formation of ZnO

$$Zn + \tfrac{1}{2}O_2 = ZnO \tag{16.4}$$

so that the voltage would be determined by the Gibbs free energy of formation of ZnO from zinc and oxygen.

$$E = \frac{\Delta G_f(ZnO)}{zF} \tag{16.5}$$

where $z = 2$, and F is the Faraday constant, 96.5 kJ/V equivalent. The value of $\Delta G_f(ZnO)$ at 298 K is 320.5 kJ/mol, so the equilibrium voltage E is 1.66 V at that temperature.

However, these cells operate in air, rather than pure oxygen. Therefore the chemical potential of oxygen is lower, and the electrical potential of the positive electrode is reduced. The chemical potential of oxygen in the positive electrode can be expressed as

$$\mu(O_2) = \mu^0(O_2) + RT\ln p(O_2) \tag{16.6}$$

where $\mu^0(O_2)$ is the chemical potential of oxygen in its standard state, a pressure of 1 atm at the temperature in question, and $p(O_2)$ is the actual oxygen pressure at the electrode.

In air the oxygen partial pressure is approximately 0.21 atm, so that the cell voltage is reduced by

$$\Delta E = \frac{RT}{zF}\ln(0.21) \tag{16.7}$$

The result is that the equilibrium voltage of the Zn/O_2 cell when air is the reactant on the positive side should be reduced by 0.02 V. Thus a Zn/air cell should have an open circuit voltage of 1.64 V. If the oxygen pressure is maintained at a constant value, the voltage will be independent of the state of charge, i.e., will have the characteristics of an infinite plateau in a battery discharge curve.

The value of the maximum theoretical specific energy can be calculated from this information using the weights of the reactants. As discussed in Chap. 9, the value of the MTSE is given by

$$MTSE = 26,805 \ (zE)/W_t \quad Wh/kg \tag{16.8}$$

The value of the reactant weight, W_t, is the weight of a mol of Zn (65.38 g) plus the weight of 1/2 mol of oxygen (8 g), or a total of 73.38 g per mol of reaction. The value of z is 2, the number of elementary charges involved in the virtual cell reaction.

Using this value and a cell voltage of 1.64 V for the case of air at the positive electrode, the MTSE is 1198 Wh/kg. If pure oxygen were used, it would be 1213 Wh/kg.

But there is a problem. The measured open circuit voltage of commercial Zn/air cells is about 1.5 V, not 1.64 V. The reason for this has to do, again, with what is actually going on at the positive electrode. The normal assumption is that the positive electrode reactant is oxygen, and therefore the potential should be that of pure oxygen at the partial pressure of air.

Experiments have shown the presence of peroxide ions at the positive electrode in alkaline aqueous cells. Instead of a conversion of oxygen from O_2 in the gas phase to O^{2-} ions in the electrolyte, there is an intermediate step, due to the presence of peroxide ions.

In peroxide ions, O^-, oxygen is at an intermediate charge state between neutral oxygen, O^0, in oxygen molecules in the gas, and oxide ions, O^{2-}, in the KOH electrolyte. In such aqueous systems this can be written as two steps in series

$$O_2 + H_2O + 2e^- = HO_2^- + OH^- \tag{16.9}$$

and

$$HO_2^- + H_2O + 2e^- = 3OH^- \tag{16.10}$$

The result is that the electrical potential in the positive electrode is determined by the presence of hydrogen peroxide, which is formed by the reaction of oxygen with the KOH electrolyte.

This is also the case with aqueous electrolyte hydrogen/oxygen fuel cells, where the open circuit voltage is determined by the presence of peroxide, rather than oxide, ions [13–15]. This is shown in Fig. 16.8. On the other hand, high temperature proton or oxide ion-conducting fuel cells have open circuit voltages that correspond to the assumption that the positive electrode reactant is oxygen.

Zn/O_2 cells are sold with a removable sealing material that prevents access of air to the positive electrode structure so that there is no self-discharge before they are used.

Their specific energy is very large, about 30 times the value of the maximum theoretical specific energy of a typical Pb/PbO_2 cell, so it is obvious why there is an interest in finding a way to make this system reversible. To do so, three general problems must be solved; the rechargeability of the zinc oxide product, the reversibility of the air electrode, and the sensitivity of the KOH electrolyte to contamination from CO_2 in the ambient air. CO_2 reacts with hydroxides to form solid carbonates, which can block the ionic transport through the electrolyte.

Development efforts toward the alleviation or avoidance of these problems have been undertaken in a number of laboratories, but they have not yet led to large-scale applications.

A large effort undertaken with the support of the German Post some years ago ran into several problems, the major one being the need for centralized chemical

Fig. 16.8 Experimental data on the voltage of aqueous fuel cells, showing the influence of the presence of peroxide ions at the positive electrode. After [6]

regeneration of the zinc oxide product back into useful zinc electrodes. Transportation to and from such facilities can put a major load on rail or highway transportation systems.

16.7 Li/CF$_x$ Cells

Lithium reacts with poly(carbon monofluoride), CF$_x$, at ambient temperatures.

The value of x in CF$_x$ can vary from about 0.9 to 1.2, depending upon its synthesis parameters.

These cells are generally used in situations in which low to moderate rates are required.

Elemental lithium is used as the negative electrode reactant, and the electrolyte is typically a solution of LiBF$_4$ in propylene carbonate. The reactant in the positive electrode is powdered CF$_x$. Although this material has a lamellar structure that can be thought of as analogous to graphite, it actually consists of an infinite array of cyclohexane "boats" [16] instead of thin graphene sheets. Lithium does not readily move between these layers, and therefore the electrode reaction mechanism does not involve insertion, as in the case of lamellar graphite.

Instead, a polyphase reaction occurs during discharge that can be written as

$$xLi + CF_x = xLiF + C \tag{16.11}$$

Since this is a simple displacement reaction, the voltage remains constant, at 2.75 V, during discharge.

Because the reactants have low weights, the maximum theoretical specific energy, the MTSE, of these cells is very high, 1940 Wh/kg. This would be attractive for use for implantable medical applications. However, because the voltage remains constant, it is difficult to determine when the capacity is almost consumed. An indication that a power source is soon going to reach its end-of-life is especially important when it is used for such purposes, and this problem is currently receiving a considerable amount of attention.

16.8   Reserve Batteries

16.8.1   Introduction

The discussion of positive electrodes in lithium batteries thus far has assumed that the reactants are either solids or gases. However, this is not necessary, and there are two types of primary batteries that have been available commercially for a number of years in which the reactant is a liquid, the Li/SO_2 and $Li/SOCl_2$ (thionyl chloride) systems. They both have very high specific energies. But because of safety considerations they are not in general use, and are being produced primarily for military and space purposes.

These are examples of *reserve batteries*, in which some method is used to prevent their operation until the energy is needed. There are two general ways in which this can be done.

One is to prevent the electrolyte from contacting one or both of the electrodes. This prevents self-discharge as well as the operation of any unwanted side reactions. One way to do this is to contain the electrolyte in a glass container that can be broken when cell operation is desired. A second method involves the use of an electrolyte that does not conduct current until it melts at an elevated temperature. When battery operation is desired, the electrolyte is heated to above its melting point. Examples of both of these strategies are discussed below.

16.8.2   The Li/SO_2 System

These cells are generally constructed with large surface area carbon electrodes on the positive side, and x-ray experiments have shown that $Li_2S_2O_4$ is formed there upon discharge. The discharge curve is very flat, at 3.0 V, as shown in Fig. 16.9.

Fig. 16.9   Schematic discharge curve for a Li/SO$_2$ cell

Table 16.1   Gibbs free energies of formation of phases in the Li-S-O system at 25 °C

Phase	Gibbs free energy of formation (kJ/mol)
Li$_2$O	−562.1
SO$_2$	−300.1
Li$_2$S	−439.1
Li$_2$S$_2$O$_4$	−1179.2

As discussed earlier, this type of behavior indicates that the cell operates by a reconstitution reaction. It should be possible to calculate the voltage by consideration of the thermodynamic properties of the phases involved in this system at ambient temperature. These are shown in Table 16.1.

From this information the stable tie lines in the ternary phase stability diagram for this system can be determined, as described earlier. The reaction equations relevant to each of the sub-triangles can also be identified, and their potentials calculated. The resulting diagram is shown in Fig. 16.10.

It can be seen that the Li$_2$S$_2$O$_4$–SO$_2$–O sub-triangle has a potential of 3.0 V versus lithium. Since the SO$_2$–Li$_2$S$_2$O$_4$ tie line on the edge of that triangle points at the lithium corner, no oxygen is formed by the reaction of lithium with SO$_2$ to produce Li$_2$S$_2$O$_4$.

The formal reaction for this cell is therefore

$$2Li + 2SO_2 = Li_2S_2O_4 \tag{16.12}$$

The theoretical specific energy of this cell can be calculated to be 4080 kWh/kg, a high value.

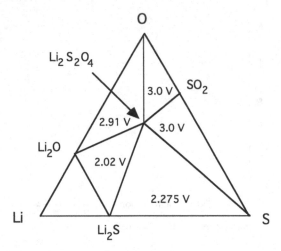

Fig. 16.10 Phase stability diagram for the ternary Li-S-O system at ambient temperature

16.8.3 The Li/SOCl₂ System

The lithium/thionyl batteries react at a somewhat higher constant voltage plateau, at 3.66 V.

The formal reaction is known to be

$$4Li + 2SOCl_2 = 4LiCl + S + SO_2 \tag{16.13}$$

This involves the Li-S-Cl-O quaternary system. In order to visualize the behavior of this system in a manner similar to that for the Li/SO₂ cell above, a tetrahedral figure would have to be drawn, and the constant voltage plateaus related to each of the sub-tetrahedra calculated. While straightforward, this is a bit too complicated to be included here, however.

The theoretical specific energy of this cell can be calculated to 7250 kWh/kg, which is a very high value.

16.8.4 Li/FeS₂ Elevated Temperature Batteries

A good deal of effort went into the development of a high temperature system that uses FeS₂ as the positive electrode reactant, but had either Li-Al or Li-Si alloys, rather than lithium metal, on the negative side. The electrolyte was a molten Li-K halide salt that has a eutectic temperature of 320 °C. These cells operated at temperatures over 400 °C, and the open circuit voltage was about 1.9 V, with most of the capacity obtained at 1.7 V. The initial development was aimed at their use to power electric vehicles, where their favorable high power operation is attractive. However, the appearance of other alternatives, such as ambient temperature lithium-ion systems, caused this work to be discontinued in the late 1990s.

Because the molten salt electrolyte is only conductive at elevated temperatures, such cells can be stored at ambient temperature and used as *reserve batteries*. Upon heating the electrolyte melts and the cell becomes operable. As mentioned above, this type of reserve batteries, often called *thermal batteries*, have been used for military applications in which a long shelf life is very important.

References

1. Scarr RF, Hunter JC, Slezak PJ (2002) Alkaline—Manganese Dioxide Batteries, in Handbook of Batteries, 3rd. edition, ed. D. Linden and T.B. Reddy, McGraw-Hill, p. 10.1
2. Ruetschi P (1984) J Electrochem Soc 131:2737
3. Ruetschi P, Giovanoli R (1988) J Electrochem Soc 135:2663
4. Coleman JJ (1946) Trans Electrochem Soc 90:545
5. Barin I (1995) Thermochemical Data of Pure Substances, 3rd edn. VCH, Weinheim, Published Online 24 Apr 2008. ISBN 9783527619829783527619825
6. Pourbaix M (1966) Atlas of Electrochemical Equilibria. Pergamon Press, Oxford, UK
7. Stotz S, Wagner C (1966) Ber Bunsenges Physik Chem 70; 781
8. Netz A, Chu WF, Thangadurai V, Huggins RA, Weppner W (1999) Ionics 5:426
9. Huggins RA (2006) J Power Sources 153:365
10. Liang CC, Boltser ME, Murphy RM (1982), US Patent 4,391,729
11. Liang CC, Boltser ME, Murphy RM (1982), US Patent 4,310,609
12. Takeuchi ES, Thiebolt WC III (1988) J Electrochem Soc 135:2691
13. Berl WG (1943) Trans Electrochem Soc 83:253
14. Winsel AW (1963) Advanced Energy Conversion, vol 3. Pergamon Press, Oxford
15. Kordesch KV, Berger C (1968) Handbook of Fuel Cell Technology. Prentice-Hall, Inc., New York, p 361
16. Ebert LB, Brauman JI, Huggins RA (1974) J Amer Chem Soc 96:7841

Chapter 17
Lead-Acid Batteries

17.1 Introduction

Over many years, the most common use of the word "battery" was in connection with the rechargeable energy source that was used to start automobiles. These were almost always what are generally called *Pb-acid batteries*, and were often a source of aggravation. A considerable amount of progress has been made in recent years, so that the SLI (starting-lighting-ignition) batteries now used in autos are actually quite reliable, assuming that they are not abused. Different types of Pb-acid batteries are used for a number of other applications, both mobile and stationary, and ranging from quite small to very large. The greatest fraction of the total battery market worldwide is now based upon this technology.

There are several reasons for the widespread use of lead-acid batteries, such as their relatively low cost, ease of manufacture, and favorable electrochemical characteristics, such as high output current and good cycle life under controlled conditions.

Pb-acid cells were first introduced by G. Planté in 1860 [1], who constructed them using coiled lead strips separated by linen cloth and immersed in sulfuric acid. By initially passing a dc current between the two lead strips, an oxide grew on the one on the positive side, forming a layer of lead dioxide. This caused the development of a voltage between them, and it was soon found that charge could be passed reversibly through this configuration, so that it could act to store electrical energy.

Significant improvements have been made over the years. One of the most important was the invention of the pasted plate electrode by C. Fauré in 1881 [2]. This involved the replacement of solid metal negative electrodes by a paste of fine particles held in a lead, or lead alloy grid. By doing this, the reaction surface area is greatly increased.

Another significant improvement has been the development of sealed cells during the last several decades. This is sometimes called *valve-regulated lead-acid* technology. These matters are discussed in the following sections.

© Springer International Publishing Switzerland 2016
R.A. Huggins, *Energy Storage*, DOI 10.1007/978-3-319-21239-5_17

There are two general types of applications that are commonly considered for Pb-acid cells, and they impose quite different requirements. One type involves keeping the cell essentially fully charged so that it maintains a constant output voltage. This is sometimes called *float charging*. Such cells are generally stationary, and are expected to have high reliability, long life, a low self-discharge rate, and a good cycling efficiency with low loss under cycling and overcharge. They are often used in telecommunication and large computer systems, railroad signaling systems, and to supply standby power as uninterruptible power sources (UPS). They are generally not optimized for energy density or specific energy, but are attractive because of their low cost.

The other general type is targeted toward applications that may involve deeper discharging, such as in load leveling systems and traction applications. In these cases, the specific energy and/or specific power can be very important, in addition to the cost, cycle efficiency, and lifetime. Periodic, rather than continuous, charging is more common in these cases.

Although hydride/"nickel" and lithium-ion cells are generally used in smaller portable applications, some sealed Pb-acid cells are now also used for such applications where the lowest cost is particularly important.

17.2   Basic Chemistry of the Pb-Acid System

The pb-acid cell is often described as having a negative electrode of finely divided elemental lead, and a positive electrode of powdered lead dioxide in an aqueous electrolyte. If this were strictly true and there were no other important species present, the cell reaction would simply involve the formation of lead dioxide from lead and oxygen:

$$Pb + O_2 = PbO_2 \tag{17.1}$$

The open-circuit voltage of such a cell would be determined from the Gibbs free energy of formation of PbO_2, $\Delta G_f^0(PbO_2)$, by

$$E = -\left(\frac{\Delta G_f^0(PbO_2)}{zF}\right) \tag{17.2}$$

in which z is 4, the number of charges involved in reaction (Eq. (17.1)), and F is the Faraday constant. The value of $\Delta G_f^0(PbO_2)$ has been measured as -215.4 kJ/mol [3]. Thus the cell voltage would be 0.56 V. However, this is far from what is actually observed, so that the reaction that determines the cell voltage must be quite different from Eq. (16.1).

Instead of the reaction in Eq. (17.1), the overall chemical process involved in the discharge of Pb-acid cells is generally described [4], in accordance with the *double-sulfate theory* [5–8] as

$$Pb + PbO_2 + 2H_2SO_4 = 2PbSO_4 + 2H_2O \qquad (17.3)$$

This reaction proceeds to the right-hand side during discharge, and toward the left side when the cell is recharged. This has been demonstrated by observations of morphological changes in both the negative [9, 10] and positive electrodes [11].

Using values of the standard Gibbs free energy of formation, ΔG_f^0, of the phases in this reaction, it has been shown that the equilibrium voltage of this reaction under standard conditions is 2.041 [12].

17.2.1 Calculation of the MTSE

It is interesting to calculate the maximum theoretical specific energy of Pb-acid cells. As discussed in Chap. 9, this can be expressed as

$$MTSE = 26,805 \left({}^{xE}\!/_{W} \right) \qquad (17.4)$$

in which x is the number of elementary charges, E the average cell voltage, and W the sum of the atomic weights of either the reactants or the products. In this case, x is 2, E is 2.05 V, and W is 642.52 g. Inserting these values, the maximum theoretical specific energy, calculated from these reactions, is 171 Wh/kg. This is fallacious, however, for it is necessary to have additional water present in order for the cell to operate. This increases the weight, and thus reduces the specific energy. But in addition, other passive components add significant amounts of weight, as is always the case in practical batteries. Values of the practical specific energy of lead-acid batteries are currently in the range of 25–40 Wh/kg. Higher values are typical for those optimized for energy, and lower values for those designed to provide more power.

17.2.2 Variation of the Cell Voltage with the State of Charge

From Eq. (17.3) it is obvious that the electrolyte changes, the amount of sulfuric acid decreases, and the amount of water present increases, as the cell becomes discharged. This causes a change in the electrolyte density. It is about 40 % by weight H_2SO_4 at full charge, but only 16 % when the cell is fully discharged. The corresponding values of equilibrium open circuit voltage are 2.15 V and 1.98 V at 25 °C. These density and voltage variations are illustrated in Fig. 17.1. Whereas it may take some time to reach the equilibrium voltage because of temporary structural and compositional inhomogeneities in the electrodes, the electrolyte density can be readily measured, and is often used to indicate the state of charge of the cell.

Fig. 17.1 Variations of the electrolyte density and open-circuit voltage in Pb-acid cells as functions of the state of charge

17.3   Potentials of the Individual Electrodes

It is clear that the cell voltage in pb-acid cells is significantly greater than the theoretical stability range of water, which is 1.23 V under equilibrium conditions. This is often attributed to (unspecified) kinetic factors in the literature.

However, this really means that the electrolytic stability window is extended by the presence of at least one additional ionically-conducting phase in series with the aqueous electrolyte [13].

Whereas it is easy to measure the cell voltage with a voltmeter, such a measurement does not give any information about the potentials of either electrode, just the difference between them. To get information about the individual potentials it is necessary to use reference electrodes.

This was done by Ruetschi [12, 14, 15], who used this information to determine the potential-determining microstructure in each electrode. He found that the surface of the lead in the negative electrode reactant is covered by a completely formed membrane layer of $PbSO_4$. He described this layer as perm selective, for it is essentially impermeable to the SO_4^{-2}, HSO_4^{-}, and Pb^{+2} ionic species in its vicinity, whereas H^+ ions can pass through it. This phase can thus be considered to be a selective ionic conductor for H^+ ions that extends the electrolytic stability window of the system. As a result, the negative electrode potential is determined by the Pb, $PbSO_4$ equilibrium, which he found to be −0.97 V relative to the Hg/Hg_2SO_4 reference electrode potential. This reference potential is + 0.65 V relative to the standard hydrogen reference potential, the SHE. PbO cannot play a role in this electrode potential, for it is only stable at potentials above −0.40 V vs. the Hg/Hg_2SO_4 reference.

Likewise, Ruetschi found that the positive electrode microstructure consists of three phases. $PbSO_4$ and PbO_2 are on top of the underlying lead structure. Again, there is a perm-selective layer of $PbSO_4$ on top of the PbO_2, which is an

Fig. 17.2 Schematic structures of the potential-determining portions of the electrodes in Pb-acid cells

Table 17.1 Potentials relevant to the Pb-acid battery

Equilibrium	Potential vs. Hg/Hg_2SO_4 reference/V	Potential vs. SHE reference/V
Hg/Hg_2SO_4	0.00	+0.65
$PbSO_4/PbO_2$ (unit activities)	+1.03	+1.68
$PbSO_4/PbO_2$ (5 m H_2SO_4)	+1.08	+1.73
$PbSO_4/PbO_2$ (1 m H_2SO_4)	+0.91	+1.56
Ag/Ag_2SO_4	+0.04	+0.69
$Pb/PbSO_4$	−1.01	−0.36
Lowest potential at which PbO is stable	−0.40	+0.25

electronic conductor. Thus the potential is determined by the $PbSO_4$, PbO_2 equilibrium in that case.

Based upon this quantitative work on their potentials and local corrosion films, the potential-determining parts of the Pb-acid battery system can be understood by considering the compositions of the two electrodes, as schematically illustrated in Fig. 17.2.

It was found that the PbO_2, $PbSO_4$ positive electrode potential depends upon the acid concentration in the same way as the voltage of the total Pb-acid cell. This is consistent with the Gibbs phase rule discussed earlier, It can be written as

$$F = C - P + 2 \tag{17.5}$$

For a fixed temperature and total pressure, it becomes simply

$$F = C - P \tag{17.6}$$

And since the PbO_2, $PbSO_4$ system has three components, Pb, S, and O, and only two phases, there is one degree of freedom left. Thus the potential is composition-dependent.

Table 17.1 shows the values of the relevant potentials in the lead-acid system, including two different liquid electrolyte compositions.

17.4   Relation to the Mechanisms of the Electrochemical Reactions in the Electrodes

The electrochemical reaction at the negative electrode is generally expressed as

$$Pb + HSO_4^- = PbSO_4 + H' + 2e^- \tag{17.7}$$

and that at the positive electrode as

$$PbO_2 + 3H^+ + HSO_4^- + 2e^- = PbSO_4 + 2H_2O \tag{17.8}$$

It can be seen that both of these reactions involve H^+ ions. These are the species that, as protons, move in and out of the electrodes by transport through the $PbSO_4$ solid electrolyte surface layers. As they move into and out of the local aqueous environment they cause the pH to vary, changing the solubility of the $PbSO_4$ in solution in the negative electrode structure, and that of both PbO_2 and $PbSO_4$ in the positive electrode structure. These solid phases are caused to precipitate and/or dissolve as the cell reaction takes place. This is generally called a dissolution—precipitation mechanism.

Thus the overall reaction in this type of battery is a composite of ionic transport of protons through a dense solid electrolyte layer of $PbSO_4$ that causes changes in the local pH, and thus of the solubility of $PbSO_4$ and PbO_2, in the adjacent liquid electrolyte. This results in their dissolution or precipitation within the multiphase electrode structure. Although the electrode potentials are determined by the two-phase equilibria under the $PbSO_4$ layer, the electrode capacity is determined by the amounts of the precipitate phases that react within the liquid electrolyte portion of the electrodes.

17.5   Construction of the Electrodes

Although descriptions of Pb-acid cells always say that the negative electrodes are primarily lead, and the positive electrodes primarily PbO_2, during manufacture they are both initially made from the same material, a paste consisting of a mixture of PbO and Pb_3O_4 [16]. It can be considered to be lead powder that is 70–85 % oxidized, and is traditionally called "leady oxide." Measured amounts of water and a H_2SO_4 solution are added, along with small polymer fibers to influence the mechanical properties, under carefully controlled temperature conditions. This results in the formation of basic lead sulfates, $3\ PbO \cdot PbSO_4 \cdot H_2O$ or $4\ PbO \cdot PbSO_4$. Various other materials are sometimes added to this mix. An example is the use of lignin, a component of wood, as a spacer, or "expander," in the paste that is used in negative electrodes [17]. Its presence reduces the tendency to form large Pb_2SO_4 crystals upon cycling those electrodes.

The paste is inserted into the electrodes by spreading it into an open grid structure. In addition to mechanically holding the paste material in place, the grid, which has better electronic conductivity than the fine-particle paste, also serves to carry the current throughout the total electrode structure.

Following this process, during which the paste is inserted into both electrode structures, generally before their final insertion into the battery case, a process called *formation* is undertaken. This process converts the materials in the two electrodes into the different compositions and structures required for the fully charged state of the cell.

This forming process is equivalent to an initial electrochemical charge. It is carried out under carefully controlled conditions, typically at a very low current density, so that the total structure in the two sets of electrodes is converted into the desired chemical species without disruption of the physical state of the paste-impregnated electrodes. This is essentially what Plante [1] did, for he made his battery using sheet lead electrodes, and cycling them in dilute H_2SO_4.

The variation of the potentials of the two electrodes during the formation process is shown in Fig. 17.3.

In an increasing number of cases, formation is followed by a carefully controlled drying process, and the batteries are supplied to the user in the dry state. The acid electrolyte is inserted into the cell at the time of the first use.

Fig. 17.3 Variation of the potentials of positive (*top*) and negative (*bottom*) electrodes during low rate formation process. After [12]

17.5.1   Volume Changes and Shedding

As is the case with a number of other battery systems, significant volume changes can occur in the electrodes as the result of the reactions that take place during charging and discharging. $PbSO_4$ has a substantially greater volume than both PbO_2 and elemental lead. The conversion of PbO_2 to $PbSO_4$ results in an increase of 92 %, whereas the change in volume from Pb to $PbSO_4$ is 164 %.

These volume changes can cause material to fall to the bottom of the cell, sometimes causing electrical shorting between adjacent electrode plates. In the past, when battery construction was different, it was sometimes found that apparently dead batteries could be "cured" by simply extracting this electrically conductive material from the bottom of the cells.

Whereas the electrodes in these batteries are generally flat, and assembled in stacks, a different configuration is also sometimes employed, in which the active electrode reactants are enclosed in porous tubes of an inert material, which serve to contain the active material, and reduce the shedding problem. Some manufacturers now encase each electrode in a porous plastic bag to prevent the results of shedding from shorting out the electrodes.

17.6   Alloys Used in Electrode Grids

The grid is generally considered to be the most critical passive component of lead-acid cells. It has two functions. One is to physically contain the active materials in the electrodes, and the other is to conduct electrons to and from the active materials. Both (relatively) pure lead and several lead alloys have been used in the manufacture of the grids in lead-acid batteries. There are two basic considerations, their mechanical, and their corrosion, properties.

Pure lead is quite soft, and although this might be an advantage in a mechanical manufacturing process, most grids are currently manufactured by casting liquid lead alloys.

Several lead alloys were developed in order to increase the mechanical strength of grids without significantly changing their electrochemical properties. Lead-antimony alloys were initially preferred. The phase diagram for this system is shown in Fig. 17.4. Compositions not far from the eutectic, which is 17 atom% antimony, are quite fluid, making it relatively easy to cast grids with relatively complex shapes. After freezing, the solid contains two phases, relatively pure lead containing a precipitate of finely divided antimony particles. The fine precipitate particles act to increase the mechanical strength of the lead.

Partly because of concern regarding the health issues related to the use of antimony—SbH_3 gas, which can form in the presence of moisture, is poisonous—attention was given to reductions in the antimony content from up to 11 wt% down to some 6–9 wt%. But with less Sb, they are not so readily cast, have

Fig. 17.4 Lead-antimony phase diagram

reduced mechanical strength and are less resistant to corrosion—especially if less than 6 % Sb.

Several other alloys were introduced subsequently. One direction that was followed by a number of manufacturers was the use of lead-calcium alloys, particularly in batteries to be used in float applications.

The phase diagram for this alloy system is quite different. Instead of having a eutectic reaction and appreciable solid solubility, as in the lead-antimony system, there is a peritectic reaction at low calcium contents at a temperature quite close to the melting point of pure lead. This is shown in Fig. 17.5. Since the solid solubility of calcium in lead is quite small and rather temperature dependent, it is possible to form fine precipitate particles of the phase $CaPb_3$. The microstructure of these alloys changes gradually after they are cast, resulting in age hardening, that increases the resistance to mechanical deformation by creep.

In addition, alloys in this system have good corrosion resistance at the potentials of the positive electrode. Their use in negative electrodes is generally thought to result in reduced hydrogen gas evolution.

A number of ternary alloys have also been explored. The presence of a small (0.5 %) amount of As increases the rate of age hardening, and provides better creep resistance, which is important for positive plates during deep discharging. Small amounts of Sn (2.5 %) increase the fluidity, making the grids easier to cast, and also give better cycle life.

Fig. 17.5 Lead-rich region of lead-calcium phase diagram

17.7   Alternative Grid Materials and Designs

There have been several other approaches to the design and materials used in electrode grids. One of these involves strengthening the lead by the inclusion of fine glass fibers, or polymers.

An alternative that has also been explored somewhat is the use of partially reduced titanium oxides for the construction of grids for positive electrodes [18]. This material, which is primarily TiO_x, where x is between 1.75 and 1.8, is called by the trade name *Ebonex*.

17.8   Development of Sealed Pb-Acid batteries

For more than 100 years Pb-acid batteries were designed as "flooded" open cells, so that the hydrogen and oxygen products that are developed upon overcharge could escape into the atmosphere. To compensate for these losses, water (preferably distilled) had to be periodically added to the electrolyte.

The technology has now changed significantly, and most common batteries do not require water replenishment. In addition, the electrolyte is immobilized, so that these products are essentially "spill-proof," and can be used in any physical orientation, upright, on the side, or even upside-down [19].

There are two general approaches that are used. One of these is generally called *"gel" technology*, and was first developed by Sonnenschein in Germany [20]. The other is known as *"glass mat" technology*, and was initially developed by Gates Energy Products in the United States [21]. Both are now described by the general term: *valve-regulated lead-acid (VRLA) cells*. This name is related to the fact that a small pressure valve must be present in such sealed cells for security purposes. It reversibly opens if the interior gas pressure exceeds about 0.5 bar above atmospheric.

In the case of the gel technology, the addition of fumed silica, a very fine amorphous form of silicon dioxide that has a very high surface area, to the sulfuric acid electrolyte causes it to thicken, or harden, into a gel. Upon the loss of some water, this mechanically stable structure develops cracks and fissures that can allow the passage of gaseous oxygen across the cell from the positive electrode to the negative electrode upon overcharge.

The other approach involves the use of a highly porous microfiber glass mat between the electrodes. This mat functions as a mechanical separator, and also as a container for the electrolyte, which adsorbs on the surface of the very fine—e.g., 1–2 μm diameter—glass fibers. If the mat is only partially filled with the liquid electrolyte, there is also space in this structure for gas to move between the positive and negative electrodes.

In both cases, the cells are designed to be positive electrode-limited. This means that the capacity of the positive electrode is less than that of the negative electrode. The cells operate by means of an *internal oxygen cycle*, or *oxygen recombination cycle*. When the positive electrode reaches the limit of its capacity, further charging causes the decomposition of water and the formation of neutral oxygen gas:

$$H_2O = 2H' + 1/2O_2 + 2e^- \qquad (17.9)$$

This gas travels through the gel or glass mat electrolyte to the negative electrode, where it reacts with hydrogen in the negative electrode to again form water, which can be written as

$$O_2 + 4H^+ + 4e^- = 2H_2O \qquad (17.10)$$

The result is that the cell suffers self-discharge as the negative electrode loses capacity.

The latter reaction is exothermic, whereas the oxygen formation reaction is endothermic.

Upon charging, part of the electrical energy is consumed by the oxygen-recombination cycle and converted into heat.

17.9 Additional Design Variations

There have been several new approaches to the design of Pb-acid cells in addition to the standard parallel flat plate and tubular configurations. One of these is the bipolar concept, which involves the construction of a stack of cells that are connected in series. To do this, it is necessary to have an electronically- conducting bipolar plate that acts as a separator between the electrodes in adjacent cells. The negative electrode of one cell is on one side of the bipolar plate, and the positive electrode of the adjacent cell is on the other side. An example of such a configuration is shown in Fig. 17.6.

It is necessary to have seals to separate the electrolytes in adjacent cells in order to prevent current flow between them. It would also be advantageous to get rid of one of the current collectors, with one layer serving as the positive electrode for one cell, and the negative electrode for the other.

The simplest case is to have a metal sheet or foil act in this way. However, this requires that this metal be stable in contact with these two electrode materials and the potentials at which they both exist, e.g., reducing on one side, and oxidizing on the other side. Alternatively, one could have an electronically conducting nonmetal serve this function. One example could be graphite. Other materials might also be considered, such as oxides, nitrides, and borides, but they also have to meet the same requirements.

Another approach would be to use a bimetallic sheet, fabricated with one material on one side, and a different one on the other side. Such double-layer sheets could be produced by electrodeposition, sputtering, or other such processes. Simply rolling the two materials together might well be the best, and least expensive, method for modest to large scale applications.

A further approach would be to put metallically-conducting layers on both sides of a mechanical support material—perhaps a polymer or ceramic. These two conducting layers could be electronically connected by the use of holes through

Fig. 17.6 Simple bipolar arrangement

Fig. 17.7 Simple model of mechanically-supported three-layer bipolar plate

Fig. 17.8 Schematic representation of the Bolder thin-layer wound system

the support material. This might be represented schematically as shown below. In this case, one conductor is both on the top and in the holes, and the other is on the other side. This is shown schematically in Fig. 17.7.

Another design variant would be to enhance the power output, rather than the voltage. One approach to this was developed by Bolder Technology [22]. It involves the use of a spiral-wound thin layer concept that essentially places a large number of local cells in parallel. This is represented schematically in Fig. 17.8. But the construction is unique. Thin film electrodes, separated by a separator are wound into a spiral. They protrude out of opposite ends of a cylin-drical can, and are electrically connected by melting and freezing caps on the two ends of the container. This provides a very large contact area, and produces a configuration with very low internal impedance. Such a design does not store much energy, but can operate at very high power for short times. In one application, starting power was supplied for internal combustion engines by a cell that is only 1/8 to 1/10 times the size and weight of conventional lead-acid cell designs.

17.9.1   Other Improvements

In addition to these innovative design changes, a number of improvements have been made in lead-acid cell components. One area involves the improvement of the mechanical properties, or the reduction in weight, of grid materials. One approach involves the use of mechanically expanded metal, rather than cast, grids. Another has been the development of extruded lead-covered glass fibers for grid structures.

Also, both polymer fibers and graphite particles have been introduced into the active materials in some cases to increase the mechanical strength or electronic conductivity.

17.10   Rapid Diffusion of Hydrogen in PbO$_2$

Experiments [23, 24] have shown that the chemical diffusion of hydrogen in PbO_2 is very fast, with a diffusion coefficient in the range of 0.4 to about 5×10^{-7} cm^{2}s^{-1}, varying with the potential, and thus with the hydrogen content. These values are some six orders of magnitude greater than hydrogen diffusion in MnO_2, which is the positive electrode reactant in the common alkaline Zn/MnO_2 cells. This very rapid diffusion explains why Pb-acid cells can provide such very high values of initial current, which is useful when they are used as starter batteries. The quantity of this proton-related charge is relatively small, however, only about 1 % of the total capacity of the positive electrode. Thus this effect does not last very long during a starting operation.

References

1. Planté G, Acad CR (1860) Sci Paris 50:640
2. Faure CA, Acad CR (1881) Sci Paris 92:951
3. I. Barin, Thermochemical Data of Pure Substances, 3rd Edition, VCH 1995, Published Online 24 Apr 2008, ISBN 9783527619825
4. P. Ruetschi, J. Power Sources 2, 3 (1977/78)
5. Gladstone JH, Tribe A (1882) Nature 25:221

6. Dolezalek F (1901) *Die Theorie des Bleiakkumulators*, Halle. Paris
7. W.H. Beck and W.F.K. Wynne-Jones, Trans. Faraday Soc. 50, 136, 147, 927 (1954)
8. Duisand JA, Giauque WF (1968) J Phys Chem 72:562
9. Weininger JL (1974) J Electrochem Soc 121:1454
10. Weininger JL, Secor FW (1974) J Electrochem Soc 121:1541
11. Euler KJ (1970) Bull Schweiz Elektrotech Ver 61:1054
12. Ruetschi P (1973) J Electrochem Soc 120:331
13. Huggins RA (2009) Advanced Batteries: Materials Science Aspects. Springer, Chapter 16
14. Ruetschi P, Angstadt RT (1964) J Electrochem Soc 111:1323
15. Ruetschi P (2003) J Power Sources 113:363
16. D. A. J. Rand, P. T. Moseley, J. Garche and C. D. Parker, eds. *Valve-regulated Lead–acid Batteries*, Elsevier (2004)
17. T.A. Willard, US Patents 1,432,508 and 1.505,990 (1920)
18. K. Ellis, A. Hill, J. Hill, A. Loyns and T.Partington, J. Power sources 136, 366 (2004)
19. D. A. J. Rand, P. T. Moseley, J. Garche and C. D. Parker, eds. *Valve-regulated Lead–acid Batteries*, Elsevier (2004)
20. Jache O (1966) US Patent 3:257,237
21. McClelland DH, Devitt JL (1975) US Patent 3:862,861
22. http://www.boldertmf.com
23. Münzberg R, Pohl JP, Phys Z (1985) Chem 146:97
24. Papazov GP, Pohl JP, Rickert H (1978) Power Sources 7:37

Chapter 18
Negative Electrodes in Other Rechargeable Aqueous Systems

18.1 Introduction

This chapter discusses two examples of negative electrodes that are used in several aqueous electrolyte battery systems, the "cadmium" electrode and metal hydride electrodes.

It will be seen that these operate in quite different ways. In the first case, the "cadmium" electrode is actually a two-phase system, with elemental cadmium in equilibrium with another solid, its hydroxide. And in the second, hydrogen is exchanged between a solid metal hydride and hydrogen-containing ionic species in the electrolyte.

18.2 The "Cadmium" Electrode

18.2.1 Introduction

Cadmium/nickel, Ni/Cd, or NiCad, cells have been important products for many years. They have alkaline electrolytes and use the "*nickel*" positive electrode, H_xNiO_2, which is discussed in the next chapter. Because they have both higher capacity and a reduced problem with environmental pollution—cadmium is considered to be environmentally hazardous—batteries with metal hydride, rather than cadmium, negative electrodes are gradually taking a larger part of the market. They are discussed later.

© Springer International Publishing Switzerland 2016
R.A. Huggins, *Energy Storage*, DOI 10.1007/978-3-319-21239-5_18

18.2.2   Thermodynamic Relationships in the H-Cd-O System

As in the case of the H-Zn-O system described in the last chapter, the first step in understanding what determines the potential of the cadmium electrode involves the use of available thermodynamic data to determine the relevant ternary phase stability diagram, for the driving forces of electrochemical reactions are the related reactions between electrically neutral species.

In this case, the key issue is the value of the Gibbs free energy of formation of CdO, which has been found to be −229.3 kJ/mol at 25 °C. This is less than the value for the formation of water, so a tie line between water and cadmium must be more stable than a tie line between CdO and hydrogen. Also, Cd(OH)$_2$ is a stable phase between water and CdO, because its Gibbs free energy of formation at 25°C is −473.8 kJ/mol, whereas the sum of the others is −466.4 kJ/mol. From these data the ternary phase stability diagram shown in Fig. 18.1 can be drawn. It is clear that it is different from the one for the H-Zn-O system discussed elsewhere.

It is seen that Cd(OH)$_2$ is also stable when Cd is in contact with water. The potential of the cadmium electrode is determined by the potential of the sub-triangle that has water, Cd(OH)$_2$, and cadmium at its corners.

Since there are three phases as well as three components, Cd, hydrogen, and oxygen, present, there are no degrees of freedom, according to the Gibbs phase rule, as discussed earlier. Therefore, the cadmium reaction should occur at a constant potential, independent of the state of charge. This is what is experimentally found.

The potential of all compositions in this triangle is determined by the reaction

$$\frac{1}{2}O_2 + Cd + H_2O = Cd(OH)_2 \tag{18.1}$$

and from the Gibbs free energies of formation of the relevant phases it is found that its value is −1.226 V relative to that of pure oxygen, as is shown in Fig. 17.2.

Fig. 18.1 The H-Cd-O ternary phase stability diagram, showing the potentials of the compositions in the sub-triangles versus pure oxygen, in volts

The discharge of the cadmium electrode can be written as an electrochemical reaction:

$$Cd + 2(OH)^- = Cd(OH)_2 + 2e^- \tag{18.2}$$

This shows that there is consumption of water from the electrolyte during discharge, as can also be seen in the neutral species reaction in Eq. 18.1. This consumption of water must be considered in the determination of the electrolyte composition.

18.2.3 Comments on the Mechanism of Operation of the Cadmium Electrode

There is another matter to be considered in the behavior of the cadmium electrode, since discharge involves the formation of a layer of $Cd(OH)_2$ on top of the Cd. This would require a mechanism to either transport Cd^{2+} or OH^{--} ions through the growing $Cd(OH)_2$ layer, both of which seem unlikely. This reaction is generally thought to involve the formation of an intermediate species that is soluble in the KOH electrolyte. The most likely intermediate species is evidently $Cd(OH)_3^-$.

The kinetics of the cadmium electrode are sufficiently rapid that the potential changes relatively little on either charge or discharge. Typical values are a deviation of 60 mV during charge, and 15 mV during discharge at the C/2, or 2-h, rate. In addition, there are small potential overshoots at the beginning in both directions if the full capacity had been employed in the previous step. This is, of course, what would be expected if the microstructure started with only one phase, and the second phase has to be nucleated. This is shown in Fig. 18.2 [1].

One of the questions that had arisen in earlier considerations of the mechanism of this electrode was the possibility of the formation of CdO. X-ray investigations have found no evidence for its presence. Thus if this phase were present it would have to be either as extremely thin layers or be amorphous.

However, this question can be readily answered by consideration of the potential of the reaction in the triangle with Cd, CdO and $Cd(OH)_2$ at its corners. This can be determined simply by the reaction along its edge:

$$\tfrac{1}{2}O_2 + Cd = CdO \tag{18.3}$$

From the Gibbs free energy of formation of CdO this is found to be -1.188 V relative to the potential of oxygen. This is 38 mV positive of the equilibrium potential of the main reaction. Since it is not expected that the electrode potential would deviate so far during operation of these electrodes, the formation of CdO is unlikely.

Fig. 18.2 The charge–discharge behavior of a sintered-plate cadmium electrode, measured at about the C/2 rate in 2M KOH at 25 °C

18.3 Metal Hydride Electrodes

18.3.1 Introduction

Metal hydrides are currently used as the negative electrode reactant in large numbers of reversible commercial batteries with aqueous electrolytes, generally in combination with "nickel" positive electrodes.

There are several families of metal hydrides, and the electrochemical properties of some of these materials are comparable to those of cadmium. Developmental efforts have led to the production of small consumer batteries with comparable kinetics, but with up to twice the energy content per unit volume of comparable small "normal" Cd/Ni cells. Typical values for AA size cells are shown in Table 18.1. For this reason, as well as because of the poisonous nature of cadmium, hydride cells are taking a larger and larger portion of this market.

18.3.2 Comments on the Development of Commercial Metal Hydride Electrode Batteries

Although there had been research activities earlier in several laboratories, work on the commercialization of small metal hydride electrode cells began in Japan's Government Industry Research Institute laboratory in Osaka (GIRIO) in 1975. By

Table 18.1 Typical capacities of AA size cells used in many small electronic devices

Type of cell	mWh/cm^3
"Normal" Cd/Ni	110
"High-capacity" Cd/Ni	150
Hydride/Ni	200

1991 there were a number of major producers in Japan, and the annual production rate had reached about 1 million cells. Activities were also underway in other countries. Those early cells had specific capacity values of about 54 Wh/kg and specific powers of about 200 W/kg.

The production rate grew rapidly, reaching an annual rate of about 100 million in 1993, and over 1 billion cells in 2005. The properties of these small consumer cells also improved greatly. By 2006 the specific capacity had reached 100 Wh/kg, and the specific power 1200 W/kg. The energy density values also improved, so that they are now up to 420 Wh/l.

They are generally designed with excess negative electrode capacity, i.e., N/P > 1. This is increased for higher power applications.

The metal hydrides used in small consumer cells are multicomponent metallic alloys, typically containing about 30 % rare earths. Prior to this development, the largest commercial use of rare earth materials was for specialty magnets. The major source of these materials is in China, where they are inexpensive and very abundant. Rare earths are also available in large quantities in the USA and South Africa.

In addition to the large current production of small consumer batteries, development efforts have been aimed at the production of larger cells with capacities of 30–100 A h at 12 V. The primary force that is driving this move toward larger cells is their use in hybrid electric vehicles. In order to meet the high power requirements, the specific capacity if these cells has to be sacrificed somewhat, down to about 45–60 Wh/kg.

18.3.3 Hydride Materials Currently Being Used

There are two major families of hydrides currently being produced that can be roughly described as AB$_5$ and AB$_2$ alloys.

The AB$_5$ alloys are based upon the pioneering work in the Philips laboratory that started with the serendipitous discovery of the reaction of gaseous hydrogen with LaNi$_5$. The basic crystal structure is of the layered hexagonal CaCu$_5$ type. Alternate layers contain both lanthanum and nickel, and only nickel. This structure is illustrated in Fig. 18.3.

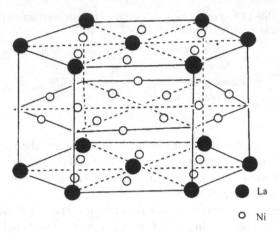

Fig. 18.3 Schematic view of the layered hexagonal lattice of LaNi$_5$ [1]

The reaction of hydrogen with materials of this type can be written as

$$3H_2 + LaNi_5 = LaNi_5H_6 \qquad (18.4)$$

The hydrogen atoms reside in tetrahedral interstitial sites between the host atoms. It has been found that hydrogen can occupy interstitial sites in suitable alloys in which the *"holes"* have spherical radii of at least 0.4 Å. In larger interstitial positions hydrogen atoms are often *"off center."*

Developmental work has involved the partial or complete replacement of the lanthanum with other metals, predominantly with Mischmetall (Mm), a mixture of rare earth elements, and zirconium. A typical composition of the relatively inexpensive Mischmetall is 45–58 % Ce, 20–27 % La, 13–20 % Nd, and 3–8 % Pr.

In addition, it has been found advantageous to replace some or all of the nickel with other elements, such as aluminum, manganese and cobalt. Furthermore, it is possible to change the A/B ratio. One major producer uses a composition that has a higher A/B ratio than 5, in the direction of A$_2$B$_7$, for example. These materials show relatively flat two-phase discharge voltage plateaus, indicating a reconstitution reaction. Various compositional factors influence the pressure (cell voltage) and the hydrogen (charge) capacity of the electrode, as well as the cycle life. There has also been a lot of developmental work on preparative methods and the influence of microstructure upon the kinetic and cycle life properties of small cells with these materials.

18.3.3.1   Disproportionation and Activation

Another reaction between hydrogen and these alloys can also take place, particularly at elevated temperatures. It can be written as

$$H_2 + LaNi_5 = LaH_2 + 5Ni \qquad (18.5)$$

and is called *disproportionation*. At 298 K the Gibbs free energies of formation of LaNi$_5$ and LaH$_2$ are -67 kJ/mol and -171 kJ/mol, respectively, so there is a significant driving force for this to occur, at least on the surface. Experiments have shown that the surface tends to contain regions that are rich in lanthanum, combined with oxygen. In addition, there are clusters of nickel. Because of the presence of these nickel islands, which are permeable to hydrogen, hydrogen can get into the interior of the alloy.

It is often found that a cyclic activation process is necessary in order to get full reaction of hydrogen with the total alloy. As hydrogen works its way into the interior there is a local volume expansion that often causes cracking and the formation of new fresh surfaces that are not covered with oxygen. This cracking can cause the bulk material to be converted into a powder, and is called *decrepitation*.

18.3.4 Pressure—Composition Relation

If the particle size is small and there are no surface contamination or activation problems, the LaNi$_5$ alloy reacts readily with hydrogen at a few atmospheres pressure. This is illustrated in Fig. 18.4.

This flat curve is an indication that this is a reconstitution, rather than insertion, reaction.

Fig. 18.4 Schematic pressure–composition isotherm for the reaction of LaNi$_5$ with hydrogen

Fig. 18.5 Relation between the logarithm of the plateau pressure and the volume of the crystal structure's unit cell. After [3]

There is a slight difference in the potential when hydrogen is added from that when hydrogen is removed. This hysteresis is probably related to the mechanical work that must occur due to the volume change in the reaction.

The pressure plateaus for the alloys that are used in batteries are a bit lower, so that the electrochemical potential remains somewhat positive of that for the evolution of hydrogen on the negative electrode.

It has been found that the logarithm of the potential at which this reaction occurs depends linearly upon the lattice parameter of the host material for this family of alloys. This is shown in Fig. 18.5.

In order to reduce the blocking of the surface by oxygen, as well as to help hold the particles together, thin layers of either copper or nickel are sometimes put on their surfaces by the use of electroless plating methods [3]. PVDF or a similar material is also often used as a binder.

18.3.5   The Influence of Temperature

The equilibrium pressure over all metal hydride materials increases at higher temperatures. This is shown schematically in Fig. 18.6.

The relation between the potential plateau pressure and the temperature is generally expressed in terms of the Van't Hoff equation

Fig. 18.6 Schematic variation of the equilibrium pressure of a metal hydride system with temperature

$$\ln p(H_2) = \left(\frac{\Delta H}{RT}\right) - \left(\frac{\Delta S}{R}\right) \tag{18.6}$$

This can readily be derived from the general relations

$$\Delta G = RT\ln p(H_2) \tag{18.7}$$

and

$$\Delta G = \Delta H - T\Delta S \tag{18.8}$$

This relationship is shown in Fig. 18.7 for LaNi$_5$ and a commercial Mischmetall-containing alloy [2]. It can be seen that the pressure is lower, and thus the electrical potential is higher, in the case of the practical alloy.

This type of representation is often use to compare metal hydride systems that are of interest for the storage of hydrogen from the gas phase. Figure 18.8 is an example of such a plot [3].

It can be seen that the range of temperature and pressure that can be considered for the storage of hydrogen gas is much greater than that which is of interest for the use in aqueous electrolyte battery systems.

Higher pressure in gas systems is equivalent to a lower potential in an electrochemical cell, as can be readily seen from the Nernst equation

$$E = -\left(\frac{RT}{zF}\right)\ln p(H_2) \tag{18.9}$$

The reaction potential must be above that for the evolution of hydrogen, and if it is too high, the cell voltage is reduced. As a result, the range of materials is quite constrained, and a considerable amount of effort has been invested in making minor modifications by changes in the alloy composition.

Fig. 18.7   Van't Hoff plot for $LaNi_5H_x$ and two compositions of a $MmNi_{3.55}Co_{0.75}Mn_{0.4}$ $Al_{0.3}H_x$ alloy. After [2]

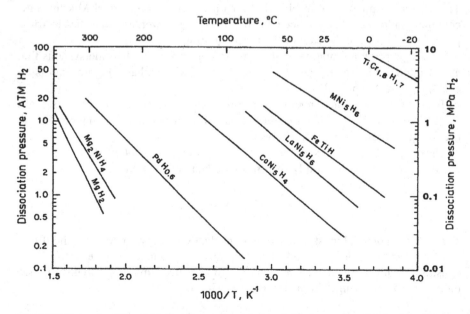

Fig. 18.8   Example of Van't Hoff plot showing data for a wide range of materials. After [3]

18.3.6 AB₂ Alloys

The other major group of materials that are now being used as battery electrodes are the AB_2 alloys. There are two general types of AB_2 structures, sometimes called Friauf-Laves phases; the C14, or $MgZn_2$, type in which the B atoms are in a close packed hexagonal array, and the C15, or $MgCu_2$, type in which the B atoms are arranged in a close packed cubic array. Many materials can be prepared with these, or closely related, structures. The A atoms are generally either Ti or Zr. The B elements can be V, Ni, Cr, Mn, Fe, Co, Mo, Cu, and Zn. Some examples are listed in Table 18.2.

It has generally been found that the C14-type structure is more suitable for hydrogen storage applications. A typical composition can be written as (Ti,Zr) $(V,Ni,Cr)_2$.

18.3.7 General Comparison of These Two Structural Types

Both of these systems can provide a high charge storage density. Present data indicate that this can be slightly (5–10 %) higher for some of the AB_2 than the AB_5 materials.

However, there is a significant difference in the electrochemical characteristics of these two families of alloys. As illustrated in Fig. 18.4, the hydrogen pressure is essentially independent of the composition over a wide range in the AB_5 case. Thus the cell potential is independent of the state of charge, characteristic of a reconstitution reaction. On the other hand, the hydrogen activity generally varies appreciably with the state of charge in the AB_2 alloys, giving charge-dependent cell voltages. The fact that the cell voltage decreases substantially during discharge of cells with the AB_2 alloy hydrides can be considered to be a significant disadvantage for the use of these materials in batteries.

A serious issue, particularly in the AB_2 materials, is the question of oxidation, and subsequent corrosion, particularly of the B metals. This can lead to drastic reductions in the capacity and cycle life, as well as causing a time-dependent increase in gaseous hydrogen pressure in the cell. Because of this, special pre-etching treatments have been developed to reduce this problem. The vanadium content of the surface is evidently particularly important.

Table 18.2 Structures of AB₂-phase materials

Material	C14 structure (hexagonal)	C15 structure (cubic)
TiMn₂	X	
ZrMn₂	X	
ZrV₂		X
TiCr₂	X	X
ZrCr₂	X	X
ZrMo₂		X

18.3.8 Other Alloys That Have Not Been Used in Commercial Batteries

An alloy of the AB_2 type based upon Ti and Mn, with some V, is currently being used for the storage of gaseous hydrogen, rather than in batteries. One commercial application is in fuel cell-propelled submarines manufactured in Germany. However, its hydrogen activity range is too high to be applicable to use in batteries.

A number of years ago there was a development program in the Battelle laboratory in Switzerland funded by Daimler Benz aimed at the use of titanium–nickel materials of the general compositions A_2B (Ti_2Ni) and AB (TiNi) for use in automobile starter batteries. This early work showed that if electrodes have the right composition and microstructure and are properly prepared, they can perform quite well for many cycles. However, this development was never commercialized.

An interesting side issue is the fact that the TiNi phase, which is very stable in KOH, is one of the materials that is known to be ferroelastic, and to have mechanical memory characteristics. Its mechanical deformation takes place by the formation and translation of twin boundaries, rather than by dislocation motion. As a result, it is highly ductile, yet extremely resistant to fracture. Thus it would be interesting to consider the use of minor amounts of this phase as a metallic binder in hydride, or other, electrodes. It should be able to accommodate the repeated microscopic mechanical deformation that typically occurs within the electrode structure upon cycling without fracturing.

18.3.9 Microencapsulation of Hydride Particles

A method was developed some years ago in which a metallic coating of either copper or nickel is deposited by electro-less methods upon the hydride particles before the mechanical formation of the hydride electrode [4]. This ductile layer helps the formation of electrodes by pressing, acts as a binder, contributes to the electronic conductivity, and thus improves electrode kinetics, and helps against overcharge. It evidently also increases the cycle life. Since both copper and nickel do not corrode in KOH, this layer also acts to prevent oxidation and corrosion.

18.3.10 Other Binders

In addition to the copper or nickel metallic binders, some Japanese cells use PTFE, silicon rubber, or SEBS rubber as a binder. It has been found that this can greatly influence the utilization at high (up to 5C) rates. With the rubber binders (e.g. 3 wt% PTFE), flexible thin-sheet electrodes can be made that make fabrication of small spiral cells easier.

18.3.11 Inclusion of a Solid Electrolyte in the Negative Electrode of Hydride Cells

An interesting development was the work in Japan on the use of a proton-conducting solid electrolyte in the negative electrodes of hydride cells. This material is tetramethyl ammonium hydroxide pentahydrate, $(CH_3)_4NOH \cdot 5 H_2O$, which has been called TMAH5. It is a clathrate hydrate, and melts (at about 70 °C) rather than decomposes, when it is heated. Thus it can be melted to impregnate a pre-formed porous electrode to act as an internal electrolyte. This is typically not true for other solid electrolytes, and can be advantageous in increasing the electrode–electrolyte contact area.

TMAH5 has a conductivity of about 5×10^{-3} S/cm^2 at ambient temperatures. While this value is higher than the conductivity of almost all other known proton-conducting solid electrolytes, it is less than that of the normal KOH aqueous electrolyte. Thus if this solid electrolyte were to be used, one would have to be concerned with the development of fine-scale geometries. This could surely be done, but it would probably involve the use of screen printing or tape casting fabrication methods, rather than conventional electrode fabrication procedures.

Both hydride/H_xNiO_2 cells and hydride/MnO_2 cells have been produced using this solid electrolyte. Because of the lower potential of the MnO_2 positive electrode relative to the "nickel" electrode, the latter cells have lower voltages.

18.3.12 Maximum Theoretical Capacities of Various Metal Hydrides

The maximum theoretical specific capacities of various hydride negative electrode materials are listed in Table 18.3. Values are shown for both the hydrogen charged and uncharged weight bases. They include two AB_5 type alloys that are being used by major producers, as well as the basic $LaNi_5$ alloy and two AB_2 materials.

As would be expected, small commercial cells have practical values that are less than the theoretical maximum values presented in the last few pages. Hydride electrodes generally have specific capacities of 320–385 mAh/g. For comparison, the H_xNiO_2 "nickel" positive electrodes typically have practical capacities about 240 mAh/g.

Table 18.3 Specific capacities of several AB_5 and AB_2 alloys

Material	Uncharged, mAh/g	H_2 charged, mAh/g
$LaNi_5H_6$	371.90	366.81
$MmNi_{3.5}Co_{0.7}Al_{0.8}H_6$	393.93	388.23
$(LaNd)(NiCoSi)_5H_4$	248.80	246.51
$TiMn_2$	509.67	500.16
$(Ti,Zr)(V,Ni)_2$	448.78	441.39

References

1. Milner PC, Thomas UB (1967) Adv Electrochem Electrochem Eng 5:1
2. Reilly JJ (1999) Handbook of Battery Materials. Besenhard JO (ed). Wiley-VCH, p. 209
3. Sandrock GD, Huston EL (1981) Chemtech 11:754
4. Sakai T, Yoshinaga H, Miyamura H, Kuriyama N, Ishikawa H (1992) J Alloys Compounds 180:37

Chapter 19
Positive Electrodes in Other Aqueous Systems

19.1 Introduction

This chapter discusses three topics relating to positive electrodes in aqueous electrolyte battery systems, the manganese dioxide electrode, the nickel electrode and the so-called *memory effect* that is found in batteries that have "nickel" positive electrodes.

The first of these deals with a very common material, MnO_2, that is used in the familiar "alkaline" cells that are found in a very large number of small portable electronic devices. This electrode operates by a simple proton insertion reaction.

MnO_2 can have a number of different crystal structures, and it has been known for many years that they exhibit very different electrochemical behavior. It is now recognized that the properties of the most useful version can be explained by the presence of excess protons in the structure, whose charge compensates for that of the Mn^{4+} cation vacancies that result from the electrolytic synthesis method.

The "nickel" electrode is discussed in the following section. This electrode is also ubiquitous, as it is used in several types of common batteries. Actually, this electrode is not metallic nickel at all, but a two-phase mixture of nickel hydroxide and nickel oxy-hydroxide. It is reversible, and also operates by the insertion and deletion of protons. The mechanism involves proton transport through one of the phases that acts as a solid electrolyte. The result is the translation of a two-phase interface at essentially constant potential.

The third topic in this group is a discussion of what has been a vexing problem for consumers. It occurs in batteries that have nickel positive electrodes. The mechanism that results in the appearance of this problem is now understood. In addition, the reason for the success of the commonly used solution to it can be understood.

© Springer International Publishing Switzerland 2016

R.A. Huggins, *Energy Storage*, DOI 10.1007/978-3-319-21239-5_19

19.2   Manganese Dioxide Electrodes in Aqueous Systems

19.2.1   *Introduction*

Manganese dioxide, MnO_2, is the reactant that is used on the positive side of the very common *alkaline* cells that have zinc as the negative electrode material. There are several versions of MnO_2, some of which are much better for this purpose than others. Thus this matter is more complicated than it might seem at first.

MnO_2 is polymorphic, with several different crystal structures. The form found in mineral deposits has the rutile (beta) structure, and is called *pyrolusite*. It is relatively inactive as a positive electrode reactant in KOH electrolytes. It can be given various chemical treatments to make it more reactive, however. One of these produces a modification containing some additional cations that is called *birnessite*. Manganese dioxide can also be produced chemically, and then generally has the delta structure. The material that is currently much more widely used in batteries is produced electrolytically, and is called *EMD*. It has the gamma (*ramsdellite*) structure.

The reason for the differences in the electrochemical behavior of the several morphological forms of manganese dioxide presented a quandary for a number of years. It was known, however, that the electrochemically active materials contain about 4 % water in their structures that can be removed by heating to elevated temperatures (100–400 °C), but the location and form of that water remained a mystery. This problem was solved by Ruetschi, who introduced a cation vacancy model for MnO_2 [1, 2].

The basic crystal structure of the various forms of MnO_2 contains Mn^{4+} ions in octahedral holes within hexagonally (almost) close packed layers of oxide ions. That means that each Mn^{4+} ion has six oxygen neighbors, and these MnO_6 octahedra are arranged in the structure to share edges and corners. Differences in the edge- and corner-sharing arrangements result in the various polymorphic structures.

If some of the Mn^{4+} ions are missing (cation vacancies), their missing positive charge has to be compensated by something else in the crystal structure. The Ruetschi model proposed that this charge balance is accomplished by the local presence of four protons. These protons would be bound to the neighboring oxide ions, forming a set of four OH^- ions. This local configuration is sometimes called a *Ruetschi defect*. There is very little volume change, as OH^- ions have essentially the same size as O^{2-} ions, and these species play the central role in determining the size of the crystal structure.

On the other hand, reduction of the MnO_2 occurs by the introduction of additional protons during discharge, as first proposed by Coleman [3], and does produce a volume change. The charge of these added mobile protons is balanced by a reduction in the charge of some of the manganese ions present from Mn^{4+} to Mn^{3+}. Mn^{3+} ions are larger than Mn^{4+} ions, and this change in volume during reduction has been observed experimentally.

The presence of protons (or OH^- ions) related to the manganese ion vacancies facilitates the transport of additional protons as the material is discharged. This is why these materials are very electrochemically reactive.

19.2.2   The Open Circuit Potential

The EMD is produced by oxidation of an aqueous solution of manganous sulfate at the positive electrode of an electrolytic cell. This means that the MnO_2 that is produced is in contact with water.

The phase relations, and the related ternary phase stability diagram, for the H–Mn–O system can be determined by use of available thermodynamic information [4, 5], as discussed in previous chapters. From this information it becomes obvious which neutral species reactions determine the potential ranges of the various phases present, and their values.

Following this approach, it is found that the lower end of the stability range of MnO_2 is at a potential that is 1.014 V vs. one atmosphere of H_2. The upper end is well above the potential at which oxygen evolves by the decomposition of water.

Under equilibrium conditions all oxides exist over a range of chemical composition, being more metal-rich at lower potentials, and more oxygen-rich at higher potentials. In the higher potential case, an increased oxygen content can result from either the presence of cation (Mn) vacancies or oxygen interstitials. In materials with the rutile, and related, structures that have close-packed oxygen lattices the excess energy involved in the formation of interstitial oxygens is much greater than that for the formation of cation vacancies. As a result, it is quite reasonable to assume that cation vacancies are present in the EMD MnO_2 that is formed at the positive electrode during electrolysis.

Due to the current that flows during the electrolytic process the potential of the MnO_2 that is formed is actually higher than the equilibrium potential for the decomposition of water. A number of other oxides with potentials above the stability range of water have been shown to oxidize water. Oxygen gas is evolved, and they become reduced by the insertion or protons. Therefore, it is quite reasonable to expect that EMD MnO_2 would have Mn vacancies, and that there would also be protons present, as discussed by Ruetschi [1, 2].

When such positive oxides oxidize water and absorb hydrogen as protons and electrons their potentials decrease to the oxidation limit of water, 1.23 V vs. H_2 at 25 °C. This is the value of the open circuit potential of MnO_2 electrodes in Zn/MnO_2 cells.

This water oxidation phenomenon that results in the insertion of protons into MnO_2 is different from the insertion of protons by the absorption of water into the crystal structure of materials that initially contain oxygen vacancies, originally discussed by Stotz and Wagner [6]. It has been shown that both mechanisms can be present in some materials [7, 8].

Fig. 19.1 Schematic discharge curve of Zn/MnO$_2$ cell

19.2.3   Variation of the Potential During Discharge

As mentioned above, this electrode operates by the addition of protons into its crystal structure. This is a single-phase insertion reaction, and therefore the potential varies with the composition, as discussed earlier.

If all of the initially present Mn^{4+} ions are converted to Mn^{3+} ions, the overall composition can be expressed as HMnO$_2$, or MnOOH.

It is also possible to introduce further protons, so that the composition moves in the direction of Mn(OH)$_2$. In this case, however, there is a significant change in the crystal structure, so that the mechanism involves the translation of a two-phase interface between MnOOH and Mn(OH)$_2$. This is analogous to the main reaction involved in the operation of the nickel electrode, as will be discussed later.

The sequence of these two types of reactions during discharge of a MnO$_2$ electrode is illustrated in Fig. 19.1.

The second, two-phase, reaction occurs at such a low cell voltage that the energy that is available is generally not used. Such cells are normally considered to only be useful down to about 1.2 V.

19.3   The "Nickel" Electrode

19.3.1   Introduction

The nickel electrode is widely used in battery technology, e.g., on the positive side of so-called Cd/Ni, Zn/Ni, Fe/Ni, H$_2$/Ni, and metal hydride/Ni cells, in some cases for a very long time. It has relatively rapid kinetics and exhibits unusually good cycling behavior. This is directly related to its mechanism of operation, which involves a solid state insertion reaction involving two ternary phases, Ni(OH)$_2$ and NiOOH, with no soluble product. While the attractive properties of this electrode have led to many investigations, there are still a number of aspects of its operation that are not fully understood. This chapter focuses primarily upon the microstructural mechanism of this two-phase insertion reaction and the thermodynamic features of the ternary Ni–O–H system that determine the observed potentials.

19.3.2 Structural Aspects of the Ni(OH)₂ and NiOOH Phases

The nanostructure of this electrode can be most simply described as a layer type configuration in which slabs of NiO_2 are separated by *galleries* in which various mobile guest species can reside. The structure of the NiO_2 layers consists of parallel sheets of hexagonally close-packed O^{2-} ions between which nickel ions occupy essentially all of the octahedral positions.

As will be described below, the mechanism of operation of this electrode involves the transition between Ni(OH)₂ and NiOOH upon oxidation, and the reverse upon reduction. Both of these phases are vario-stoichiometric (have ranges of stoichiometry). One can thus also describe their compositions in terms of the value of x in the general formula H_xNiO_2.

In the case of stoichiometric β Ni(OH)₂, the equilibrium crystal structure, which is isomorphous with *brucite*, Mg(OH)₂, has galleries that contain a proton concentration such that one can consider it as consisting of nickel-bonded layers of OH^- ions instead of O^{2-} ions. The nominal stoichiometry could thus be written as NiO_2H_2. Stoichiometric NiOOH has half as many protons in the galleries, and thus can be thought of as having an ordered mixture of O^{2-} and OH^- ions. Its nominal composition would then be H_1NiO_2.

When it is initially prepared, Ni(OH)₂ is often in the α modification, with a substantial amount of hydrogen-bonded water in the galleries. This structure is, however, not stable, and it gradually loses this water and converts to the equilibrium β Ni(OH)₂ structure, in which the galleries are free of water and contain only protons.

The equilibrium form of NiOOH, likewise called the β form, also has only protons in the galleries. However, there is also a γ modification of the NiOOH phase that contains water, as well as other species from the electrolyte, in the galleries. This γ modification forms at high charge rates or during prolonged overcharge in the alkali electrolyte. In both cases the potential is quite positive. It can also be formed by electrochemical oxidation of the α Ni(OH)₂ phase.

One can understand the transition of the β NiOOH to the γ modification at high potentials under overcharge conditions qualitatively in terms of the structural instability of the H_xNiO_2-type phase when the proton concentration is reduced substantially. Under those conditions, the bonding between adjacent slabs will be primarily of the relatively weak van der Waals type. This allows the entry of species from the electrolyte into the gallery space. This type of behavior is commonly found in other insertion reaction materials, such as TiS_2, mentioned in Chap. 9, if the interslab forces are weak and the electrolyte species are compatible.

The general relations between these various phases is generally described in terms of the scheme presented by Bode and co-workers [9].

A number of very good papers were published by the Delmas group in Bordeaux [10–14] that were aimed at the stabilization of the α Ni(OH)₂ phase by the presence of cobalt so that one might be able to cycle between the α Ni(OH)₂ and γ NiOOH

Table 19.1 Interslab distances for a number of phases related to the "nickel" electrode

Phase	Spacing (Å)
β-Ni(OH)$_2$	4.6
β-NiOOH	4.7
NaNiO$_2$	5.2
Na$_y$(H$_2$O)$_z$NiO$_2$	5.5
Na$_y$(H$_2$O)$_z$CoO$_2$	5.5
γ-H$_x$Na$_y$(H$_2$O)$_z$NiO$_2$	7.0
γ-H$_x$K$_y$(H$_2$O)$_z$NiO$_2$	7.0
γ'-H$_x$Na$_y$(H$_2$O)$_{2z}$NiO$_2$	9.9

phases. Since both of these phases have water, as well as other species, in the galleries, they have faster kinetics than the proton-conducting γ phases, although the potential is less positive. An important feature of their work has been the synthesis of sodium analogs by solid-state preparation methods and the use of solid-state ion-exchange techniques (*chimie douce*, or *soft chemistry*) to replace the sodium with other species [15].

The available information concerning the interslab spacing, the critical feature of the crystallographic structure of these phases in the nickel electrode, is presented in Table 19.1. It is readily seen that the crystallographic changes involved in the β Ni(OH)$_2$–β NiOOH reaction are very small, as they have almost the same value of interslab spacing. This is surely an important consideration in connection with the very good cycle life that is generally experienced with these electrodes. It can also be seen that the structural change involved in the a Ni(OH)$_2$–γ NiOOH transformation is somewhat larger. There are also differences in the slab stacking sequence in these various phases, but that factor will not be considered here.

Both the α and β versions of the Ni(OH)$_2$ phase are predominantly ionic, rather than electronic, conductors, and have a pale green color. The NiOOH phase, on the other hand, is a good electronic conductor, and both the β and γ versions are black.

19.3.3   Mechanism of Operation

The normal cycling reaction of commercial cells containing this electrode involves back and forth conversion between the β Ni(OH)$_2$ structure and the β NiOOH structure. It has been well established that these are separate, although vario-stoichiometric, phases, rather than end members of a continuous solid solution. The experimental evidence for this conclusion involves both x-ray measurements that show no gradual variation in lattice parameters with the extent of reaction [16], as well as similar IR observations [17] that indicate only changes in the amounts of the two separate phases as the electrode is charged or discharged.

Although the electrode potential when this two-phase structure is present is appreciably above the potential at which water is oxidized to form oxygen gas, as recognized long ago by Conway [18], gaseous oxygen evolution cannot happen if the solid electrolyte $Ni(OH)_2$ separates the water from the electronic conductor NiOOH. Oxygen evolution can only occur when the electronically conducting NiOOH phase is present on the surface in contact with the aqueous electrolyte.

Therefore, as a first approximation, one can describe the microstructural changes occurring in the electrode in terms of the translation of the $Ni(OH)_2$/NiOOH interface. When the electrode is fully reduced, its structure consists of only $Ni(OH)_2$, whereas upon full oxidation, only NiOOH is present. This is shown schematically in Fig. 19.2. The crystallographic transition between the $Ni(OH)_2$ and NiOOH structures, with their different proton concentrations in the galleries, is shown schematically in Fig. 19.3.

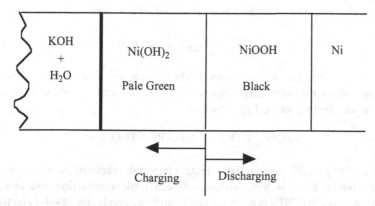

Fig. 19.2 Schematic drawing of the microstructure of the nickel electrode. The major phases present are $Ni(OH)_2$, which is a proton-conducting solid electrolyte, and NiOOH, a proton-conducting mixed conductor. The electrochemical reaction takes place by the translation of the $Ni(OH)_2$/NiOOH interface and the transport of protons through the $Ni(OH)_2$ phase

Fig. 19.3   Schematic drawing of the crystallographic transition between the $Ni(OH)_2$ and NiOOH structures, showing the step in the proton concentration in the galleries

It has long been known [19] that the NiOOH forms first at the interface between the Ni(OH)$_2$ and the underlying electronic conductor, rather than at the electrolyte/ Ni(OH)$_2$ interface. Other authors (e.g., [20, 21]) have observed the motion of the color boundary during charge and discharge of such electrodes.

19.3.4 Relations Between Electrochemical and Structural Features

It is useful to consider the operation of this electrode in terms of the net reaction in which hydrogen is either added to or deleted from the layer structure. In the case of oxidation, this can be written as a neutral chemical reaction:

$$Ni(OH)_2 - {}^1\!/_2 H_2 = NiOOH \tag{19.1}$$

However, in electrochemical cells this oxidation reaction takes place electrochemically, and since this normally involves an alkaline electrolyte, it is generally written in the electrochemical literature as

$$Ni(OH)_2 + OH^- = NiOOH + H_2O + e^- \tag{19.2}$$

However, the general rule is that electrochemical reactions take place at the boundary where there is a transition between ionic conduction and electronic conduction. Since Ni(OH)$_2$ is predominantly an ionic conductor (a solid electrolyte) the electrochemical reaction occurs at the Ni(OH)$_2$/NiOOH interface, where neither H$_2$O nor OH$^-$ are present. The electrochemical reaction should therefore more properly be written as

$$Ni(OH)_2 - H^+ = NiOOH + e^- \tag{19.3}$$

In order for the reaction to proceed, the protons must be transported away from the interface through the galleries in the Ni(OH)$_2$ phase and into the electrolyte. However, in the alkaline aqueous electrolyte environment hydrogen is not present as either H$^+$ or H$_2$. Instead, hydrogen is transferred between the electrolyte and the Ni(OH)$_2$ phase by the interaction of neutral H$_2$O molecules and OH$^-$ ions in the electrolyte with the H$^+$ ions at the electrolyte/Ni(OH)$_2$ interface. Thus the reaction at the electrolyte/Ni(OH)$_2$ interface must be electrically neutral and can be written as

$$OH^-_{\text{(electrolyte)}} + H^+_{\left(Ni(OH)_2\right)} = H_2O \tag{19.4}$$

The equilibrium coulometric titration curve shows that under highly reducing conditions, when only the pale green Ni(OH)$_2$ phase is present throughout, there is a relatively steep potential-composition dependence. However, the fact that this part of the titration curve is not vertical indicates that there is a range of composition in this phase. It was shown some time ago that up to about 0.25 electrons (and thus 0.25 protons) per mole can be deleted from the $Ni(OH)_2$ phase before the onset of the two-phase $Ni(OH)_2/NiOOH$ equilibrium [21]. Translated to the crystallographic picture, this means that the proton concentration in the phase nominally called $Ni(OH)_2$ can deviate significantly from the stoichiometric value, up to a proton vacancy fraction of some 12.5 %. The proton-deficient composition limit for the $Ni(OH)_2$ phase can thus be expressed as $H_{(1.75}NiO_2)$.

When both phases are present, there is a relatively long constant-potential plateau in the limit of negligible current density. This extends from the proton deficient concentration limit in the $Ni(OH)_2$ phase ($H_{(1.75}NiO_2)$) to the maximum proton concentration in the $NiOOH$ phase. According to Barnard et al. [21] this is when about 0.75 electrons (or protons) per mole are deleted from the electrode. This is equivalent to a composition of $H_{1.25}NiO_2$. Under more oxidizing conditions, when further protons are deleted, the potential of the $NiOOH$ phase becomes more positive.

The apparent length of the constant potential two-phase plateau that is observed experimentally depends upon when the $NiOOH$ phase reaches the electrolyte/electrode interface, and thus upon the thickness of the $Ni(OH)_2$ phase and the geometrical shape of the $Ni(OH)_2/NiOOH$ interface. The morphology of this interface, which is often not flat [22], is dependent upon several factors. As will be discussed subsequently, a flat interface is inherently unstable during the oxidation reaction. On the other hand, the interface will tend toward a smooth shape when it translates in the reduction direction. In both cases, it will be shown that the current density is a critical parameter.

Under more oxidizing conditions, when only the $NiOOH$ phase is present, the electrode is black and electronically conducting. This phase has wide ranges of both composition and potential. As mentioned above, the upper limit of proton concentration has been found to be approximately $H_{1.25}NiO_2$ for the β modification. Upon further oxidation in the $NiOOH$ single-phase regime the gallery proton concentration is reduced. It is generally found that the proton concentration can be substantially lower for the γ modification than in the β case. These can thus be far from the nominal composition of $NiOOH$.

19.3.5   Self-Discharge

Since the $NiOOH$ phase is a good mixed-conductor, with a high mobility of both ionic and electronic species, equilibrium with the adjacent electrolyte is readily attained. In the absence of current through the external circuit, there will be a chemical reaction at the $NiOOH$ surface with water in the electrolyte that results in the addition of hydrogen to the electrode. This causes a shift in the direction of a less positive potential. This increase in the hydrogen content and decrease of the potential thus results effectively in a gradual self-discharge of the electrode.

The electrochemical literature generally assumes that this self-discharge reaction involves the generation of oxygen, since the potential of the electrode is more positive than that necessary for the evolution of oxygen by the decomposition of water, as mentioned above.

There are two possible oxygen evolution reactions involving species in the electrolyte:

$$2NiOOH + H_2O = 2Ni(OH)_2 + \tfrac{1}{2}O_2 \tag{19.5}$$

which can also be written as

$$H_2O = \tfrac{1}{2}O_2 \text{(into electrolyte or gas)} + H_2 \text{(into electrode)} \tag{19.6}$$

and

$$2OH^- = H_2O + \tfrac{1}{2}O_2 + 2e^- \tag{19.7}$$

However, the latter does not provide any hydrogen to the electrode, and thus cannot contribute to self-discharge. Instead, it is the electrochemical oxygen evolution reaction, involving passage of current through the outer circuit, as mentioned later.

The rate of the self-discharge reaction can be simply measured for any value of electrode potential in the single-phase NiOOH regime, where the potential is state-of-charge dependent by using a potentiostat to hold the potential at a constant value, and measuring the anodic current through the external circuit that is required to maintain that value of the potential (and thus also the corresponding proton concentration in the electrode). This is the opposite of the self-discharge process, and can be written as

$$2Ni(OH)_2 + OH^- = 2NiOOH + H_2O + e^- \tag{19.8}$$

Measurements of the self-discharge current as a function of potential in the NiOOH regime for the case of electrodes produced by two different commercial manufacturers are shown in Fig. 19.4. The differences between the two curves are not important, as they are related to differences between the microstructures of the two electrodes.

If anodic current is passed through the NiOOH electrode, part will be used to counteract the self-discharge mentioned above. If the magnitude of the current is greater than the self-discharge current, additional protons will be removed from the electrode's crystal structure, making the potential more positive. This results in an increased rate of self-discharge. Thus a steady state will evolve in which the applied current will be just balanced by the rate of self-discharge and the proton concentration in the galleries will reach a new steady (lower) value.

Fig. 19.4 Self-discharge current as a function of potential in the NiOOH regime measured on electrodes produced by two different commercial manufacturers

19.3.6   Overcharge

If the applied current density exceeds that which can be accommodated by the kinetics of the compositional change and the self-discharge process, another mechanism must come into play. This is the direct generation of oxygen gas at the electrolyte/NiOOH interface by the decomposition of water in the electrolyte. This can be described by the reaction

$$2\,OH^- = H_2O + \tfrac{1}{2}O_2 + 2\,e^- \tag{19.9}$$

in which the electrons go into the current collector

The relationship between the potential of the "nickel" electrode and the amount of hydrogen that is deleted when it is charged (oxidized) is shown schematically in Fig. 19.5.

19.3.7   Relation to Thermodynamic Information

The available thermodynamic data relating to the various phases in the Ni–O–H system can be used to produce a ternary phase stability diagram. From this information, one can also readily calculate the potentials of the various possible stable phase combinations. This general methodology [23–26] has been used with great success to understand the stability windows of a number of electrolytes, as well as the potential-composition behavior of many electrode materials in lithium, sodium, oxide, and other systems.

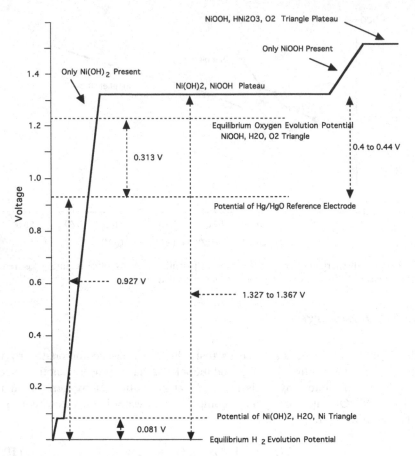

Fig. 19.5 Variation of the potential as hydrogen is removed from the nickel electrode during oxidation when the cell is being charged

With this information, the microstructural model discussed above, and the available information about the stoichiometric ranges of the important phases, one should be able to explain the observed electrochemical behavior of the nickel electrode.

Unfortunately, reliable thermodynamic information for this system is rather scarce. The data that have been used are included in Table 19.2, taken mostly from the compilation in [27]. Unfortunately, no recognition was given to the question of stoichiometry or the ordered/disorder state of crystal perfection, or even to the differences between the α and β structures of $Ni(OH)_2$ and the β and γ structures of NiOOH. Therefore, the calculated potentials can only be considered semiquantitative at present.

The results of these calculations are shown in the partial ternary phase stability diagram of Fig. 19.6, in which all phases are assumed to have their

Table 19.2 Thermodynamic data

Phase	ΔG_f° (25 °C)
NiO	−211.5
NiOOH	−329.4
Ni(OH)$_2$	−458.6
H$_2$O	−237.14

Fig. 19.6 Partial Gibbs triangle. The main charge-discharge reaction takes place along the thick line at 1.339 V vs. H$_2$. The overall composition moves along that line upon further charging

nominal compositions. The two-phase tie lines and three-phase triangles indicate the phases that are stable in the presence of each other at ambient temperatures. Also shown are the potentials of the various relevant three-phase equilibria versus the hydrogen evolution potential at one atmosphere. The composition of the electrode during operation on the main plateau follows along the heavy line that points toward the hydrogen corner of the diagram and lies on the edge of the triangle in which all compositions have a potential of 1.339 V vs. hydrogen.

Parts of this figure are incomplete, for the data available did not indicate what happens when additional hydrogen is removed from HNiO$_2$ and the potential exceeds 1.339 V at the time that it was first written [28]. It was obvious, of course, that the composition follows further along the arrow, but the species at the corners of the sub-triangle that has the observed higher potential could not be identified. Subsequent information, that led to an explanation of the so-called *memory effect*, is discussed later in this chapter.

One important question is whether there is a stable tie line between NiOOH and H$_2$O. The alternative is a tie line between Ni(OH)$_2$ and oxygen. Only one of these can be stable, as tie lines cannot cross. The Gibbs free energy change involved in the determining reaction can be calculated as

$$2NiOOH + H_2O = 2Ni(OH)_2 + \tfrac{1}{2}O_2 \tag{19.10}$$

This is found to be -21.26 kJ at 25 °C, which means that the NiOOH–H_2O tie line is not stable, and that there is a stable three-phase equilibrium involving Ni(OH)$_2$, NiOOH, and oxygen. A situation in which both Ni(OH)$_2$ and NiOOH are in contact with water can only be metastable.

It is possible to calculate the potential of the Ni(OH)$_2$, NiOOH, O_2 triangle relative to the one atmosphere hydrogen evolution potential from the reaction

$$\frac{1}{2}H_2 + NiOOH = Ni(OH)_2 \qquad (19.11)$$

that is the primary reaction of the nickel electrode, as discussed above, since this reaction occurs along one of the sides of this three-phase equilibrium triangle. From the data in Table 19.2 it can be determined that the Gibbs free energy change ΔG_r° accompanying this reaction is -129.2 kJ per mol. From the relation

$$\Delta G_r^0 = -zFE \qquad (19.12)$$

since z is unity for this reaction, the equilibrium potential is 1.34 V positive of the hydrogen evolution potential.

Since the potential of a Hg/HgO reference electrode is 0.93 V positive of the reversible hydrogen evolution potential (RHE), this calculation predicts that the equilibrium value of the two-phase constant potential plateau for the main reaction of the nickel electrode should occur at about 0.41 V positive of the *Hg/HgO* reference. This is also about 0.11 V more positive than the equilibrium potential of oxygen evolution from water.

This result can be compared with the experimental information from Barnard et al. [21] on the potentials of both the *activated* (highly disordered) and *deactivated* (more highly ordered) β Ni(OH)$_2$–β NiOOH reaction. Their data fell in the range 0.44–0.47 V positive of the *Hg/HgO* electrode potential. They found the comparable values for the a Ni(OH)$_2$–γ NiOOH reaction to be in the range 0.39–0.44 V relative to the Hg/HgO reference. Despite the lack of definition of the structures to which the thermodynamic data relate, this should be considered to be a quite good correlation.

Further oxidation causes the electrode composition to move along the arrow further away from the hydrogen corner of the ternary diagram, and leads to an electrode structure in which the Ni(OH)$_2$ phase is no longer stable, as is found experimentally. The potential moves to more positive values as the stoichiometry of the NiOOH phase changes, and if no other reaction interferes, should eventually arrive at another, higher, plateau in which the lower proton concentration limit NiOOH is in equilibrium with some other phase or phases.

Another complicating fact is that electrolyte enters the β NiOOH at high potentials, converting it to the γ modification. As mentioned earlier, the water-containing α Ni(OH)$_2$ and γ NiOOH phases are not stable, and during normal cycling are gradually converted to the corresponding β phases that have only protons in their galleries. When these metastable phases are present the electrode potential of the reaction plateau is less positive, as is characteristic of insertion structures with larger interslab spacings. Correspondingly, the apparent capacity of the electrode prior to rapid oxygen evolution is greater. These several factors are discussed further in the next chapter.

19.4   Cause of the Memory Effect in "Nickel" Electrodes

19.4.1   Introduction

It is often found that batteries with nickel positive electrodes, e.g., Cd/Ni, Hydride/ Ni, Zn/Ni, Fe/Ni, and H_2/Ni cells, have a so-called *memory effect*, in which the available capacity apparently decreases if they are used under conditions in which they are repeatedly only partially discharged before recharging. In many cases these batteries are kept connected to their chargers for long periods of time. It is also widely known that this problem can be "cured" by subjecting them to a slow, deep discharge.

The phenomena that take place in such electrodes have been studied by many investigators over many years, but no rational and consistent explanation of the *memory effect* related to nickel electrodes emerged until recently. Although it has important implications for the practical use of such cells, some of the major reviews in this area don't even mention this problem, and others give it little attention and/or no explanation.

In studying this apparent loss of capacity, a number of investigators have shown that a second plateau appears at a lower potential during discharge of nickel electrodes [29–45]. Importantly, it is found that under low current conditions the total length of the two plateaus remains constant. As the capacity on the lower one, sometimes called *residual capacity*, becomes greater, the capacity of the higher one shrinks. The relative lengths of the two plateaus vary with the conditions of prior charging. This is shown schematically in Fig. 19.7.

Since the capacity of the lower plateau is at about 0.78 V positive of the reversible hydrogen electrode potential, it is generally not useful for most of the applications for which nickel electrode batteries are employed. The user does not see this capacity, but instead, sees only the dwindling capacity on the upper plateau upon discharge. Thus it is quite obvious that the appearance of this lower plateau and reduction in the length of the upper plateau is an important component of the memory effect.

It is also found that this lower plateau and the memory effect both disappear if the cell is deeply discharged. Thus the existence of the lower plateau, and its disappearance, are both obviously related to the *curing* of the memory effect.

These phenomena can now be explained on the basis of available thermodynamic and structural information by using the ternary Gibbs phase stability diagram for the H–Ni–O system as a thinking tool [46, 47].

Fig. 19.7 Schematic representation of the two discharge plateaus. With increasing overcharge the length of the upper one decreases and the lower one increases

19.4.2   Mechanistic Features of the Operation of the "Nickel" Electrode

The microscopic mechanism of the basic operation of these electrodes was discussed earlier in this chapter. However, it is important for this discussion, and will be briefly reviewed here. It involves an insertion reaction that results in the translation of a two-phase interface between H_2NiO_2 and $HNiO_2$, both of which are vario-stoichiometric (have ranges of stoichiometry). The H_2NiO_2 is in contact with the alkaline electrolyte, and the $HNiO_2$ is in contact with the metallic current collector. The outer layer of the H_2NiO_2 phase is pale green and is predominantly an ionic conductor, allowing the transport of protons to and from the two-phase $H_2NiO_2/HNiO_2$ boundary. $HNiO_2$ is a good electronic conductor, and is black. The electrochemical reaction takes place at that ionic/electronic two-phase interface. This boundary is displaced as the reaction proceeds, and the motion of the color boundary has been experimentally observed. When the electrode is fully reduced, its structure consists of only H_2NiO_2, whereas oxidation causes the interface to translate in the opposite direction until only $HNiO_2$ is present. Although these are both ternary phases, the only compositional change involves the amount of hydrogen present, and the structure of the host "NiO_2" does not change. Thus this is a pseudo-binary reaction, although it takes place in a ternary system, and the potential is independent of the overall composition, i.e., the state of charge.

Once the H_2NiO_2 has been completely consumed, and the $HNiO_2$ phase comes into contact with the aqueous electrolyte it is possible to obtain further oxidation. This involves a change in the hydrogen content of the $HNiO_2$ phase. The variation of the composition of this single phase results in an increase in the potential from this two-phase plateau to higher values, as is expected from the Gibbs phase rule.

After the low-hydrogen limit of the composition of the $HNiO_2$ phase is reached, further oxidation can still take place. Another potential plateau is observed, and oxygen evolution occurs. This is often called *overcharging*, and obviously involves another process.

A number of authors have shown that the length of the lower plateau observed upon discharging is a function of the amount of the γ NiOOH phase formed during overcharging [38]. However, other authors [12] have shown that it is possible to prevent the formation of the γ phase during overcharging by using a dilute electrolyte. Yet the lower discharge potential plateau still appears. There is also evidence that the γ phase can disappear upon extensive overcharging, but the lower discharge plateau is still observed [38].

Neutron diffraction studies [43], which see only crystalline structures, showed a gradual transition between the γ and β NiOOH structures upon discharge, with no discontinuity at the transition between the upper and lower discharge plateaus. There was no evidence of a change in the compositions of either of the two phases, just a variation in their amounts, which changed continuously along both discharge plateaus. These authors attributed the presence of the lower plateau to undefined "technical parameters."

Several other authors have explained the presence of the lower discharge plateau in terms of the formation of some type of barrier layer [30, 36], and there is evidence for the formation of β H_2NiO_2, which is not electronically conducting, on the lower plateau [41]. This can, of course, be interpreted as a barrier.

These studies all seem to assume that the oxygen that is formed during operation upon the upper plateau during charging comes only from decomposition of the aqueous electrolyte. However, something else is obviously happening that leads to the formation of the lower plateau that is observed upon discharge. It must also relate to a change in the amounts, compositions or structure of the solid phase, or phases, present.

Although the electrochemical behavior of the nickel electrode upon the lower potential plateau can be understood in terms of a pseudo-binary insertion/extraction hydrogen reaction, the evolution of oxygen and the formation of the second discharge plateau indicate that the assumption that the oxygen comes from (only) the electrolysis of the aqueous electrolyte during overcharge cannot be correct. In order to understand this behavior, recognition must be given to the fact that the evolution of oxygen indicates that at this potential this electrode should be treated in terms of the ternary H–Ni–O system, rather than as a simple binary phase reaction.

Use of the Gibbs triangle as a *thinking tool* to understand the basic reactions in the H–Ni–O system has been discussed in several places [48–50]. The major features of the lower-potential portion of this system can be readily determined from available thermodynamic information. A major part of the Gibbs phase stability triangle for this system is again shown in Fig. 19.8, copied from the discussion earlier in this chapter.

Since the two phases H_2NiO_2 and $HNiO_2$ are on a tie line that points to the hydrogen corner, neither hydrogen insertion nor deletion involve any change in the Ni/O_2 ratio, and this can be considered to be a pseudo-binary reaction. The tie line between those two phases is one side of a triangle that has pure oxygen at its other corner. This means that both of these phases are stable in oxygen, as is well known.

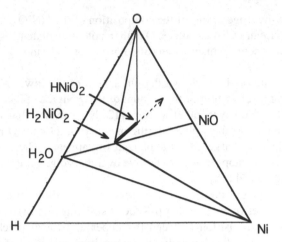

Fig. 19.8 Partial Gibbs triangle. The main charge-discharge reaction takes place along the *thick line*. The overall composition moves along the *dashed line* upon further charging

As the result of the Gibbs phase rule, movement of the overall composition along this tie line occurs at a constant potential plateau. It was shown earlier that the potential of this plateau is 1.34 V versus pure hydrogen at 25 °C.

Thus the equilibrium electrode potential of the basic H_2NiO_2–$HNiO_2$ reaction is not only composition-independent, but also more positive than the potential of the decomposition of water, as is experimentally observed. Also, because the H_2NiO_2 that is between the $HNiO_2$ and the water is a solid electrolyte, there is little or no oxygen evolution.

As additional hydrogen is removed the potential moves up the curve where only *HNiO$_2$* is present. When the overall composition exceeds the stability range of that phase it moves further from the hydrogen corner and enters another region in the phase diagram, as indicated by the dashed line in Fig. 19.2.

19.4.3 Overcharging Phenomena

The potential then moves along the upper charging (or overcharging) plateau. Since all of the area within a Gibbs triangle must be divided into sub-triangles, the overall composition must be moving into a new sub-triangle. One corner of this new triangle must be $HNiO_2$, and another must be oxygen. This is consistent with the observation that oxygen is evolved at this higher charging potential. The question is then, what is the composition of the phase that is at the third corner?

If gaseous oxygen is evolved from the electrode, not just from decomposition of the water, the third-corner composition must be below (i.e., have less oxygen) than all compositions along the dashed line.

One possibility might be the phase Ni_3O_4, another could be NiO. However, neither of these phases, which readily crystallize, has been observed. There must be another phase with a reduced ratio of oxygen to nickel.

Fig. 19.9 Gibbs triangle showing the presence of the HNi_2O_3 phase

Although evidently not generally recognized by workers interested in the nickel electrode, it has been found [51, 52] that a phase with a composition close to HNi_2O_3 can be formed under conditions comparable to those during charging the electrode on the upper voltage plateau. This phase can form as an amorphous product during the oxidation of $HNiO_2$. It's crystal structure and composition were determined after hydrothermal crystallization. In addition, the mean nickel oxidation state was found by active oxygen analysis to be only 2.65.

The composition HNi_2O_3 lies on a line connecting $HNiO_2$ and NiO. This would then lead to a sub-triangle as shown in Fig. 19.9, which meets the requirement that there be another phase in equilibrium with both $HNiO_2$ and oxygen that has a reduced ratio of oxygen to nickel.

The gradual formation of amorphous HNi_2O_3 during oxygen evolution upon the upper overcharging plateau, and its influence upon behavior during discharge, is the key element in the memory effect puzzle.

As overcharge continues, oxygen is evolved, and more and more of the HNi_2O_3 phase forms. Thus the overall composition of the solid gradually shifts along the line connecting $HNiO_2$ and HNi_2O_3.

Upon discharge, the overall composition moves in the direction of the hydrogen corner of the Gibbs triangle. This is indicated by the dashed line in Fig. 19.10.

It is seen that the HNi_2O_3 portion of the total solid moves into a different sub-triangle that has H_2NiO_2, HNi_2O_3 and NiO at its corners. From the available thermodynamic data one can calculate that the potential in this sub-triangle is 0.78 V versus hydrogen. That is essentially the same as experimentally found for the lower discharge plateau. The larger the amount of HNi_2O_3 that has been formed during overcharging, the longer the corresponding lower discharge plateau will be. The upper discharge plateau becomes correspondingly shorter.

After traversing this triangle, the overall composition of what had been HNi_2O_3 moves into another sub-triangle that has H_2NiO_2, NiO, and Ni at its corners. The HNi_2O_3 disappears, and the major product is H_2NiO_2. The potential in this sub-triangle can be calculated to be 0.19 V versus hydrogen.

Fig. 19.10 Composition path during the discharge of the HNi_2O_3 formed during overcharge

Fig. 19.11 Calculated values of the potential in the various sub-triangles in the H–Ni–O ternary system versus hydrogen. It is seen how the composition path during discharge of HNi_2O_3 leads to the observation of the lower discharge plateau at about 0.78 V, and the disappearance of that phase when the potential moves to a much lower value

If the electrode is now recharged, its potential does not go back up to the 0.8 V plateau, since HNi_2O_3 is no longer present, but goes to the potential for the oxidation of its major component, H_2NiO_2, The overall composition again moves away from the hydrogen corner, and the H_2NiO_2 loses hydrogen and gets converted to $HNiO_2$. This is the standard charging cycle low potential plateau. This also means that the lower reduction plateau is no longer active, for the HNi_2O_3 has disappeared, and the *memory effect* has been *cured*.

The calculated potentials in the various sub-triangles of the H–Ni–O system are shown in Fig. 19.11.

The reactions in the H–Ni–O system obviously have very rapid kinetics, for this electrode can be both charged and discharged at high rates. Therefore, it is quite reasonable to expect the phases present to be at or near their equilibrium amounts and compositions. This is indicated by the very good correlation between experimental results and the information obtained by the use of ternary phase stability diagrams based upon the available thermodynamic data.

19.4.3.1 Conclusions

The basic mechanisms that are involved in causing the memory effect have been identified. The key element is the formation of an amorphous HNi_2O_3 phase upon overcharging into the potential range where oxygen is evolved. Upon subsequent reduction, the presence of this phase produces the potential plateau at about 0.8 V versus hydrogen, reducing the available capacity at the normal higher reduction potential. The more the overcharge, the more HNi_2O_3 that is formed, and the longer the lower plateau. If the electrode undergoes further reduction this phase disappears, and the potential drops to a much lower value. Subsequent charging of the electrode brings the composition back to the initial state, and the *memory effect is cured*.

This model provides an understanding of the main features of the memory effect, and also explains the several confusing and apparently contradictory observations in the literature. It is expected that further experimental work will address this matter. Additional confirmation of the presence of HNi_2O_3 in the microstructures of overcharged electrodes would be especially useful.

The implication from this mechanism is that the major reason for the memory effect, a decrease in the capacity at the normal discharge potential, is related to extensive overcharging, rather than to the use of shallow discharge cycles.

References

1. Ruetschi P (1984) Cation-Vacancy Model for MnO_2. J Electrochem Soc 131:2737
2. Ruetschi P, Giovanoli R (1988) J Electrochem Soc 135:2663
3. Coleman JJ (1946) Trans Electrochem Soc 90:545
4. Barin (1995) Thermochemical Data of Pure Substances, 3rd edn. VCH, ISBN 9783527619825
5. Pourbaix M (1966) Atlas of Electrochemical Equilibria. Pergamon, Oxford
6. Stotz S, Wagner C (1966) Ber Bunsenges Physik Chem 70: 781
7. Netz A, Chu WF, Thangadurai V, Huggins RA, Weppner W (1999) Ionics 5:426
8. Huggins RA (2006) J Power Sources 153:365
9. Bode H, Dehmelt K, Witte J (1966) Electrochim Acta 11:1079
10. Oliva P et al (1982) J Power Sources 8:229
11. Faure C et al (1991) J Power Sources 35:249
12. Faure C, Delmas C, Willmann P (1991) J Power Sources 35:263
13. Faure C, Delmas C, Fouassier M (1991) J Power Sources 35:279

14. Delmas C (1991) In: Nazri GA, Shriver DF, Huggins RA, Balkanski M (eds) Solid State Ionics II. Materials Research Society, p 335
15. Delmas C et al (1988) Solid State Ionics 28–30:1132
16. Briggs GWD, Jones E, Wynne-Jones WFK (1955) Trans Faraday Soc 51:394
17. Kober FP (1965) J Electrochem Soc 112:1064
18. Conway BE, Bourgault PL (1959) Can J Chem 37:292
19. Kuchinskii EM, Erschler BV (1940) J Phys Chem 14:985
20. Briggs GWD, Fleischmann M (1971) Trans Faraday Soc 67:2397
21. Barnard R, Randell CF, Tye FL (1980) J Appl Electrochem 10:109
22. Crocker RW, Muller RH (1992) Presented at the Meeting of the Electrochemical Society, Toronto
23. Huggins RA (1979) In: Vashishta PP, Mundy JN, Shenoy GK (ed) Fast Ion Transport in Solids. North-Holland, p 53
24. Weppner W, Huggins RA (1980) Solid State Ionics 1:3
25. Godshall NA, Raistrick ID, Huggins RA (1980) Mater Res Bull 15:561
26. Godshall NA, Raistrick ID, Huggins RA (1984) J Electrochem Soc 131:543
27. Balej J, Divisek J (1992) Presented at the Meeting of the Bunsengesellschaft, Wien.
28. Huggins RA, Wohlfahrt-Mehrens M, Jörissen L (1993) Presented at Symposium on Intercalation Chemistry and Intercalation Electrodes. Meeting of the Electrochemical Society in Hawaii, Spring 1993
29. Milner PC, Thomas UB (1967) In: Tobias CW (ed) Advances in Electrochemistry and Electrochemical Engineering. p 1
30. Barnard R, Crickmore GT, Lee JA, Tye FL (1980) J Appl Electrochem 10:61
31. Klapste B, Mickja K, Mrha J, Vondrak J (1982) J Power Sources 8:351
32. Zimmerman AH, Effa PK (1984) J Electrochem Soc 131:709
33. Lim HS, Verzwyvelt SA (1988) J Power Sources 22:213
34. Vaidyanathan H (1988) J Power Sources 22:221
35. McBreen J (1990) Mod Aspect Electrochem 21:29
36. Zimmerman AH (1990) In: Corrigan DA, Zimmerman AH (ed) Nickel Hydroxide Electrode. Electrochem Soc Proc 90-4: 311
37. Zimmerman AH (1994) Proc IECEC 4:63
38. Wilde P (1996) PhD Thesis. University of Ulm, Germany
39. Sac-Epee N, Palacín MR, Beaudoin B, Delahaye-Vidal A, Jamin T, Chabre Y, Tarascon J-M (1997) J Electrochem Soc 144:3896
40. Sac-Epee N, Palacìn MR, Delahaye-Vidal A, Chabre Y, Tarâscon J-M (1998) J Electrochem Soc 145:1434
41. Leger C, Tessier C, Ménétrier M, Denage C, Delmas C (1999) J Electrochem Soc 146:924
42. Fourgeot F, Deabate S, Henn F, Costa M (2000) Ionics 6:364
43. Deabate S, Fourgeot F, Henn F (2000) Ionics 6:415
44. Barde F, Palacin MR, Chabre Y, Isnard O, Tarascon J-M (2004) Chem Mater 16:3936
45. Huggins RA (2006) Solid State Ionics 177:2643
46. Huggins RA (2007) J Power Sources 165:640
47. Huggins RA, Wohlfahrt-Mehrens M, Jörissen L (1992) Presented at Meeting of the Electrochemical Society, Hawaii
48. Huggins RA, Wohlfahrt-Mehrens M, Jörissen L (1993) In: Nazri GA, Tarascon JM, Armand (eds) Solid State Ionics III. Mater Res Soc Proc 293: 57
49. Huggins RA, Prinz H, Wohlfahrt-Mehrens M, Jörissen L, Witschel W (1994) Solid State Ionics. 70/71: 417
50. Greaves C, Thomas MA, Turner M (1983) Power Source 9:163
51. Greaves C, Malsbury AM, Thomas MA (1986) Solid State Ionics. 18/19: 763
52. Malsbury AM, Greaves C (1987) J Solid State Chem 71:418

Chapter 20
Negative Electrodes in Lithium Systems

20.1 Introduction

A great deal of attention is currently being given to the development and use of batteries in which lithium plays an important role. Looked at very simply, there are two major reasons for this. One is that lithium is a very electropositive element, and its employment in electrochemical cells can lead to larger voltages than are possible with the other, less electropositive alkali metals. The second positive aspect of lithium systems is the possibility of major reductions in weight, at least partly due to the light weight of elemental lithium and many of its alloys and compounds.

Although there are now a number of lithium-based batteries available commercially, there is still a large amount of research and development effort under way. There are two general targets, the achievement of significant improvements in performance and safety, and a great reduction in costs. Since this technology has not matured and stabilized, the discussion here focuses upon phenomena and components, rather than complete systems. This chapter deals with negative electrodes in lithium systems. Positive electrode phenomena and materials are treated in the next chapter.

Early work on the commercial development of rechargeable lithium batteries to operate at or near ambient temperatures involved the use of elemental lithium as the negative electrode reactant. As discussed below, this leads to significant problems. Negative electrodes currently employed on the negative side of lithium cells involving a solid solution of lithium in one of the forms of carbon.

Lithium cells that operate at temperatures above the melting point of lithium must necessarily use alloys instead of elemental lithium. These are generally binary or ternary metallic phases.

There is also increasing current interest in the possibility of the use of metallic alloys instead of carbons at ambient temperatures, with the goal of reducing the electrode volume, as well as achieving significantly increased capacity.

© Springer International Publishing Switzerland 2016   361
R.A. Huggins, *Energy Storage*, DOI 10.1007/978-3-319-21239-5_20

There are differences in principle between the behavior of elemental and binary phase materials as electrodes. It is the purpose of this chapter to elucidate these principles, as well as to present some examples. Ternary systems are discussed elsewhere.

20.2 Elemental Lithium Electrodes

It is obvious that elemental lithium has the lowest potential, as well as the lowest weight per unit charge, of any possible lithium reservoir material in an electrochemical cell. Materials with lower lithium activities have higher potentials, leading to lower cell voltages, and they also carry along extra elements as dead weight.

There are problems with the use of elemental lithium, however. These are due to phenomena that occur during the recharging of all electrodes composed of simple metallic elements. In the particular case of lithium, however, this is not just a matter of increasing electrode impedance and reduced capacity, as are typically found with other electrode materials. In addition, severe safety problems can ensue. Some of these phenomena will be discussed in the following sections.

In the case of an electrochemical cell in which an elemental metal serves as the negative electrode the process of recharging may seem to be very simple, for it merely involves the electrodeposition of the metal from the electrolyte onto the surface of the electrode. This is not the case, however.

In order to achieve good rechargeability, a consistent geometry must be maintained on both the macroscopic and microscopic scales. Both electrical disconnection of the electroactive species and electronic short circuits must also be avoided. In addition, thermal runaway must not occur.

Phenomena related to the inherent microstructural and macrostructural instability of a growth interface and related thermal problems will now be briefly reviewed.

20.2.1 Deposition at Unwanted Locations

In the absence of a significant nucleation barrier, deposition will tend to occur anywhere at which the electric potential is such that the element's chemical potential is at, or above, that corresponding to unit activity. This means that electrodeposition may take place upon current collectors and other parts of an electrochemical cell that are at the same electrical potential as the negative electrode, as well as upon the electrode structure where it is actually desired. This was a significant problem during the period in which attempts were being made to use pure (molten) lithium as the negative electrode in high-temperature molten halide salt electrolyte cells. Another problem with these high-temperature cells was the fact that alkali metals dissolve in their halides at elevated temperatures. This leads to electronic conduction and self-discharge.

20.2.2 Shape Change

Another difficulty is the *shape change* phenomenon, in which the location of the electrodeposit is not the same as that where the discharge (deplating) process took place. Thus, upon cycling the electrode metal gets preferentially transferred to new locations. For the most part, this is a problem of current distribution and hydrodynamics, rather than being a materials issue. Therefore, it will not be discussed further here.

20.2.3 Dendrites

An additional type of problem relates to the inherent instability of a flat interface on a microscopic scale during electrodeposition, even in the case of a chemically clean surface. It has been shown that there can be an electrochemical analog of the constitutional supercooling that occurs ahead of a growth interface during thermally-driven solidification [1].

This will be the case if the current density is such that solute depletion in the electrolyte near the electrode surface causes the local gradient of the element's chemical potential in the electrolyte immediately adjacent to the solid surface to be positive. Under such a condition there will be a tendency for any protuberance upon the surface to grow at a faster rate than the rest of the interface. This leads to exaggerated surface roughness, and eventually to the formation of either dendrites or filaments. In more extreme cases, it leads to the nucleation of solid particles in the liquid electrolyte ahead of the growing solid interface.

This is also related to the inverse phenomenon, the formation of a flat interface during electropolishing, as well as the problem of morphology development during the growth of an oxide layer upon a solid solution alloy [2, 3]. Another analogous situation is present during the crystallization of the solute phase from liquid metal solutions.

The protuberances upon a clean growing interface can grow far ahead of the general interface, often developing into dendrites. A general characteristic of dendrites is a tree-and-branches type of morphology, which has very distinct geometric and crystallographic characteristics, due to the orientation dependence of either the surface energy or the growth velocity.

20.2.4 Filamentary Growth

A different phenomenon that is often mistakenly confused with dendrite formation is the result of the presence of a reaction product layer upon the growth interface if the electrode and electrolyte are not stable in the presence of each other.

The properties of these layers can have an important effect upon the behavior of the electrode. In some cases they may be useful solid electrolytes, and allow electro-deposition by ionic transport through them. Such layers upon negative electrodes in lithium systems have been given the name *SEI*, and will be discussed in a later chapter. But in other cases reaction product layers may be ionically blocking, and thus significantly increase the interfacial impedance.

Interfacial layers often have defects in their structure that can lead to local variations in their properties. Regions of reduced impedance can cause the formation of deleterious filamentary growths upon recharge of the electrode. This is an endemic problem with the use of organic solvent electrolytes in contact with lithium electrodes at ambient temperatures.

When a protrusion grows ahead of the main interface the protective reaction product layer will typically be locally less thick. This means that the local impedance to the passage of ionic current is reduced, resulting in a higher current density and more rapid growth in that location. This behavior can be exaggerated if the blocking layer is somewhat soluble in the electrolyte, with a greater solubility at elevated temperatures. When this is the case, the higher local current leads to a higher local temperature, and a greater solubility. The result is then a locally thinner blocking layer, and an even higher local current.

Furthermore, the current distribution near the tip of a protrusion that is well ahead of the main interface develops a 3-dimensional character, leading to even faster growth than the main electrode surface, where the mass transport is essentially 1-dimensional. Especially in relatively low concentration solutions, this leads to a runaway type of process, so that the protrusions consume most of the solute, and grow farther and farther ahead of the main, or bulk, interface.

This phenomenon can result in the metal deposit having a hairy or spongy character. During a subsequent discharge step, the protrusions often get disconnected from the underlying metal, so that they cannot participate in the electrochemical reaction, and the rechargeable capacity of the electrode is reduced.

This unstable growth is a major problem with the rechargeability of elementary negative electrodes in a number of electrochemical systems, and constitutes an important limitation upon the development of rechargeable lithium batteries using elemental lithium as the negative electrode reactant.

20.2.5   Thermal Runaway

The organic solvent electrolytes that are typically used in lithium batteries are not stable in the presence of high lithium activities. This is a common problem when using elemental lithium negative electrodes in contact with electrolytes containing organic cationic groups, regardless of whether the electrolyte is an organic liquid or a polymer [4].

They react with lithium and form either crystalline or amorphous product layers upon the surface of the electrode structure. These reactions are exothermic and cause local heating. Experiments using an *accelerating rate calorimeter* have shown that this problem increases dramatically as cells are cycled, presumably due to an increase in the surface area of the lithium due to morphological instability during repetitive recharging [5]. This is a fundamental difficulty with elemental lithium electrodes, and has led to serious safety problems.

The exothermic formation of reaction product films also occurs when carbon or alloy electrodes are used that operate at potentials at which the electrolyte reacts with lithium. However, if their morphology is constant the surface area does not change substantially, so that it can lead to heating, but typically does not lead to thermal runaway at the negative electrode.

20.3 Alternatives to the Use of Elemental Lithium

Because of these safety and cycle life problems with the use of elemental lithium, essentially all commercial rechargeable lithium batteries now use lithium–carbon alloys as negative electrode reactants today.

A considerable amount of research attention is now also being given to the possibility of the use of metallic lithium alloys instead of the carbons, because of the expectation that this may lead to significant increases in capacity. The large volume changes that accompany increased capacity present a significant problem, however. These matters, as well as the possibility of the use of novel micro- or nano-structures to alleviate this difficulty, are briefly discussed later in this chapter.

20.4 Lithium–Carbon Alloys

20.4.1 Introduction

Lithium–carbons are currently used as the negative electrode reactant in the very common small rechargeable lithium batteries used in consumer electronic devices. As will be seen in this chapter, a wide range of structures, and therefore of properties, is possible in this family, depending upon how the carbon is produced. The choices made by the different manufacturers are not all the same. Several good reviews of the materials science aspects of this topic can be found in the literature [6, 7].

The crystal structure of pure graphite is shown schematically in Fig. 20.1. It consists of parallel sheets containing interconnected hexagons of carbon, called *graphene* layers or sheets. They are stacked with alternate layers on top of one another. This is described as A-B-A-B-A stacking.

Graphite is amphoteric, and either cations or anions can be inserted into it between the graphene layers. When cations are inserted, the host graphite structure

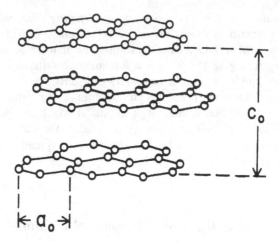

Fig. 20.1 Model of a portion of the crystal structure of graphite

takes on a negative charge. Cation examples are Li^+, K^+, Rb^+, Cs^+. When anions are inserted, the host graphite structure takes on a positive charge, and anion examples are Br^-, SO_4^{2-}, or SbF_6^-.

The insertion of alkali metals into carbon was first demonstrated in 1926 [8], and the chemical synthesis of lithium–carbons was demonstrated in 1955 [9]. X-ray photoemission spectroscopy experiments showed that the inserted lithium gives up its electron to the carbon, and thus the structure can be viewed as Li^+ ions contained between the carbon layers of the graphite structure [10]. A general review of the early work on the insertion of species into graphite can be found in [11].

Insertion often is found to occur in "stages," with nonrandom filling of positions between the layers of the host crystal structure. This ordering can occur in individual layers, and also in the filling of the stack of layers.

The possibility of the use of graphite as a reactant in the negative electrode of electrochemical cells containing lithium was first investigated than some 30 years ago [12]. The experiments were, however, unsuccessful. Swelling and defoliation occurred due to co-intercalation of species from the organic solvent electrolytes that were used at that time.

This problem has been subsequently solved by the use of other liquid electrolytes.

Attention was again brought to this possibility by a conference paper that was presented in 1983 [13] that showed that lithium can be reversibly inserted into graphite at room temperatures when using a polymeric electrolyte. Although not publicly known at that time, two patents relating to the use of the insertion of lithium into graphite as a reversible negative electrode in lithium systems, at both elevated [14] and ambient [15] temperatures, had already been submitted by Bell Laboratories. Royalties paid for the use of these patents have become very large.

This situation changed abruptly as the result of the successful development by SONY in 1990 of commercial rechargeable batteries containing negative electrodes based upon materials of this family and their commercial introduction as the power source in camcorders [16, 17].

There has been a large amount of work on the understanding and development of graphites and related carbon-containing materials for use as negative electrode materials in lithium batteries since that time.

Lithium–carbon materials are, in principle, no different from other lithium-containing metallic alloys. However, since this topic is treated in more detail later, only a few points that specifically relate to carbonaceous materials are discussed here.

One is that the behavior of these materials is very dependent upon the details of both the nanostructure and the microstructure. Therefore, the composition and the thermal and mechanical treatment of the electrode materials all play important roles in determining the resulting thermodynamic and kinetic properties. Materials with a more graphitic structure have properties that are much different from those with less well-organized structures. The materials that are used by the various commercial producers are not all the same, as they reflect the different choices that they have made for their specific products. However, the major producers of small consumer lithium batteries generally now use relatively graphitic carbons.

An important consideration in the use of carbonaceous materials as negative electrodes in lithium cells is the common observation of a considerable loss of capacity during the first charge-discharge cycle due to irreversible lithium absorption into the structure, as will be seen later. This has the distinct disadvantage that it requires that an additional amount of lithium be initially present in the cell. If this irreversible lithium is supplied from the positive electrode, an extra amount of the positive electrode reactant material must be put into the cell during its fabrication. As the positive electrode reactant materials often have relatively low specific capacities, e.g., around 140 mAh/g, this irreversible capacity in the negative electrode leads to a requirement for an appreciable amount of extra reactant material weight and volume in the total cell.

20.4.2 Ideal Structure of Graphite Saturated with Lithium

Lithium can be inserted into the graphite structure up to a maximum concentration of one Li per six carbons, or LiC_6. One of the major influences of the presence of lithium is the graphite crystal structure is that the stacking of graphene layers is changed by the insertion of lithium. It changes from A-B-A-B-A stacking to A-A-A-A-A stacking. This is illustrated schematically in Fig. 20.2.

The distribution of lithium ions within the gallery space between the graphene layers is illustrated schematically in Fig. 20.3.

A-B-A-B-A Stacking

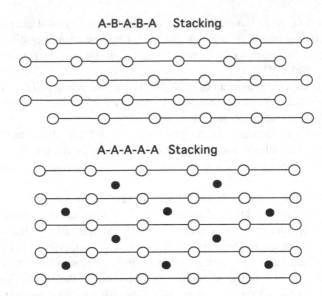

A-A-A-A-A Stacking

Fig. 20.2 Difference between the A-B-A-B-A and A-A-A-A-A stacking of the graphene layers when lithium is inserted. The *black circles* are the lithium ions

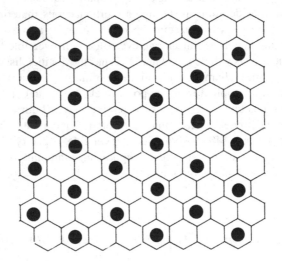

Fig. 20.3 Schematic representation of the lithium distribution in the gallery space in relation to the carbon hexagonal network in the adjacent graphene layers

20.4.3 Variations in the Structure of Graphite

There is actually a wide range of lithium–carbon structures, and most such materials do not actually have the ideal graphite structure. The ones that are closest are made synthetically by vapor transport, and are called highly ordered pyrolytic graphite (HOPG). This is a slow and very expensive process. The graphites that are used commercially range from natural graphite to materials formed by the pyrolysis of various polymer or hydrocarbon precursors. They are often divided into two general types, designated as *soft, or graphitizing, carbons*, and *hard carbons* [18].

At modest temperatures and pressures there is a strong tendency for carbon atoms to be arranged in a planar graphene-type configuration, rather than a 3-dimensional structure such as that in diamond.

Soft carbons are generally produced by the pyrolysis of liquid materials such as petroleum pitch, which is the residue from the distillation of petroleum fractions.

The carbon atoms in their structure are initially arranged in small graphene-type groups, but there is generally a significant amount of imperfection in their two-dimensional honeycomb networks, as well as randomness in the way that the layers are vertically stacked upon each other. In addition there is little coordination in the rotational orientation of nearby graphene layers. The term *turbostratic* is generally used to describe this general type of 3-dimensional disorder in carbons [18].

The three types of initial disorder, in-plane defects, inter-plane stacking defects, and rotational misorientation, gradually become healed as the temperature is raised: the first two earlier than the rotational disorder between adjacent layers, for that requires more thermal energy.

The microstructure of such materials that have been heated to intermediate temperatures is shown schematically in Fig. 20.4.

At this intermediate stage, the structure contains many small three-dimensional subgrains.

In addition to containing some internal imperfections, they differ from their neighbors in both vertical and horizontal orientations. They are separated by subgrain walls (boundaries) that have surface energy. This subgrain wall surface energy gradually gets reduced as the individual subgrains grow in size and the overall graphitic structure becomes more perfect.

The *hard carbons*, that are typically produced by the pyrolysis of solid materials, such as chars or glassy carbon, initially have a significant amount of initial cross-linking, related to the structure of their precursors. In addition, they can have a

Fig. 20.4 Schematic drawing of the microstructure of graphite after heating to intermediate temperatures

substantial amount of nano-porosity. As a result, it is more difficult to make these structural rearrangements and turbostratic disorder is more persistent. The result is the requirement for more thermal energy, i.e., higher temperatures.

The structure that results from the pyrolysis of carbonaceous precursors depends greatly upon the maximum temperature that is reached. Heating initially amorphous, or *soft*, carbons to the range of 1,000–2,000 °C produces microstructures in which graphene sheets form and begin to grow, with diameters up to about 15 nm, and they become assembled into small stacks of 50–100 sheets. These subgrains initially have a turbostratic arrangement, but their alignment into larger ordered, i.e., graphitic, regions gradually takes place as the temperature is increased from 2000 to 3000 °C.

20.4.4   Structural Aspects of Lithium Insertion into Graphitic Carbons

One of the important features in the interaction of lithium with graphitic materials is the phenomenon of *staging*. Lithium that enters the graphite structure is not distributed uniformly between all the graphene layers at ambient temperatures. Instead, it resides in certain interlayer *galleries*, but not others, depending upon the total amount of lithium present.

The distribution is described by the number of graphene layers between those that have the lithium guest ions present. A stage 1 structure has lithium between all of the graphene layers, a stage 2 structure has an empty gallery between each occupied gallery, and a stage 4 structure has four graphene layers between each gallery containing lithium. This is discussed a bit more later in this chapter. This is obviously a simplification, for in any real material there will be regions with predominately one structure, and other regions with another.

The phenomenon of nonrandom gallery occupation is found in a number of other materials, and can be attributed to a catalytic effect, in which the ions that initially enter a gallery pry open the van der Waals-bonded interlayer space, making it easier for following ions to enter.

However, the situation is a bit more complicated, for there must be communication between nearby galleries in order for the structure to adopt the ordered stage structure. This is related to the inter-tunnel communication in the *hollandite* structure described in Chap. 13, but will not be further discussed here.

20.4.5   Electrochemical Behavior of Lithium in Graphite

The electrochemical behavior of lithium in carbon materials is highly variable, depending upon the details of the graphitic structure. Materials with a more perfect

Fig. 20.5 Typical discharge curve of a lithium battery negative electrode

graphitic structure react with lithium at more negative potentials, whereas those with less well organized structures typically operate over much wider potential ranges, resulting in cell voltages that are both lower and more state-of-charge dependent.

In a number of cases, the carbons that are used in commercial batteries have been heated to temperatures over bout 2400 °C, where they become quite well graphitized. Capacities typically range from 300 to 350 mAh/g, whereas the maximum theoretical value (for LiC_6) is 372 mAh/g.

A typical discharge curve under operating conditions, with currents as large as 2–4 mA/cm^2, is shown in Fig. 20.5.

This behavior is not far from what is found under near equilibrium conditions, as shown in Fig. 20.6. It can be seen that there is a difference between the data during charge, when lithium is being added, and discharge, when lithium is being deleted. This displacement (hysteresis) between the charge and discharge curves is at least partly due to the mechanical energy involved in the structural changes.

It can be seen that these data show plateaus, indicating the presence of three ranges of composition within which reconstitution reactions take place. As the composition changes along these plateaus the relative amounts of material with the two end compositions varies. This means that there will be regions, or domains, where the graphene layer stacking is of one type, and regions in which it has the other. The relative volumes of these two domains varies as the overall composition traverses these *two phase regions*. The differences in stacking results in differences in interlayer spacing, and therefore considerable amount of distortion of the structure. Such a model was presented some time ago by Daumas and Herold [20].

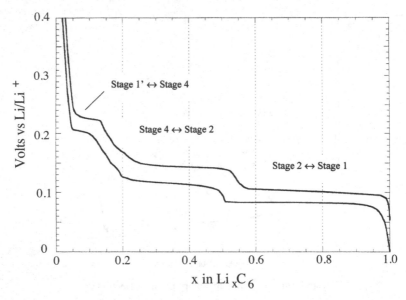

Fig. 20.6 Potential versus composition during lithiation and delithiation of a graphite electrode at the C/50 rate at ambient temperature. After [19]

20.4.6   Electrochemical Behavior of Lithium in Amorphous Carbons

The electrochemical behavior is quite different when the carbon has not been heated so high, and the structure is not so well ordered. There is a wide range of possible sites in which the lithium can reside, with different local structures, and therefore different energies. The result is that the potential varies gradually, rather than showing the steps characteristic of more ordered structures. This is shown in Fig. 20.7. It can be seen that, in addition to varying with the state of charge, the potential is significantly greater than is found in the graphitic materials. This means that the cell voltages are correspondingly lower.

It can be seen that there was some capacity loss on the first cycle. The capacity upon the first charging (that is not useful) was greater than the capacity in the subsequent discharge cycle. The source of this phenomenon is not yet understood, but there must be some lithium that is *trapped* in the structure and does not come out during discharge. Because of this extra (useless) capacity during the initially charging of this negative electrode it is necessary to put extra capacity in the positive electrode. This is unfortunate, for the specific capacity of the positive electrodes in such systems is less than that in the negative electrodes. As a result, a significant amount of extra weight and volume is necessary.

Fig. 20.7 Typical data for the reaction of lithium with an amorphous carbon

20.4.7 *Lithium in Hydrogen-Containing Carbons*

It is often found that there is a considerable amount of hydrogen initially present in various carbons, depending upon the nature of the precursor. This gradually disappears as the temperature is raised.

If the precursor is heated to only 500–700 °C, there is still a lot of hydrogen present in the structure. It has been found experimentally that this can lead to a very large capacity for lithium that is proportional to the amount of hydrogen present [21–23]. There is a loss in this capacity upon cycling, perhaps due to the gradual loss of hydrogen in the structure.

The large capacity may be due to lithium binding to hydrogen-terminated edges of small graphene fragments. The local configuration would then be analogous to that in the organolithium molecule $C_2H_2L_2$. This is consistent with the experimental observation of the dependence of the lithium capacity upon the amount of hydrogen present. This would also result in a change in the local bonding of the host carbon atom from sp^2 to sp^3.

In addition to a large capacity, experiments have shown a very large hysteresis with these materials [23]. Hysteresis is generally considered to be a disadvantage, as the discharge potential is raised, reducing the cell voltage.

Hysteresis is characteristic of reactions that involve a lot of mechanical energy as the result of shape and volume changes. However, in this case it is more likely due to the energy involved in the change of the bonding of the nearby carbon atoms [23].

The results of experiments performed on one example of a hydrogen-containing material are shown in Fig. 20.8. It can be seen that there was a very large capacity loss on the first cycle. The capacity upon the first charging (that is not useful) was much greater than the capacities in subsequent cycles. As mentioned above, this extra lithium must be supplied by the positive electrode. The

Fig. 20.8 Charge–discharge curves for a material containing hydrogen. After [7]

source of this phenomenon is not yet understood, but there must be a lot of lithium that is *trapped* in the structure and does not come out during the first, and subsequent, discharges.

20.5   Metallic Lithium Alloys

20.5.1   Introduction

Attention has been given to the use of lithium alloys as an alternative to elemental lithium for some time. Groups working on batteries with molten salt electrolytes that operate at temperatures of 400–450 °C, well above the melting point of lithium, were especially interested in this possibility. Two major directions evolved. One involved the use of lithium-aluminum alloys [24, 25], whereas another was concerned with lithium–silicon alloys [26–28].

Whereas this approach can avoid the problems related to lithium melting, as well as the others mentioned above, there are always at least two disadvantages related to the use of alloys. Because they reduce the activity of the lithium they necessarily reduce the cell voltage. In addition, the presence of additional species that are not directly involved in the electrochemical reaction always brings additional weight, and often volume. Thus the maximum theoretical values of the specific energy are always reduced compared to what might be attained with pure lithium. The energy density is also often reduced. But lithium has a large specific volume, so that this is not always the case.

In practical cases, however, the excess weight and volume due to the use of alloys may not be very far from those required with pure lithium electrodes, for it is generally necessary to have a large amount of excess lithium in rechargeable cells in order to make up for the capacity loss related to the dendrite or filament growth problem upon cycling.

Lithium alloys have been used for a number of years in the high temperature "thermal batteries" that are produced commercially for military purposes. These devices are designed to be stored for long times at ambient temperatures before use, where their self-discharge kinetic behavior is very slow. They must be heated to elevated temperatures when their energy output is desired. An example is the Li alloy/FeS_2 battery system that employs a chloride molten salt electrolyte. In order to operate, the temperature must be raised to over the melting point of the electrolyte. This type of cell typically uses either Li–Si or Li–Al alloys in the negative electrode.

The first use of lithium alloys as negative electrodes in commercial batteries to operate at ambient temperatures was the employment of Wood's metal alloys in lithium-conducting button type cells by Matsushita in Japan. Development work on the use of these alloys started in 1983 [29], and they became commercially available somewhat later.

20.5.2 Equilibrium Thermodynamic Properties of Binary Lithium Alloys

Useful starting points when considering lithium alloys as electrode reactants are their phase diagrams and equilibrium thermodynamic properties. In some cases this information is available, so that predictions can be made of their potentials and capacities. In other cases, experimental measurements are required. Relevant principles were discussed in earlier chapters, and will not be repeated here.

Elevated temperature data for a number of phases in the Li–Al, Li–Bi, Li–Cd, Li–Ga, Li–In, Li–Pb, Li–Sb, Li–Si, and Li–Sn binary lithium alloy systems, made using a LiCl–KCl molten salt electrolyte, are listed in Table 20.1.

20.5.3 Experiments at Ambient Temperature

Experiments have also been performed to determine the equilibrium values of the electrochemical potentials and capacities in a smaller number of binary lithium systems at ambient temperatures [30, 31]. Because of slower kinetics at lower temperatures, these experiments took longer to perform. Data are presented in Table 20.2.

Table 20.1 Plateau potentials and composition ranges of a number of binary lithium alloys at 400 °C

Voltage vs Li/Li$^+$	System	Range of y
0.910	Li$_y$Sb	0–2.0
0.875	Li$_y$Sb	2.0–3.0
0.760	Li$_y$Bi	0.6–1.0
0.750	Li$_y$Bi	1.0–2.82
0.570	Li$_y$Sn	0.57–1.0
0.455	Li$_y$Sn	1.0–2.33
0.430	Li$_y$Sn	2.33–2.5
0.387	Li$_y$Sn	2.5–2.6
0.283	Li$_y$Sn	2.6–3.5
0.170	Li$_y$Sn	3.5–4.4
0.565	Li$_y$Ga	0.15–0.82
0.122	Li$_y$Ga	1.28–1.48
0.09	Li$_y$Ga	1.53–1.93
0.558	Li$_y$Cd	0.12–0.21
0.373	Li$_y$Cd	0.33–0.45
0.058	Li$_y$Cd	1.65–2.33
0.507	Li$_y$Pb	0–1.0
0.375	Li$_y$Pb	1.1–2.67
0.271	Li$_y$Pb	2.67–3.0
0.237	Li$_y$Pb	3.0–3.5
0.089	Li$_y$Pb	3.8–4.4
0.495	Li$_y$In	0.22–0.86
0.145	Li$_y$In	1.74–1.92
0.080	Li$_y$In	2.08–2.67
0.332	Li$_y$Si	0–2.0
0.283	Li$_y$Si	2.0–2.67
0.156	Li$_y$Si	2.67–3.25
0.047	Li$_y$Si	3.25–4.4
0.300	Li$_y$Al	0.08–0.9

20.5.4   Liquid Binary Alloys

Although the discussion here has involved solid lithium alloys, similar considerations apply to those based on sodium or other species. In addition, it is not necessary that the active material be solid. The same principles hold for liquids.

An example was discussed in Chap. 3 relating to the so-called sodium–sulfur battery that operates at about 300 °C. In this case, both of the electrodes are liquids, and the electrolyte is a solid sodium ion conductor. This configuration can thus be described as an L/S/L system. It is the inverse of conventional systems with solid electrodes and a liquid electrolyte, S/L/S systems.

Table 20.2 Plateau potentials and composition ranges of lithium alloys at ambient temperatures under equilibrium conditions

Voltage vs. Li/Li$^+$	System	Range of y
0.956	Li$_y$Sb	1.0–2.0
0.948	Li$_y$Sb	2.0–3.0
0.828	Li$_y$Bi	0–1.0
0.810	Li$_y$Bi	1–3.0
0.680	Li$_y$Cd	0–0.3
0.352	Li$_y$Cd	0.3–0.6
0.055	Li$_y$Cd	1.5–2.9
0.660	Li$_y$Sn	0.4–0.7
0.530	Li$_y$Sn	0.7–2.33
0.485	Li$_y$Sn	2.33–2.63
0.420	Li$_y$Sn	2.6–3.5
0.380	Li$_y$Sn	3.5–4.4
0.601	Li$_y$Pb	0–1.0
0.449	Li$_y$Pb	1.0–3.0
0.374	Li$_y$Pb	3.0–3.2
0.292	Li$_y$Pb	3.2–4.5
0.256	Li$_y$Zn	0.4–0.5
0.219	Li$_y$Zn	0.5–0.67
0.157	Li$_y$Zn	0.67–1.0
0.005	Li$_y$Zn	1.0–1.5

20.5.5 Mixed-Conductor Matrix Electrodes

In order to be able to achieve appreciable macroscopic current densities while maintaining low local microscopic charge and particle flux densities, many battery electrodes that are used in conjunction with liquid electrolytes are produced with porous microstructures containing very fine particles of the solid reactant materials. This high reactant surface area porous structure is permeated with the electrolyte.

This porous fine-particle approach has several characteristic disadvantages, however. Among these are difficulties in producing uniform and reproducible microstructures, and limited mechanical strength when the structure is highly porous. In addition, they often suffer Ostwald ripening, sintering, or other time-dependent changes in both microstructure and properties during cyclic operation.

Furthermore, it is often necessary to have an additional material present in order to improve the electronic transport within an electrode. Various highly dispersed carbons are often used for this purpose.

A quite different approach was introduced some years ago [32–34] in which it was demonstrated that a rather dense solid electrode can be fabricated that has a composite microstructure in which particles of the reactant phase or phases are finely dispersed within a solid electronically-conducting matrix in which the electroactive species is also mobile, i.e., within a mixed conductor. There is thus a large internal reactant/mixed-conducting matrix interfacial area. The electroactive species is transported through the solid matrix to this interfacial region, where it undergoes the chemical part of the electrode reaction. Since the matrix material is also an electronic conductor, it can also act as the electrode's current collector. The electrochemical part of the reaction takes place on the outer surface of the composite electrode.

When such an electrode is discharged by deletion of the electroactive species, the residual particles of the reactant phase remain as relics in the microstructure. This provides fixed permanent locations for the reaction to take place during following cycles, when the electroactive species again enters the structure. Thus this type of configuration has the additional advantage that it can provide a mechanism for the achievement of true microstructural reversibility.

In order for this concept to be applicable, the matrix and the reactant phases must be thermodynamically stable in contact with each other. One can evaluate this possibility if one has information about the relevant phase diagrams as well as the titration curves of the component binary systems. The stability window of the matrix phase must span the reaction potential of the reactant material. It has been shown that one can evaluate the possibility that these conditions are met from knowledge of the binary titration curves.

Since there is generally a common component, these two binaries can also be treated as a ternary system. Although ternary systems are not explicitly discussed here, it can be simply stated that the two materials must lie at corners of the same constant-potential tie triangle in the relevant isothermal ternary phase diagram in order to not interact. The potential of the tie triangle determines the electrode reaction potential, of course. An additional requirement is that the reactant material must have two phases present in the tie triangle, but the matrix phase only one.

The kinetic requirements for a successful application of this concept are readily understandable. The primary issue is the rate at which the electroactive species can reach the matrix/reactant interfaces. The critical parameter is the chemical diffusion coefficient of the electroactive species in the matrix phase. This can be determined by various techniques, as discussed in later chapters.

The first example that was demonstrated was the use of the phase with the nominal composition $Li_{13}Sn_5$ as the matrix, in conjunction with reactant phases in the lithium-silicon system at temperatures near 400 °C. This is an especially favorable case, due to the very high chemical diffusion coefficient of lithium in the $Li_{13}Sn_5$ phase.

The relation between the potential-composition data for these two systems under equilibrium conditions is shown in Fig. 20.9 [32]. It is seen that the phase $Li_{2.6}Sn$ ($Li_{13}Sn_5$) is stable over a potential range that includes the upper two-phase reconstitution reaction plateau in the lithium–silicon system. Therefore, lithium can react with Si to form the phase $Li_{1.71}Si$ ($Li_{12}Si_7$) inside an all-solid composite electrode containing the $Li_{2.6}Sn$ phase, which acts as a lithium-transporting, but electrochemically inert matrix.

Figure 20.10 shows the relatively small polarization that is observed during the charge and discharge of this electrode, even at relatively high current densities [32].

Fig. 20.9 Composition dependence of the potential in the Li–Sn and Li–Si systems. After [32]

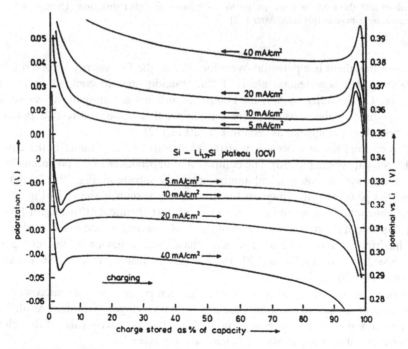

Fig. 20.10 Charge and discharge curves of the Li–Si alloy in the matrix of the electrochemically inert mixed-conducting Li–Sn alloy at different current densities. After [32]

Fig. 20.11 Charge and discharge curves of the Li–Si and Li–Sn composite if the capacity is limited so that the reaction does not go to completion in either direction. There is no large nucleation overshoot in this case. After [32]

It is seen that there is a potential overshoot due to the free energy involved in the nucleation of a new second phase if the reaction goes to completion in each direction. On the other hand, if the composition is not driven quite so far, so that there is some of the reactant phase remaining, this nucleation-related potential overshoot does not appear, as seen in Fig. 20.11 [32].

This concept has also been demonstrated at ambient temperature in the case of the Li–Sn–Cd system [35, 36]. The composition-dependence of the potentials in the two binary systems at ambient temperatures is shown in Fig. 20.12, and the calculated phase stability diagram for this ternary system is shown in Fig. 20.13. It was shown that the phase $Li_{4.4}Sn$, which has fast chemical diffusion for lithium [37], is stable at the potentials of two of the Li–Cd reconstitution reaction plateaus, and therefore can be used as a matrix phase. The behavior of this composite electrode, in which Li reacts with the Cd phases inside of the Li-Sn phase, is shown in Fig. 20.14.

In order to achieve good reversibility, the composite electrode microstructure must have the ability to accommodate any volume changes that might result from the reaction that takes place internally. This can be taken care of by clever microstructural design and alloy fabrication techniques.

Fig. 20.12 Potential versus composition for Li–Sn and Li–Cd systems at ambient temperature. After [36]

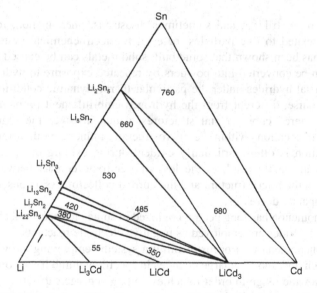

Fig. 20.13 Calculated phase stability diagram for the Li–Cd–Sn system at ambient temperature. *Numbers* are voltages vs. Li. After [36]

20.5.6 Decrepitation

A phenomenon called *decrepitation*, that is also sometimes called *crumbling*, can occur in materials that undergo significant volume changes upon the insertion of guest species. These dimensional changes cause mechanical strain in the microstructure, often resulting in the fracture of particles in an electrode into smaller pieces.

Fig. 20.14 Charge–discharge curve of the Li–Cd system with a fast mixed-conducting phase in the lithium–tin system at ambient temperature. After [37]

This can be a striking, and sometimes disastrous, phenomenon, for it is not specifically related to fine particles, or even to electrochemical systems. As an example, it has been shown that some bulk solid metals can be caused to fracture, and can even be converted into powders by repeated exposure to hydrogen gas if they form metal hydrides under the particular thermodynamic conditions present. This is, of course, different from the hydrogen embrittlement problem in metals with body-centered cubic crystal structures, which involves the segregation of hydrogen to dislocations within the microstructure, influencing their mobility.

Decrepitation is often particularly evident during cycling of electrochemical systems. It can readily result in the loss of electronic contact between reactive constituents in the microstructure and the current collector. As a consequence, the reversible capacity decreases.

This phenomenon has long been recognized in some electrochemical systems in which metal hydrides are employed as negative electrode reactants.

Similar phenomena also occur in lithium systems employing alloy electrodes, some of which undergo very large changes in specific volume if the composition is varied over a wide range in order to achieve a large charge capacity.

Because of its potentially large capacity, a considerable amount of attention has been given recently to the Li–Sn system, which is a fine example of this phenomenon. The phase diagram of the Li–Sn system shows that there are six intermediate phases. The thermodynamic and kinetic properties of the different phases in this system were investigated some time ago at elevated temperatures [37, 38] and also at ambient temperatures [30, 31, 35, 36]. The volume changes that occur in connection with phase changes in this alloy system are large. The phase that forms at the highest lithium concentration, $Li_{4.4}Sn$, has a specific volume that is

283 % of that of pure tin. Thus Li–Sn electrodes swell and shrink, or *breathe*, a lot as lithium is added or deleted.

Observations on metal hydrides that undergo larger volume changes have shown that this process does not continue indefinitely. Instead, it is found that there is a terminal particle size that is characteristic of a particular material. Particles with smaller sizes do not fracture further.

Experiments on lithium alloy electrodes have also shown that the electrochemical cycling behavior is significantly improved if the initial particle size is already very small [39], and it is reasonable to conclude that this is related to the terminal particle size phenomenon.

A theoretical study of the mechanism and the influence of the important parameters related to decrepitation utilized a simple one-dimensional model to calculate the conditions under which fracture will be caused to occur in a two-phase structure due to a specific volume mismatch [40]. This model predicts that there will be a terminal particle size below which further fracture will not occur. The value of this characteristic dimension is material-specific, depending upon two parameters, the magnitude of a strain parameter related to the volume mismatch and the fracture toughness of the lower-specific-volume phase. For the same value of volume mismatch, the tendency to fracture will be reduced and the terminal particle size will be larger the greater the toughness of the material. The results of this model calculation are shown in Fig. 20.15 [40].

The magnitude of the volume change depends upon the amount of lithium that has entered the alloy crystal structure, and is essentially the same for all lithium alloys. This is shown in Fig. 20.16 [41].

Fig. 20.15 Variation of the critical particle size as a function of the dilation strain for several values of the fracture toughness of the phase in tension. After [40]

Fig. 20.16 Relation between volume expansion and the amount of lithium introduced into lithium alloys. After [41]

20.5.7   Modification of the Micro- and Nano-Structure of the Electrode

Some innovated approaches have been employed to ameliorate the decrepitation problem due to the large volume changes inherent in the use of metal alloy and silicon negative electrodes in lithium systems. If that can be done, there is the possibility of a substantial improvement in the electrode capacity.

The general objective is to give the reactant particles room to "breathe," so that they do not impinge upon each other. However, this has to be done so that they are maintained in electrical contact with the current collector system. Thus they cannot be physically isolated.

One interesting direction involves the modification of the shape of the surface upon which thin films of active material are deposited [42]. When the reactant film is dense, the volume changes and related stresses parallel to the surface cause separation from the substrate and loss of electronic contact. But if the surface is rough, there are high spots and low spots that have different local values of current

Electrodeposited Active Material

Fig. 20.17 Schematic drawing of the preferential deposition of reactant material upon protrusions on the substrate surface

Electrolyte

Metal Substrate

Fig. 20.18 Schematic drawing of electrode with a large number of nanowires

density when the active material is electrodeposited. The deposition rate is greater at the higher locations, and less elsewhere. The result is that the active material is mostly deposited at the high spot locations, and grows in a generally columnar shape away from the substrate. This leaves some space between the columnar growths to allow for their volume changes during operation of the electrode. This is illustrated schematically in Fig. 20.17.

Another alternative would be to make separated conductive spots on the surface, perhaps by the use of photolithography, that become the preferred locations for the deposition of reactant. By control of the spot arrangement, the electrodeposition can result in the formation of reactant material with limited impingement, thus allowing more *breathing room* when it undergoes charge and discharge.

It has been recently shown that a very attractive potential solution to this cycling problem is the use of reactant material in the form of nanowires. This is illustrated schematically in Fig. 20.18.

The particular example has been silicon [43]. Such wires can be grown directly upon a metallic substrate, so that they are all in good electronic contact. Because there is some space between the individual wires, they can expand and contract as lithium is added or deleted without the constraints present in either thin film or powdered electrode structures. Experiments showed that such fine wires can attain essentially the theoretical capacity of the Li–Si system.

20.5.8 Formation of Amorphous Products at Ambient Temperatures

This chapter has been primarily concerned with understanding the behavior of negative electrode materials under equilibrium or near-equilibrium conditions, from which the potential and capacity limits can be determined. Actual behavior in real applications always deviates from these limiting values, of course.

It was mentioned in earlier that repeated cycling can cause crystalline materials to become amorphous. The spectrum of materials in which amorphous phases have been formed under these conditions is now quite broad, and includes some materials of potential interest as positive electrode reactants, such as some vanadium-based materials with the general formula RVO_4, which R is Al, Cr, Fe, In, or Y [44].

There have been a number of observations that the operation of negative electrode materials at very high lithium activities can result in the formation of amorphous, rather than crystalline, products. The properties of these amorphous materials are different from those of the corresponding crystalline materials. This is very different from the amorphization of positive electrode materials under cycling conditions.

One example is a group of nitride alloys with structures related to that of Li_3N, which is known to be a fast ionic conductor for lithium, but in which some of the lithium is replaced by a transition metal, such as Co, have been found to become amorphous upon the first insertion of lithium [45–48].

Experimental evidence for the electrochemical amorphization of alloys in the Li–Si, Li–Sn, or Li–Ag systems was presented by Limthongkul [49]. In the latter two cases, this was only a transient phenomenon.

Especially interesting, however, have been experiments that gave evidence for the formation of amorphous silicon during the initial lithiation of a number of silicon-containing precursors, including SiB_3, SiO, $CaSi_2$, and $NiSi_2$ [50–52]. The electrochemical behavior of these materials after the initial lithiation cycle was essentially the same as that found in Si powder that was initially amorphous. There was, however, an appreciable amount of irreversible capacity in the first cycles of these precursors, about 1 mol of Li in the case of SiB_3 and the disilicides, which was evidently due to an irreversible displacement reaction with Li to form one mol of amorphous silicon. In the case of SiO the irreversible capacity amounted to about two mols of Li, which was surely related to the irreversible formation of Li_2O as well as the amorphous silicon.

Some of these materials with amorphous Si are of considerable potential interest as negative electrode reactants in lithium systems, as their charge/discharge curves are in an attractive potential range, they have reasonable kinetics, and their reversible capacities are quite high. The materials with silicon nanowire structure appear to be particularly attractive.

20.6   Protected Lithium Aqueous Electrolyte Systems

20.6.1   Introduction

As can be seen from the discussion thus far in this chapter, the attainment of two major advantages of the use of lithium negative electrodes, the production of electrochemical cells with large voltages and low weight, has involved the use of organic electrolytes. The stability range of aqueous electrolytes is limited by the decomposition of water.

There is another alternative, however, the use of "protected electrodes." This approach has been developed by the firm PolyPlus Battery Co., and involves the use of thin solid electrolyte ion-permeable membranes to separate lithium metal or alloy electrodes from an adjacent aqueous electrolyte. This concept was first publically presented in 2004 [53–55], and is currently being developed for production.

By the use of these solid electrolyte membranes in series with an aqueous electrolyte it is possible to make lithium-based aqueous batteries with output voltages considerably higher than is possible with an aqueous electrolyte alone.

This general scheme involves surrounding the lithium metal negative electrode reactant by a protective 20–50 μm thick lithium-conducting solid electrolyte membrane, and a gel electrolyte interlayer. This membrane is made of a version of "Lisicon," $Li_{1.3}Al_{0.3}Ti_{1.7}(PO_4)_3$, which has an ionic conductivity of 7×10^{-4} S/cm at 25 °C. This is illustrated schematically in Fig. 20.19.

Fig. 20.19 Schematic representation of the PolyPlus cell configuration. After [55]

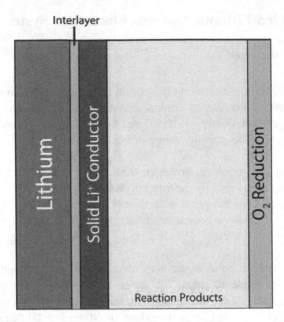

Fig. 20.20 General configuration of protected lithium. After [55]

Since the volume of the enclosed lithium changes during charging or discharging of such an electrode, the electrode shrinks, and the external seal must remain protective as the volume of the contained lithium varies. This is accomplished by the use of a flexible laminate seal material.

The theoretical specific energy of this configuration is high, about 10,000 Wh/kg, assuming that all of the lithium present can be used.

This type of configuration can be employed with several types of lithium-based batteries, Li–water primary cells, primary and secondary Li–air cells, and rechargeable Li–S systems.

The general configuration is illustrated in Fig. 20.20.

As an example, discharge curves at three different current densities for cells with protected lithium metal electrodes in a neutral aqueous electrolyte are shown in Fig. 20.21.

If there were no barrier layer separating them, lithium would react with water to form LiOH and hydrogen.

By protecting the lithium from water, the reaction product of lithium and air is lithium peroxide, Li_2O_2,

$$Li + O_2 = Li_2O_2 \tag{20.1}$$

From the Gibbs free energy of formation of Li_2O_2, it is found that this occurs at a potential of 2.96 V vs. Li in a nonaqueous solvent. The associated specific energy is 3,450 Wh/kg.

Fig. 20.21 Discharge curves of protected lithium electrodes at several current densities: (*1*) 2.0 mA/cm^2, (*2*) 1.0 mA/cm^2, (*3*) 0.5 mA/cm^2 [55]

However, in an aqueous solvent the reaction product is LiOH, from

$$4\,Li + O_2 + 2H_2O = 4LiOH \tag{20.2}$$

which occurs at a potential of 3.45 V vs. Li. The associated specific energy in this case is 3,850 Wh/kg.

References

1. Huggins RA, Elwell D (1977) J Cryst Growth 37:159
2. Wagner C (1954) J Electrochem Soc 101:225
3. Wagner C (1956) J Electrochem Soc 103:571
4. Deublein, Huggins RA (1986) Solid State Ionics. 18/19: 1110
5. von Sacken U, Nodwell E, Dahn JR (1994) Solid State Ionics 69:284
6. Winter M, Moeller K-C, Besenhard JO (2004) In: Nazri G-A, Pistoia G (eds) Lithium Batteries, Science and Technology. Kluwer Academic. p 144
7. Dahn JR, Sleigh AK, Shi H, Way BM, Weydanz WJ, Reimers JN, Zhong Q, von Sacken U (1994) In: Pistoia G (ed) Lithium Batteries. Elsevier. p 1
8. Fredenhagen K, Cadenbach G (1926) Z Anorg Allg Chem 158:249
9. Guerard D, Herold A (1975) Carbon. 13: 337
10. Wertheim GK, Van Attekum P.M.Th.M, Basu S (1980) Solid State Communications. 33:1127
11. Ebert LB (1976) In: Huggins RA (ed) Annual Review of Materials Science, vol. 6. Annual Reviews Inc. p 181
12. Besenhard JO, Fritz HP (1974) J Electroanal Chem 53:329
13. Yazami R, Touzain P (1983) J Power Sources 9:365
14. Basu S (1981) US Patent No 4,304,825, Dec 8 1981

15. Basu S (1983) US Patent No 4,423,125, Dec 27 1983
16. Nagaura T, Tozawa K (1990) Progress in Batteries and Solar Cells. JEC. 9: 209
17. Nagaura T (1991) Progress in Batteries and Solar Cells. JEC 10: 218
18. Franklin RE (1951) Proc Roy Soc (London) A209: 196
19. Yazami R. Personal Communication
20. Daumas N, Herold A (1969) CR Acad Sci C 286:373
21. Zheng T, Liu Y, Fuller EW, Tseng S, von Sacken U, Dahn JR (1995) J Electrochem Soc 142:2581
22. Zheng T, Xue JS, Dahn JR (1996) Chem Mat 8:389
23. Zheng T, McKinnon WR, Dahn JR (1996) J Electrochem Soc 143:2137
24. Yao NP, Heredy LA, Saunders RC (1971) J Electrochem Soc 118: 1039
25. Gay EC et al (1976) J Electrochem Soc 123:1591
26. Lai SC (1976) J Electrochem Soc 123:1196
27. Sharma RA, Seefurth RN (1976) J Electrochem Soc 123:1763
28. Seefurth RN, Sharma RA (1977) J Electrochem Soc 124:1207
29. Ogawa H (1984) Proceedings of 2nd International Meeting on Lithium Batteries. Elsevier Sequoia, Lausanne, p 259
30. Wang J, King P, Huggins RA (1986) Solid State Ionics 20:185
31. Wang J, Raistrick ID, Huggins RA (1986) J Electrochem Soc 133:457
32. Boukamp BA, Lesh GC, Huggins RA (1981) J Electrochem Soc 128:725
33. Boukamp BA, Lesh GC, Huggins RA (1981) In: Venkatasetty HV (ed) Proc. Symp. on Lithium Batteries. Electrochem Soc, p 467
34. Huggins RA, Boukamp BA. US Patent 4,436,796
35. Anani A, Crouch-Baker S Huggins RA (1987) In: Dey AN (ed) Proc. Symp. on Lithium Batteries. Electrochem Soc, p 382
36. Anani A, Crouch-Baker S, Huggins RA (1988) J Electrochem Soc 135:2103
37. Wen CJ, Huggins RA (1980) J Solid State Chem 35:376
38. Wen CJ, Huggins RA (1981) J Electrochem Soc 128:1181
39. Yang J, Winter M, Besenhard JO (1996) Solid State Ionics 90:281
40. Huggins RA, Nix WD (2000) Ionics 6:57
41. Timmons A (2007) PhD Dissertation. Dalhousie University
42. Fujimoto M, Fujitani S, Shima M et al. (2007) US Patent 7,195,842. March 27, 2007
43. Chan CK, Peng H, Liu G, McIlwrath K, Feng Zhang X, Huggins RA, Cui Y (2008) Nat Nanotechnol 3:31
44. Piffard Y, Leroux F, Guyomard D, Mansot J-L, Tournoux M (1997) J Power Sources 68:698
45. Nishijima M, Kagohashi T, Imanishi N, Takeda Y, Yamamoto O, Kondo S (1996) Solid State Ionics 83:107
46. Shodai T, Okada S, Tobishima S-i, Yamaki J-i (1996) Solid State Ionics 86–88:785
47. Nishijima M, Kagohashi T, Takeda Y, Imanishi N, Yamamoto O (1996) 8th International Meeting on Lithium Batteries. p 402
48. Shodai T, Okada S, Tobishima S, Yamaki J (1996) 8th International Meeting on Lithium Batteries. p 404
49. Limthongkul P (2002) PhD Thesis. Mass. Inst. of Tech
50. Klausnitzer B (2000) PhD Thesis. University of Ulm
51. Netz A (2001) PhD Thesis. University of Kiel
52. Netz A, Huggins RA, Weppner W (2002) Presented at 11th International Meeting on Lithium Batteries. Abstract No. 47
53. Visco SJ, Nimon E, Katz B, De Jonghe LC, Chu M-Y (2004) 12th International Meeting on Lithium Batteries, Nara, Japan, June 2004
54. Visco SJ, Nimon E, Katz B, Chu M-Y, De Jonghe LC (2009) Scalable Energy Storage: Beyond Li-Ion. Almaden Inst., San Jose, CA, August 2009
55. Visco SJ, Nimon E, Nimon VY, Katz B, Chu M-Y, De Jonghe LC (2015) International battery Association and Pacific Power Source Symposium. Hilton Waikoloa Village, Hawaii, Jan 2015

Chapter 21
Positive Electrodes in Lithium Systems

21.1 Introduction

Several types of lithium batteries are used in a variety of commercial products, and are produced in very large numbers. According to various reports, the sales volume in 2008 is approximately 10 billion dollars per year, and it is growing rapidly. Most of these products are now used in relatively small electronic devices, but there is also an extremely large potential market if lithium systems can be developed sufficiently to meet the requirements for hybrid, or even plug-in hybrid vehicles.

As might be expected, there is currently a great deal of interest in the possibility of the development of improved lithium batteries in both the scientific and technological communities. An important part of this activity is aimed at the improvement of the positive electrode component of lithium cells, where improvements can have large impacts upon the overall cell performance.

However, before giving attention to some of the details of positive electrodes for use in lithium systems, some comments will be made about the evolution of lithium battery systems in recent years.

Modern advanced battery technology actually began with the discovery of the high ionic conductivity of the solid phase $NaAl_{11}O_{17}$, called sodium beta alumina, by Kummer and coworkers at the Ford Motor Co. laboratory [1]. This led to the realization that ionic transport in solids can actually be very fast, and that it might lead to a variety of new technologies. Shortly thereafter, workers at Ford showed that one can use a highly conducting solid electrolyte to produce an entirely new type of battery, using molten sodium at the negative electrode and a molten solution of sodium in sulfur as the positive electrode, with the sodium-conducting solid electrolyte in between [2].

This attracted a lot of attention, and scientists and engineers from a variety of other fields began to get interested in this area, which is so different from conventional aqueous electrochemistry, in the late 1960s. This concept of a liquid electrode, solid electrolyte (L/S/L) system was quite different from conventional S/L/S

© Springer International Publishing Switzerland 2016
R.A. Huggins, *Energy Storage*, DOI 10.1007/978-3-319-21239-5_21

batteries. The development of the Na/NiCl$_2$ "Zebra" battery system, which has since turned out to be more attractive than the Na/Na$_x$S version, came along somewhat later [3–5]. This is discussed elsewhere in this text.

As might be expected, consideration was soon given to the possibility of analogous lithium systems, for it was recognized that an otherwise equivalent lithium cell should produce higher voltages than a sodium cell. In addition, lithium has a lower weight than sodium, another potential plus. There was a difficulty, however, for no lithium-conducting solid electrolyte was known that had a sufficiently high ionic conductivity to be used for this purpose.

Instead, a concept employing a lithium-conducting molten salt electrolyte, a eutectic solution of LiCl and KCl that has a melting point of 356 °C, seemed to be an attractive alternative. However, because a molten salt electrolyte is a liquid, the electrode materials had to be solids. That is, the lithium system had to be of the S/L/S type.

Elemental lithium could not be used, because of its low melting point. Instead, solid lithium alloys, primarily the Li/Si and Li/Al systems, were investigated [6], as discussed elsewhere in this text.

A number of materials were investigated as positive electrode reactants at that time, with most attention given to the use of either FeS or FeS$_2$. Upon reaction with lithium, these materials undergo *reconstitution reactions*, with the disappearance of the initial phases and the formation of new ones [7].

21.2   Insertion Reaction, Instead of Reconstitution Reaction, Electrodes

An important next step was the introduction of the concept that one can reversibly *insert* lithium into solids to produce electrodes with useful potentials and capacities. This was first demonstrated by Whittingham in 1976, who investigated the addition of lithium to the layer-structured TiS$_2$ to form Li$_x$TiS$_2$, where x went from 0 to 1 [8, 9].

Evidence that this insertion-driven solid solution redox process is quite reversible, even over many cycles, is shown in Fig. 21.1, where the charge and discharge behavior of a Li/TiS$_2$ cell is shown after 76 cycles [10].

Subsequently, the insertion of lithium into a significant number of other materials including V$_2$O$_5$, LiV$_3$O$_8$, and V$_6$O$_{13}$ was investigated in many laboratories. In all of these cases, this involved the assumption that one should assemble a battery with pure lithium negative electrodes and positive electrodes with small amounts of, or no, lithium initially. That is, the electrochemical cell is assembled in the charged state.

The fabrication method generally involved the use of glove boxes and a molten salt or organic liquid electrolyte. This precluded operation at high potentials, and the related oxidizing conditions, as discussed elsewhere.

Fig. 21.1 Charge–discharge behavior of a Li/TiS$_2$ cell after 76 cycles. After ref. [10]

That work involved the study of materials by the addition of lithium, and thus scanned their behavior at potentials lower than about 3 V vs. Li, for this is the starting potential for most electrode materials that are synthesized in air. As lithium is added and the cell is discharged, the potential of the positive electrode goes down toward that of pure lithium.

21.2.1 More Than One Type of Interstitial Site or More Than One Type of Redox Species

The variation of the potential depends upon the distribution of available interstitial places that can be occupied by the Li guest ions. If all sites are not the same in a given crystal structure, the result can be the presence of more than one plateau in the voltage-composition curve. An example of this is the equilibrium titration curve for the insertion of lithium into the V$_2$O$_5$ structure shown in Fig. 21.2 [11].

As seen later, similar voltage/composition behavior can result from the presence of more than one species that can undergo a redox reaction as the amount of inserted lithium is varied.

21.3 Cells Assembled in the Discharged State

On the other hand, if a positive electrode material initially contains lithium, and some or all of the lithium is deleted, the potential goes up, rather than down, as it does upon the insertion of lithium. Therefore, it is possible to have positive electrode materials that react with lithium at potentials above about 3 V, if they already contain lithium, and this lithium can be electrochemically extracted.

Fig. 21.2 Variation of the potential with the concentration of lithium guest species in the V_2O_5 host structure. After ref. [11]

This concept is shown schematically in Fig. 21.3 for a hypothetical material that is *amphoteric*, and can react at both high and low potentials.

This approach, involving the use of materials in which lithium is already present, was first demonstrated in Prof. Goodenough's laboratory in Oxford. The first examples of materials initially containing with lithium, and electrochemically deleting lithium from them, was the work on $Li_{1-x}CoO_2$ [12] and $Li_{1-x}NiO_2$ [13] in 1980. They showed that it is possible in this way to obtain high reaction potentials, up to over 4 V.

It was not attractive to use such materials in cells with metallic Li negative electrodes, however, and this approach did not attract any substantial interest at that time. This abruptly changed as the result of the surprise development by SONY of a lithium battery containing a carbon negative electrode and a $LiCoO_2$ positive electrode that became commercially available in 1990. These cells were initially assembled in the discharged state. They were activated by charging, whereby lithium left the positive electrode material, raising its potential, and moved to the carbon negative electrode, whose potential was concurrently reduced.

This cell can be represented as

$$Li_xC/\text{organic solvent electrolyte}/Li_{1-x}CoO_2$$

and the cell reaction can be written as

$$C + LiCoO_2 = Li_xC + Li_{1-x}CoO_2 \tag{21.1}$$

This general type of cell and related reaction are most common in commercial cells at the present time.

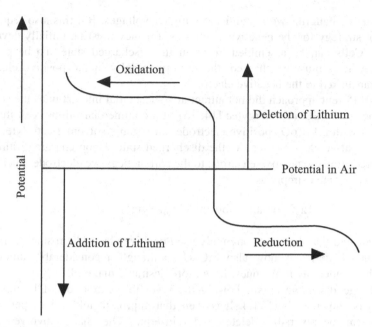

Fig. 21.3 Schematic representation of the behavior of a material that is amphoteric, i.e., that can be both electrochemically oxidized at high potentials by the deletion of lithium, and electrochemically reduced at lower potentials by the addition of lithium

21.4   Solid Positive Electrodes in Lithium Systems

21.4.1   Introduction

In almost every case, the materials that are now used as positive electrode reactants in reversible lithium batteries operate by the use of insertion reactions. This general concept has been discussed several times in this text already. The early ambient temperature lithium battery developments were based upon the observation that lithium could be readily inserted into solids with crystal structures containing available interstitial space. A number of such materials were found, the most notable being TiS_2 and V_6O_{13}. These cells utilized elemental lithium, or lithium alloys, in the negative electrode.

Although precautions had to be taken in preparing and handling the negative materials, due to their propensity to oxidize, the positive electrode materials were typically stable in air.

As the insertion of lithium causes the potential to decrease, and those positive electrodes necessarily operated at potentials lower than that of air, the voltage of such cells was limited to about 3 V.

The shift in concept to the use of air-stable positive electrode materials that already contained lithium, and their operation by the deletion of lithium, led to the possibility of batteries with significantly higher voltages. But this also required a different strategy for the negative electrodes, for they must be initially devoid of lithium. Cells can be assembled in air in the discharged state. To be put into operation, they must be charged, the lithium initially in the positive electrode being transferred to the negative electrode.

This different approach did not attract any substantial interest until the surprise development by SONY Energytec [14, 15] of a commercial lithium cell that was produced with a $LiCoO_2$ positive electrode, an organic solvent electrolyte, and a carbon negative electrode, i.e., in the discharged state. Upon charging, lithium is transferred from the positive electrode to the carbon negative electrode. Such a cell can be represented simply as

$$Li_xC/organic \ solvent \ electrolyte/Li_{1-x}CoO_2.$$

It is interested that the most commonly used positive electrode in small consumer electronics batteries is now also $LiCoO_2$, although a considerable amount of research is underway in the quest for a more desirable material.

A charge–discharge curve showing the reversible extraction of lithium from $LiCoO_2$ is shown in Fig. 21.4. It is seen that approximately 0.5 Li per mol of $LiCoO_2$ can be reversibly deleted and reinserted. The charge involved in the transfer of lithium ions is balanced by the Co^{3+}/Co^{4+} redox reaction. This process cannot go further, because the layered crystal structure becomes unstable, and there is a transformation into another structure.

Quite a number of materials are now known from which it is possible to delete lithium at high potentials. Some of these are described briefly below, but it is important to realize that this is a very active research area at the present time, and no such discussion can be expected to be complete.

Fig. 21.4 Charge–discharge behavior of Li_xCoO_2

There are a number of interesting materials that have a *face-centered cubic packing* of oxide ions, including both those with the *spinel* structure, e.g., $LiMn_2O_4$, variants containing more than one redox ion, and those with *ordered cation distributions*, which are often described as having *layered structures* (e.g., $LiCoO_2$ and $LiNiO_2$). There also are materials with *hexagonal close packed oxide ion packing*, including some with ordered *olivine-related structures* (e.g., $LiFePO_4$).

In addition, there are a number of interesting materials that have more open crystal structures, sometimes called *framework*, or *skeleton* structures. These are sometimes described as containing *polyanions*. Examples are some sulfates, molybdates, tungstates, and phosphates, as well as Nasicon, and Nasicon-related materials (e.g., $Li_3V_2(PO_4)_3$ and $LiFe_2(SO_4)_3$). In these materials lithium ions can occupy more than one type of interstitial position. Especially interesting are materials with more than one type of polyanion. In some cases the reaction potentials are related to the potentials of the redox reactions of ions in octahedral sites, which are influenced by the charge and crystallographic location of other highly charged ions on tetrahedral sites in their vicinity.

Since the reaction potentials of these positive electrode materials are related to the redox reactions that take place within them, consideration should be given to this matter.

The common values of the formal valence of a number of redox species in solids are given in Table 21.1. In some cases the capacity of a material can be enhanced by the use of more than one redox reaction. In such cases, an issue is whether this can be done without a major change in the crystal structure.

An example of the reaction of lithium with an electrode material containing two redox ions, a Li-Mn-Fe phosphate with the olivine structure, shown in Fig. 21.5 [16].

Not all redox reactions are of practical value in electrode materials, and in some cases, their potentials depend upon their environments within the crystal structure. Some experimental data are presented in Table 21.2.

When lithium or other charged mobile guest ions are inserted into the crystal structure, their electrostatic charge is balanced by a change in the oxidation state of one or more of the redox ions contained in the structure of the host material. The reaction potential of the material is determined by the potential at which this oxidation or reduction of these ions occurs in the host material. In some cases, this redox potential is rather narrowly defined, whereas in others redox occurs over

Table 21.1 Common valences of redox ions in solids

Element	Valences	Valence range	Comments
Ti	2,3,4	2	
V	2,3,4,5	3	
Cr	2,3,6	1	6 is poisonous
Mn	2,3,4,6,7	2	6,7 usable ?
Fe	2,3	1	
Co	2,3	1	
Ni	2,3,4	2	
Cu	1,2	1	

Fig. 21.5 Charge–discharge curve of the reaction of lithium with an example of a double-cation olivine material. After ref. [16]

Table 21.2 Potentials of redox reactions in a number of host materials/volts vs. lithium

Redox System	Nasicon framework phosphates	Layered close-packed oxides	Cubic close-packed spinels	Hexagonal close-packed olivines
V^{2+}/V^{3+}	1.70–1.75			
Nb^{3+}/Nb^{4+}	1.7–1.8			
Nb^{4+}/Nb^{5+}	2.2–2.5			
Ti^{3+}/Ti^{4+}	2.5–2.7		1.6	
Fe/Fe^{2+}	2.65			
Fe^{2+}/Fe^{3+}	2.7–3.0			3.4
V^{3+}/V^{4+}	3.7–3.8			
Mn^{2+}/Mn^{3+}		4.0	1.7	>4.3
Co^{2+}/Co^{3+}		4.2	1.85	>4.3
Ni^{2+}/Ni^{3+}		4.8		>4.3
Mn^{3+}/Mn^{4+}			4.0	
Fe^{3+}/Fe^{4+}	4.4			
Co^{3+}/Co^{4+}			5.0	

a range of potential, due to the variation of the configurational entropy with the guest species concentration, as well as the site distribution.

21.4.2   Influence of the Crystallographic Environment on the Potential

It has been shown that the environment in which a given redox reaction takes place can affect the value of its potential. This matter has been investigated by comparing the potentials of the same redox reactions in a number of oxides with different

Redox Ion   High Valence Host Ion

Oxygen Ion

Displacement of center
of electron cloud

Fig. 21.6 Schematic representation of the displacement of the electron cloud around an oxide ion by the charge upon nearby cations

polyanions, but with the same type of crystal structure. Some of the early references to this topic are listed here [17, 18].

These materials all have crystal structures in which the redox ion is octahedrally surrounded by oxide ions, and the oxide ions also have cations with a different charge in tetrahedral environments on the other side. The electron clouds around the oxide ions are displaced by the presence of adjacent cations with different charges. This is shown schematically in Fig. 21.6.

One of the first cathode materials with a polyanion structure to be investigated was $Fe_2(SO_4)_3$. It can apparently reversibly incorporate up to 2 Li per formula unit, has a very flat discharge curve, indicating a reconstitution reaction, at 3.6 V vs. Li/Li^+ [19, 20].

21.4.3 Oxides with Structures in Which the Oxygen Anions are in a Face-Centered Cubic Array

21.4.3.1 Materials with Layered Structures

As mentioned above, the positive electrode reactant in the SONY cells was Li_xCoO_2, whose properties were first investigated at Oxford [8]. It can be synthesized so that it is stable in air, with $x = 1$. Its crystal structure can be described in terms of a close-packed face-centered cubic arrangement of oxide ions, with the Li^+

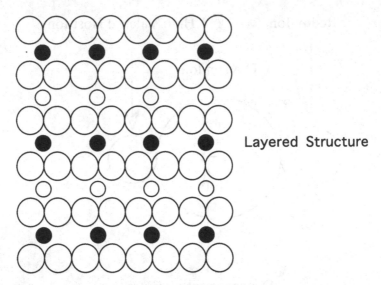

Layered Structure

Fig. 21.7 Simplified schematic drawing of a layered structure in which there is alternate occupation of the cation layers between the close-packed oxide ion layers. The solid and open small circles represent two different types of cations. The larger circles are oxide ions

and Co^{3+} cations occupying octahedrally coordinated positions in between layers of oxide ions. The cation positions are ordered such that the lithium ions and the transition metal ions occupy alternate layers between close-packed (111) planes of oxide ions. As a result, these materials are described as having layered, rather than simple cubic, structures. This is shown schematically in Fig. 21.7. However, there is a slight distortion of the cubic oxide stacking because of the difference between the bonding of the monovalent and trivalent cations.

When lithium ions move between octahedral sites within the layers of this structure they must go through nearby tetrahedral sites that lie along the jump path.

Li_xCoO_2 can be cycled many times over the range $1 > x > 0.5$, but there is a change in the structure and a loss of capacity if more $Li+$ ions are deleted.

Because it has an inherently lower cost and is somewhat less poisonous, it would be preferable to use $LiNiO_2$ instead of $LiCoO_2$. However, it has been found that Li_xNiO_2 is difficult to prepare with the right stoichiometry, as there is a tendency for nickel ions to reside on the lithium layers. This results in a loss of capacity. It was also found that $LiNiO_2$ readily loses oxygen at high potentials, destroying its layer structure, and tending to lead to safety problems because of exothermic reaction with the organic solvent electrolyte.

There have been a number of investigations of the modification of Li_xNiO_2 by the substitution of other cations for some of the Ni^{3+} ions. It has been found that the replacement of 20–30 % of the Ni^{3+} by Co^{3+} ions will impart sufficient stability [21]. Other aliovalent alternatives have also been explored, including the introduction of Mg^{2+} or Ti^{4+} ions.

In the case of $LiMnO_2$, that also has the alpha $NaFeO_2$ structure, it has been found that if more than 50 % of the lithium ions are removed during charging, conversion to the spinel structure tends to occur. About 25 % of the Mn ions move from octahedral sites in their normal layers into the alkali metal layers, and lithium is displaced into tetrahedral sites [22]. But this conversion to the spinel structure can be avoided by the replacement of half of the Mn ions by chromium [23]. In this case, the capacity (190 mAh/g) is greater than can be accounted for by a single redox reaction, such as Mn^{3+} to Mn^{4+}. This implies that the chromium ions are involved, whose oxidation state can go from Cr^{3+} to Cr^{6+}. Unfortunately, the use of chromium is not considered desirable because of the toxicity of Cr^{6+}.

The replacement of some of the manganese ions in $LiMnO_2$ by several other ions in order to prevent the conversion to the spinel structure has been investigated [24].

A number of other layer-structure materials have also been investigated. Some of them contain two or more transition metal ions at fixed ratios, often including Ni, Mn, Co, and Al. In some cases, there is evidence of ordered structures at specific compositions and well-defined reaction plateaus, at least under equilibrium or near-equilibrium conditions. This indicates reconstitution reactions between adjacent phases.

There have been several investigations of layer phases with manganese and other transition metals present. A number of these, including $LiMn_{1-y}Co_yO$, have been found to not be interesting, as they convert to the spinel structure rather readily.

However, the manganese–nickel materials, $Li_xMn_{0.5}Ni_{0.5}O_2$ and related compositions, have been found to have very good electrochemical properties, with indications of a solid solution insertion reaction in the potential range 3.5 to 4.5 V vs. Li [25–28]. It appears as though the redox reaction involves a change from Ni^{2+} to Ni^{4+}, whereas the Mn remains as Mn^{4+}. This means that there is no problem with Jahn–Teller distortions, which are related to the presence of Mn^{3+}. The stability of the manganese ions is apparently useful in stabilizing this structure.

At higher manganese concentrations these materials adopt the spinel structure and apparently react by reconstitution reactions, as discussed later in this chapter.

Success with this cation combination apparently led to considerations of compositions containing three cations, such as Mn, Ni, and Co. One of these is $LiMn_{1/3}Ni_{1/3}Co_{1/3}O_2$ [29, 30]. The presence of the cobalt ions evidently stabilizes the layer structure against conversion to the spinel structure. These materials have good electrochemical behavior, and have been studied in many laboratories, but one concern is that they evidently have limited electronic conductivity.

In these materials, as well, when they are fully lithiated, the nickel is evidently predominantly divalent, the cobalt trivalent, and the manganese tetravalent. Thus the major electrochemically active species is nickel, with the cobalt playing an active role only at high potentials. The manganese evidently does not play an active role. It does reduce the overall cost, however.

An extensive discussion of the various approaches to the optimization of the layer structure materials can be found in [31].

21.4.3.2   Materials with the Spinel Structure

The spinel class of materials, with the nominal formula AB_2O_4, has a related structure that also has a close-packed face-centered cubic arrangement of oxide ions. Although this structure is generally pictured in cubic coordinates, it also has parallel layers of oxide ions on (111) planes, and there are both octahedrally-coordinated sites and tetrahedrally-coordinated sites between the oxide ion planes. The number of octahedral sites is equal to the number of oxide ions, but there are twice as many tetrahedral sites. The octahedral sites reside in a plane intermediate between every two oxide ion planes. The tetrahedral sites are in parallel planes slightly above and below the octahedral site planes between the oxide ion planes.

In *normal spinels*, the A (typically monovalent or divalent) cations occupy 1/8 of the available tetrahedral sites, and the B (typically trivalent or quadrivalent) cations 1/2 of the B sites. In *inverse spinels*, the distribution is reversed.

The spinel structure is quite common in nature, indicating a large degree of stability. As mentioned above, there is a tendency for the materials with the layer structures to convert to the closely related spinel structure. This structure is shown schematically in Fig. 21.8.

A wide range of materials with different A and B ions can have this structure, and some of them are quite interesting for use in lithium systems. An especially important example is $Li_xMn_2O_4$. There can be both lithium insertion and deletion from the nominal composition in which $x = 1$. This material has about 10 % less capacity than Li_xCoO_2, but it has somewhat better kinetics and does not have as great a tendency to evolve oxygen.

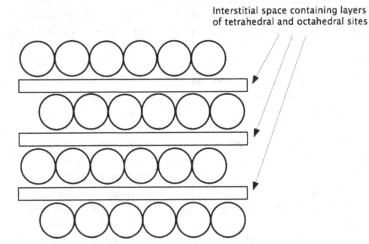

Interstitial space containing layers of tetrahedral and octahedral sites

Fig. 21.8 Schematic drawing of the spinel structure in which the cations between the close-packed (111) planes of oxide ions are distributed among both tetrahedral and octahedral sites

Fig. 21.9 Charge–discharge behavior of $Li_xMn_2O_4$ [32]

Li$_x$Mn$_2$O$_4$ can be readily synthesized with x equal to unity, and this composition can be used as a positive electrode reactant in lithium batteries. A typical charge–discharge curve is shown in Fig. 21.9.

It is seen that there are two plateaus. This is related to an ordering reaction of the lithium ions on the tetrahedral sites when x is about 0.5.

Although the Li$_x$Mn$_2$O$_4$ system, first investigated by Thackeray et al. [32, 33], has the inherent advantages of low cost, good kinetics, and being nonpoisonous, it has been found to have some problems that result in a gradual loss of capacity [34]. Thorough discussions of early work to optimize this material can be found in refs. [35, 36].

One of the problems with this material is the loss of Mn^{2+} into the organic solvent electrolyte as the result of a disproportionation reaction when the potential is low near the end of discharge.

$$2Mn^{3+} = Mn^{4+} + Mn^{2+}. \tag{21.2}$$

These ions travel to the carbon negative electrode, with the result that a layer of manganese metal is deposited that act to block lithium ion transport.

Another problem that can occur at low potentials is the local onset of Jahn–Teller distortion that can cause mechanical damage to the crystal structure. On the other hand, if the electrode potential becomes too high as the result of the extraction of too much lithium, oxygen can escape and react with the organic solvent electrolyte.

These problems are reduced by modification of the composition of the electrode by the presence of additional lithium and a reduction of the manganese [37]. This increase in stability comes at the expense of the capacity. Although the theoretical capacity of $LiMn_2O_4$ is 148 mAh/g, this modification results in a capacity of only 128 mAh/g.

There have also been a number of investigations in which various other cations have been substituted for part of the manganese ions. But in order to avoid the loss of a substantial amount of the normal capacity, it was generally thought at that time that the extent of this substitution must be limited to relatively small concentrations.

At that time the tendency was to perform experiments only up to a voltage about 4.2 V above the Li/Li$^+$ potential, as had been done for safety reasons when using Li_xCoO_2. But it was soon shown that it is possible to reach potentials up to 5.4 V vs. Li/Li$^+$ using some organic solvent electrolytes [38, 39].

Experiments on the substitution of some of the Mn^{2+} ions in Li_xMnO_2 by Cr^{3+} ions [40] showed that the capacity upon the 3.8 V plateau was decreased in proportion to the concentration of the replaced Mn ions. But when the potential was raised to higher values, it was found that this missing capacity at about 4 V reappeared at potentials about 4.9 V that was obviously related to the oxidation of the Cr ions that had replaced the manganese ions in the structure. This particular option, replacing inexpensive and nontoxic manganese with more expensive and toxic chromium is, of course, not favorable.

In both the cases of chromium substitution and nickel substitution the sum of the capacities of the higher potential plateau and the lower plateau are constant. This implies that there is a one-to-one substitution, and thus that the oxidation that occurs in connection with the chromium and nickel ions is a one-electron process. This is in contradiction to the normal expectation that these ions undergo a 3-electron (Cr^{3+} to Cr^{6+}) or a two electron (Ni^{2+} to Ni^{4+}) oxidation step.

Another example is work on lithium manganese spinels in which some of the manganese ions have been replaced by copper ions. One of these is $LiCu_xMn_{2-x}O_4$ [41–43]. Investigations of materials in which up to a quarter of the manganese ions are replaced by Cu ions have shown that a second plateau appears at 4.8 to 5.0 V vs. Li/Li$^+$ that is due to a Cu^{2+}/Cu^{3+} reaction, in addition to the normal behavior of the Li-Mn spinel in the range 3.9 to 4.3 V vs. Li/Li$^+$ that is related to the Mn^{3+}/Mn^{4+} reaction . Data for this case are shown in Fig. 21.10 [41]. Unfortunately, the overall capacity seems to be reduced when there is a substantial amount of copper present in this material [42]. When x is 0.5 the total capacity is about 70 mAh/g, with only about 25 mAh/g obtainable in the higher potential region.

The redox potentials that are observed when a number of elements are substituted into lithium manganese spinel structure materials are shown in Fig. 21.11 [44].

An especially interesting example is the spinel structure material with a composition $Li_xNi_{0.5}Mn_{1.5}O_4$. Its electrochemical behavior is different from the others, showing evidence of two reconstitution reactions, rather than solid solution behavior [45].

Fig. 21.10   Potential-composition curves for LiCu$_{0.5}$Mn$_{1.5}$O4. After ref. [40]

Fig. 21.11   Potential ranges, vs. Li, of redox potentials found as the result of the introduction of a number of cations into lithium manganese spinels. The operating potential range of lithium manganese spinel itself is also shown. After ref. [43]

The constant potential charge–discharge curve for this material in the high potential range is shown in Fig. 21.12 [45]. Careful coulometric titration experiments showed that this apparent plateau is actually composed of two reactions with a potential separation of only 20 mV.

In addition to this high potential reaction, this material also has a reconstitution reaction with a capacity of 1 Li per mol at 2.8 V vs. Li, as well as further lithium

Fig. 21.12 Charge–discharge curves for $Li_xNi_{0.5}Mn_{1.5}O_4$. After ref. [44]

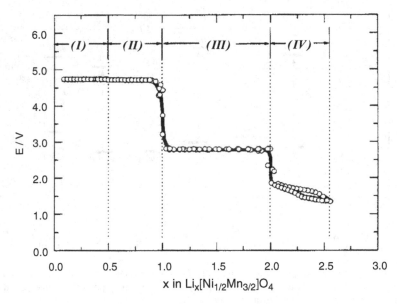

Fig. 21.13 Coulometric titration curve for the reaction of lithium with $Li_xNi_{0.5}Mn_{1.5}O_4$. After ref. [44]

uptake via a single phase reaction below 1.9 V. These features are shown in Fig. 21.13 [45]. It is not fully known what redox reactions are involved in this behavior, but it is believed that those at the higher potentials relate to nickel, and the lower ones to manganese.

21.4.3.3 Lower Potential Spinel Materials with Reconstitution Reactions

Whereas this discussion here has centered on lithium-containing materials that exhibit high potential reactions, and thus are useful as reactants in the positive electrode, attention should also be given to another related spinel structure material that has a reconstitution reaction at 1.55 V vs. Li [46, 47]. This is $Li_{1.33}Ti_{1.67}O_4$, that can also be written as $Li_x[Li_{0.33}Ti_{1.67}O_4]$ for some of the lithium ions share the octahedral sites in an ordered arrangement with the titanium ions. It also sometimes appears in the literature as $Li_4Ti_5O_{12}$.

This spinel structure material is unusual in that there is essentially no change in the lattice dimensions with variation of the amount of lithium in the crystal structure, and it has been described as undergoing a *zero-strain insertion reaction* [48]. This is an advantage in that there is almost no volume change-related hysteresis, leading to very good reversibility upon cycling.

As was mentioned earlier, this material can also be used on the negative electrode side of a battery. Although there is a substantial voltage loss compared to the use of carbons, the good kinetic behavior can make this option attractive for high power applications, where the lithium-carbons can be dangerous because their reaction potential is rather close to that of elemental lithium.

A charge–discharge curve for this interesting material is shown in Fig. 21.14.

Fig. 21.14 Charge–discharge curve for $Li_4Ti_5O_{12}$. After ref. [46]

21.4.4 Materials in Which the Oxide Ions are in a Close-Packed Hexagonal Array

Whereas in the spinel and the related layered materials such as Li_xCoO_2, Li_xNiO_2, and Li_xMnO_2 the oxide ions are in a cubic close-packed array, there are also many materials in which the oxide ions are in a hexagonal close-packed configuration. Some of these are currently of great interest for use as positive electrode reactants in lithium batteries, but are generally described as having *framework structures*. They are sometimes also called "scaffold," "skeleton," "network," or "polyanion" structures.

21.4.4.1 The Nasicon Structure

The *Nasicon structure* first attracted attention within the solid state ionics community because some materials with this structure were found to be very good solid electrolytes for sodium ions. One such composition was $Na_3Zr_2Si_2PO_{12}$.

This structure has monoclinic symmetry, and can be considered as consisting of MO_6 octahedra sharing corner oxide ions with adjacent XO_4 tetrahedra. Each octahedron is surrounded by six tetrahedral, and each tetrahedron by four octahedra. These are assembled in as a three-dimensional network of M_2X_3 groups. Between these units is three-dimensional interconnected interstitial space, through which small cations can readily move. This structure is shown schematically in Fig. 21.15.

Fig. 21.15 Schematic representation of the Nasicon structure

Fig. 21.16 Charge–discharge curve for $Li_3V_2(PO_4)_3$, that has the Nasicon structure. After ref. [50]

Unfortunately, Nasicon was found to not be thermodynamically stable versus elemental sodium, so that it did not find use as an electrolyte in the Na/Na_xS and $Na/NiCl_2$ Zebra cells, that are discussed elsewhere in this text, at that time.

However, by using M cations whose ionic charge can be varied, it is possible to make materials with this same structure that undergo redox reactions upon the insertion or deletion of lithium within the interstitial space. The result is that although Nasicon materials may not be useful for the function for which they were first investigated, they may be found to be useful for a different type of application.

As mentioned earlier, it has been found that the identity of the X ions in the tetrahedral parts of the structure influences the redox potential of the M ions in the adjacent octahedra [17, 49]. This has been called an *induction effect*.

A number of compositions with this structure have been investigated for their potential use as positive electrode reactants in lithium cells [17, 49–51]. An example is $Li_3V_2(PO_4)_3$, whose potential vs. composition data are shown in Fig. 21.16 [51]. The related differential capacity plot is shown in Fig. 21.17.

It is seen that the titration curve shows three two-phase plateaus, corresponding to the extraction of two of the lithium ions in the initial structure. The first two plateaus indicate that there are two slightly different configurations for one of the two lithium ions. The potential must be increased substantially, to over 4 V, for the deletion of the second. Experiments showed that it is possible to extract the third lithium from this material by going up to about 5 V, but that this process is not readily reversible, whereas the insertion/extraction of the first two lithium ions is highly reversible.

These phosphate materials all show significantly more thermal stability than is found in some of the other, e.g., layer- and spinel-structure, positive electrode reactants. This is becoming ever more important as concerns about the safety aspects of high-energy batteries mounts.

Fig. 21.17 Differential capacity plot corresponding to the charge–discharge data for $Li_3V_2(PO_4)_3$ shown in Fig. 21.10. After ref. [50]

21.4.4.2   Materials with the Olivine Structure

Another group of materials that have a hexagonal stacking of oxide ions are those with the Olivine structure. These materials have caused a great deal of excitement, as well as controversy, in the research community since it was first shown that they can reversibly react with lithium at ambient temperature [52]. The most interesting of these materials is $LiFePO_4$, that has the obvious advantage of being composed of safe and inexpensive materials.

The olivine structure can be described as M_2XO_4, in which the M ions are in half of the available sites of the close-packed hexagonal oxygen array. The more highly charged X ions occupy one eighth of the tetrahedral sites. Thus it is a hexagonal analog of the cubic spinel structure discussed earlier. However, unlike spinel, the two octahedral sites in olivine are crystallographically distinct, and have different sizes. This results in a preferential ordering if there are two M ions of different sizes and/or charges. Thus $LiFePO_4$ and related materials containing lithium and transition metal cations have an ordered cation distribution. The M_1 sites containing lithium are in linear chains of edge-shared octahedral that are parallel to the c-axis in the hexagonal structure in alternate a–c planes. The other (M_2) sites are in a zig-zag arrangement of corner-shared octahedral parallel to the c-axis in the other a–c planes. The result is that lithium transport is highly directional in this structure.

Experiments showed that extraction of lithium did not readily occur with olivines containing the Mn, Co, or Ni, but proceeded readily in the case of $LiFePO_4$. Deletion of lithium from $LiFePO_4$ occurs by a reconstitution reaction with a moving two-phase interface in which $FePO_4$ is formed at a potential of 3.43 V vs. Li. Although the initial experiments only showed the electrochemical removal of about 0.6 Li ions per mol, subsequent work has shown that greater values can be attained.

A reaction with one lithium ion per mol would give a theoretical specific capacity of 170 mAh/g, which is higher than that obtained with $LiCoO_2$. It has been found that the extraction/insertion of lithium in this material can be quite reversible over many cycles.

These phases have the mineralogical names triphylite and heterosite, although the latter was given to a mineral that also contains manganese. Although this reaction potential is significantly lower than those of many of the materials discussed earlier in this chapter, other properties of this class of materials makes them attractive for application in lithium-ion cells. There is active commercialization activity, as well as a measure of conflict over various related patent matters.

These materials do not tend to lose oxygen and react with the organic solvent electrolyte nearly so much as the layer structure materials, and are they are evidently much safer at elevated temperatures. As a result, they are being considered for larger format applications, such as in vehicles or load leveling, where there are safety questions with some of the other positive electrode reactant materials.

It appeared that the low electronic conductivity of these materials might limit their application, so work was undertaken in a number of laboratories aimed at the development of two-phase microstructures in which electronic conduction within the electrode structure could be enhanced by the presence of an electronic conduction, such as carbon [53]. Various versions of this process quickly became competitive and proprietary.

A different approach is to dope the material with highly charged (supervalent) metal ions, such as niobium, that could replace some of the lithium ions on the small M_1 sites in the structure, increasing the n-type electronic conductivity [54]. On the other hand, experimental evidence seems to indicate that the electronic conduction in the doped Li_xFePO_4 is p-type, not n-type [54, 55]. This could be possible if the cation doping is accompanied by a deficiency of lithium.

This interpretation has been challenged, however, based upon observations of the presence of a highly conductive iron phosphide phase, Fe_2P under certain conditions [56]. Subsequent studies of phase equilibria in the Li-Fe-P-O quaternary system [57] seem to contradict that interpretation.

Regardless of the interpretation, it has been found that the apparent electronic conductivity in these Li_xFePO_4 materials can be increased by a factor of 10^8, reaching values above 10^{-2} S cm^{-1} in this manner. These are higher than those found in some of the other positive electrode reactants, such as $LiCoO_2$ (10^{-3} S cm^{-1}) and $LiMn_2O_4$ (2 to 5×10^{-5} S cm^{-1}).

An interesting observation is that very fine scale cation-doped Li_xFePO4 has a restricted range of composition at which the two phases "LiFePO4" and "FePO4" are in equilibrium, compared to undoped and larger particle-size material [58]. Thus there is more solid solubility in each of the two end phases. This may play an important role in their increased kinetics, for in order for the moving interface reconstitution phase transformation involved in the operation of the electrode to proceed there must be diffusion of lithium through the outer phase to the interface. The rate of diffusional transport is proportional to the concentration gradient. A wider compositional range allows a greater concentration gradient, and thus faster kinetics.

These materials have been found to be able to react with lithium at very high power levels, greater than those that are typical of common hydride/H_xNiO_2 cells, and commercial applications of this material are being vigorously pursued.

21.4.5   Materials Containing Fluoride Ions

Another interesting variant has also been explored somewhat. This involves the replacement of some of the oxide ions in lithium transition metal oxides by fluoride ions. An example of this is the lithium vanadium fluorophosphate $LiVPO_4F$, that was found to have a triclinic structure analagous to the mineral tavorite, $LiFePO_4 \cdot OH$ [59]. As in the case of the Nasicon materials mentioned earlier, the relevant redox reaction in this material involves the V^{3+}/V^{4+} couple. The charge–discharge behavior of this material is shown in Fig. 21.18 [60], and the related differential capacity results are presented in Fig. 21.19.

21.4.6   Hybrid Ion Cells

An additional variant involves the use of positive electrode reactants that contain other mobile cations. An example of this were the reports of the use of $Na_3V_2(PO_4)_3F_3$ as the positive electrode reactant and either graphite [61] or $Li_{4/3}Ti_{5/3}O_4$ [62] as the

Fig. 21.18 Charge–discharge behavior of $LiVPO_4F$. After ref. [60]

Fig. 21.19 Differential capacity plot corresponding to the charge–discharge data for LiVPO$_4$F shown in Fig. 20.18. After ref. [60]

negative reactant in lithium-conducting electrolyte cells. It appears as though the mobile insertion species in the positive electrodes gradually shifts from Na$^+$ to Li$^+$. Consideration of this type of mixed-ion materials may lead to a number of interesting new materials.

21.4.7   Amorphization

It is pointed out in Chap. 13 that crystal structures can become amorphous as the result of multiple insertion/extraction reactions. A simple explanation for this phenomenon can be based upon the dimensional changes that accompany the variation in the composition. These dimensional changes are typically not uniform throughout the material, so quite significant local shear stresses can result that disturb the regularity of the atomic arrangements in the crystal structure, resulting in regions with amorphous structures. The degree of amorphization should increase with cycling, as is found experimentally.

There is also another possible cause of this effect that has to do with the particle size. As particles become very small, a significant fraction of their atoms actually reside on the surface. Thus the surface energy present becomes a more significant fraction of the total Gibbs free energy. Amorphous structures tend to have lower values of surface energy than their crystalline counterparts. As a result, it is easy to understand that there will be an increasing tendency for amorphization as particles become smaller.

21.4.8 The Oxygen Evolution Problem

It is generally considered that a high cell voltage is desirable, and the more the better, since the energy stored is proportional to the voltage, and the power is proportional to the square of the voltage. However, there are other matters to consider, as well. One of these is the evolution of oxygen from a number of the higher potential positive electrode materials.

There is a direct relationship between the electrical potential and the chemical potential of oxygen in materials containing lithium. In this connection it is useful to remember that the chemical potential has been called the *escaping tendency* in the well-known book on thermodynamics by Pitzer and Brewer.

Experiments have shown that a number of the high potential positive electrode reactant materials lose oxygen into the electrolyte. It is also generally thought that the presence of oxygen in the organic solvent electrolytes is related to thermal runaway and the safety problems that are sometimes encountered in lithium cells. As example of experimental measurements that clearly show oxygen evolution is shown in Fig. 21.20.

The relationship between the potential and the chemical potential of oxygen in electrode materials was investigated a number of years ago, but under conditions that are somewhat different from those in current ambient temperature lithium cells. Nevertheless, the principles are the same, and thus it is useful to review what was found about the thermodynamics of such systems at that time [64].

Fig. 21.20 The influence of temperature upon the derivative of the sample weight versus temperature for three different layer structure materials. After ref. [62]

As discussed in this chapter, many of the positive electrode materials in lithium batteries are ternary lithium transition metal oxides. Since there are three kinds of atoms, i.e., three components, present, compositions in these systems can be represented on an Isothermal Gibbs Triangle. As discussed in Chaps. 10 and 12, the Gibbs Phase Rule can be written as

$$F = C - P + 2 \tag{21.3}$$

where F is the number of degrees of freedom, C the number of components, and P the number of phases present. At constant temperature and overall pressure $F = 0$ when $C = P = 3$. This means that all of the intensive variables have fixed values when three phases are present in such three-component systems. Since the electrical potential is an intensive property, this means that the potential has the same value, independent of how much of each of the three phases is present.

It has already been pointed out that the *isothermal phase stability diagram*, an approximation of the Gibbs Triangle in which the phases are treated as though they have fixed, and very narrow, compositions, is a very useful thinking tool to use when considering ternary materials.

The compositions of all of the relevant phases are located on the triangular coordinates, and the possible two-phase tie lines identified. Tie lines cannot cross, and the stable ones can readily be determined from the energy balance of the appropriate reactions. The stable tie lines divide the total triangle into sub-triangles that have two phases at the ends of the tie lines along their boundaries. There are different amounts of the three corner phases at different locations inside the sub-triangles. All of these compositions have the same values of the intensive properties, including the electrical potential.

The potentials within the sub-triangles can be calculated from thermodynamic data on the electrically neutral phases at their corners. From this information it is possible to calculate the voltages versus any of the components. This means that one can also calculate the equilibrium oxygen activities and pressures for the phases in equilibrium with each other in each of the sub-triangles. As was shown earlier, one can also do the reverse, and measure the equilibrium potential at selected compositions in order to determine the thermodynamic data, including the oxygen pressure. The relation between the potential and the oxygen pressure is of special interest because of its practical implications for high voltage battery systems.

The experimental data that are available for ternary lithium–transition metal oxide systems are, however, limited to only three system and one temperature. The Li-Mn-O, Li-Fe-O, and Li-Co-O systems were studied quantitatively using molten salt electrolytes at 400 °C [64]. Because of the sensitivity of lithium to both oxygen and water, they were conducted in a helium-filled glove box. The maximum oxygen pressure that could be tolerated was limited by the formation of Li_2O in the molten salt electrolyte, which was determined to occur at an oxygen partial pressure of 10^{-25} atmospheres at 400 °C. This is equivalent to 1.82 V versus lithium at that temperature. Thus it was not possible to study materials with potentials above 1.82 V versus lithium at that temperature.

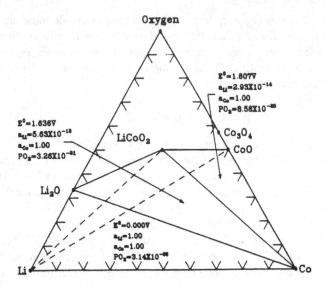

Fig. 21.21 Equilibrium data for the Li-Co-O ternary system at 400 °C. After ref. [63]

As an example, the results obtained for the Li-Co-O system under those conditions are shown in Fig. 21.21 [64], which is also included in Chap. 13.

The general equilibrium equation for a ternary sub-triangle that has two binary transition metal oxides (MO_y and MO_{y-x}) and lithium oxide (Li_2O) at its corners can be written as

$$2xLi + MO_y = xLi_2O + MO_{y-x} \tag{21.4}$$

According to Hess's law, this can be divided into two binary reactions, and the Gibbs free energy change ΔG_r is the sum of the two

$$\Delta G_r = \Delta G_r^1 + \Delta G_r^2 \tag{21.5}$$

One is the reaction

$$MO_y = (x/2)O_2 + MO_{y-x} \tag{21.6}$$

The related Gibbs free energy change is given by

$$\Delta G_r^1 = -RTlnK \tag{21.7}$$

where K is the equilibrium constant.

The other is the formation of Li_2O, that can be written as

$$2xLi + (x/2)O_2 = xLi_2O \tag{21.8}$$

for which the Gibbs free energy change is the standard Gibbs free energy of formation of Li_2O.

$$\Delta G_r^2 = x\Delta G_f^o(Li_2O) \tag{21.9}$$

The potential is related to ΔG_r by

$$E = -\Delta G_r/zF \tag{21.10}$$

that can also be written as

$$E = RT/(4F)ln(pO_2) - \Delta G_f^0(Li_2O)/(2F) \tag{21.11}$$

This can be simplified to become a linear relation between the potential E and $\ln p$ (O_2), with a slope of $RT/(4F)$ and an intercept related to the Gibbs free energy of formation of Li_2O at the temperature of interest.

Experimental data were obtained on the polyphase equilibria within the subtriangles in the Li-Mn-O, Li-Fe-O, and Li-Co-O systems by electrochemical titration of lithium into various Li_xMO_y materials to determine the equilibrium potentials and compositional ranges. The results are plotted in Fig. 21.22 [64].

Fig. 21.22 Experimental data on the relation between the potential and the oxygen pressure in phase combinations in the Li-Mn-O, Li-Fe-O, and Li-Co-O systems at 400 °C. After ref. [63]

It is seen that there is a clear correlation between the potentials and the oxygen pressure in all cases. The equation for the line through the data is

$$E = 3.34 \times 10^{-2} \log p(O_2) + 2.65 \, V \tag{21.12}$$

The data all fit this line very well, even though the materials involved had a variety of compositions and crystal structures. Thus the relation between the potential and the oxygen pressure is obviously independent of the identity and structures of the materials involved.

Extrapolation of the data in Fig. 21.22 shows that the oxygen pressure would be 1 atmosphere at a potential of 2.65 V vs. Li/Li^+ at 400 °C.

At 25 °C the Gibbs free energy of formation is -562.1 kJ/mol, so the potential at 1 atmosphere oxygen is 2.91 V vs. Li/Li^+. This is about what is observed as the initial open circuit potential in measurements on many transition metal oxide materials when they are fabricated in air.

Evaluating Eq. (21.11) for a temperature of 298 K, it becomes

$$E = 1.476 \times 10^{-2} \log p(O_2) + 2.91 \, V \tag{21.13}$$

At this temperature the slope of the potential versus oxygen pressure curve is somewhat less than at the higher temperature. But considering it the other way around, the pressure increases more rapidly as the potential is raised.

This result shows that the equilibrium oxygen pressures in the Li-M-O oxide phases increase greatly as the potential is raised. Values of the equilibrium oxygen pressure as a function of the potential are shown in Table 21.3. These data are plotted in Fig. 21.23.

It can be seen that these values become very large at high electrode potentials, and from the experimental data taken under less extreme conditions, it is obvious

Table 21.3 Values of the equilibrium oxygen pressure over oxide phases in Li-M-O systems at 298 K

E vs. Li/Li^+/V	Logarithm of equilibrium oxygen pressure/atm
1	−129
1.5	−95
2	−62
2.5	−28
3	6
3.5	40
4	73.7
4.5	107.6
5	141.4

Fig. 21.23 Dependence of the logarithm of the equilibrium oxygen pressure upon the potential in lithium–transition metal oxide systems [63]

that the critical issue is the potential, not the identity of the electrode reactant material or the crystal structure.

One can understand the tendency toward the evolution of oxygen from oxides at high potentials from a different standpoint. Considerations of the influence of the potential on the point defect structure of oxide solid electrolytes has shown that electronic holes tend to be formed at higher potentials, and excess electrons at lower potentials. The presence of holes means that some of the oxide ions have a charge of $1-$, rather than $2-$. That is, they become peroxide ions, O^-. This is an intermediate state on the way to neutral oxygens, as in the neutral oxygen gas molecule O_2.

Such ions have been found experimentally on the oxygen electrode surface where the transition between neutral oxygen molecules and oxide ions takes place at the positive electrodes of fuel cells.

21.4.9   Final Comments on This Topic

It is evident that this is a very active research area, with a number of different avenues being explored in the pursuit of higher potentials, greater capacity, longer cycle life, greater safety, and lower cost. It will be interesting to see which of these new materials, if any, actually come into commercial application.

21.5   Hydrogen and Water in Positive Electrode Materials

21.5.1   Introduction

The electrochemical insertion and deletion of hydrogen is a major feature in some important types of aqueous batteries. The use of metal hydrides as negative electrode reactants in aqueous systems is discussed in Chap. 16, and the hydrogen-driven $H_2NiO_2/HNiO_2$ phase transformation is the major reaction in the positive electrode of a number of "nickel" cells, as described in Chap. 17.

It is generally known that alkali metals react vigorously with water, with the evolution of hydrogen. In addition, a number of materials containing lithium are sensitive to air and/or water, and thus have to be handled in dry rooms or glove boxes. Yet most of the lithium-containing oxides now used as positive electrode reactants in lithium battery systems are synthesized in air, often with little heed given to this problem.

It has long been known that hydrogen (protons) can be present in oxides, including some that contain lithium, and that water (a combination of protons and extra oxide ions) can be absorbed into some selected cases. There are several different mechanisms whereby these can happen.

21.5.2   Ion Exchange

It is possible to simply exchange one type of cationic species for another of equal charge without changing the ratio of cations to anions or introducing other defects in oxides. For example, the replacement of some or all of the sodium cations present in oxides by lithium cations is discussed in several places in this text.

Especially interesting is the exchange of lithium ions by protons. One method is chemically driven ion exchange, in which there is inter-diffusion in the solid state between native ionic species and ionic species from an adjacent liquid phase. An example of this is the replacement of lithium ions in an oxide solid electrolyte or mixed-conductor by protons as the result of immersion in an acidic aqueous solution. Protons from the solution diffuse into the oxide, replacing lithium ions, which move back into the solution. The presence of anions in the solution that react with lithium ions to form stable products, such as LiCl, can provide a strong driving force. An example could be a lithium transition metal oxide, $LiMO_2$ placed in an aqueous solution of HCl. In this case the ion exchange process can be written as a simple chemical reaction

$$HCl + LiMO_2 = HMO_2 + LiCl \tag{21.14}$$

The LiCl product can either remain in solution or precipitate as a solid product.

One can also use electrochemical methods to induce ion exchange. That is, one species inside a solid electrode can be replaced in the crystalline lattice by a different species from the electrolyte electrochemically. The species that is displaced leaves the solid and moves into the electrolyte or into another phase. This electrochemically-driven displacement process is now sometimes called "extrusion" by some investigators.

21.5.3 Simple Addition Methods

Instead of exchanging with lithium, hydrogen can be simply added to a solid in the form of interstitial protons. The charge balance requirement can be accomplished by the co-addition of either electronic or ionic species, i.e., either by the introduction of extra electrons or the introduction of negatively charged ionic species, such as O^{2-} ions. If electrons are introduced, the electrical potential of the material will become more negative, with a tendency toward n-type conductivity.

Similarly, oxygen, as oxide ions, can be introduced into solids, either directly from an adjacent gas phase or by reaction with water, with the concurrent formation of gaseous hydrogen molecules. Oxide ions can generally not reside upon interstitial sites in dense oxides because of their size, and thus their introduction requires the presence of oxygen vacancies in the crystal lattice. If only negatively charged oxide ions are introduced, electroneutrality requires the simultaneous introduction of electron holes. Thus the electrical potential of the solid becomes more positive, with a tendency toward p-type conductivity.

There is another possibility, first discussed by Stotz and Wagner [65, 66]. This is the simultaneous introduction of species related to both the hydrogen component and the oxygen component of water, i.e., both protons and oxide ions. This requires, of course, mechanisms for the transport of both hydrogen and oxygen species within the crystal structure. As mentioned already, hydrogen can enter the crystal structure of many oxides as mobile interstitial protons. The transport of oxide ions, that move by vacancy motion, requires the preexistence of oxide ion vacancies. This typically involves cation doping. In this dual mechanism the electrical charge is balanced. Neither electrons nor holes are involved, so the electrical potential of the solid is not changed. The concurrent introduction of both protons and oxide ions is, of course, compositionally equivalent to the addition of water to the solid, although the species H_2O does not actually exist in the crystal structure.

21.5.4 Thermodynamics of the Lithium: Hydrogen: Oxygen System

A number of the features of the interaction between lithium, hydrogen and oxygen in solids can be understood in terms of the thermodynamics of the ternary Li-H-O

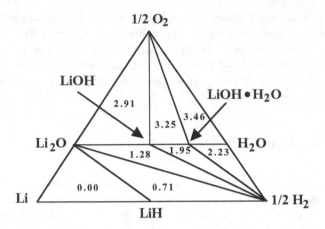

Fig. 21.24 Calculated phase stability diagram for the Li-H-O system at 298 K, assuming unit activities of all phases. The numbers within the triangles are their respective potentials vs. pure lithium. After ref. [66]

system. A useful thinking tool that can be used for this purpose is the *ternary phase stability diagram* with these three elements at the corners. This is discussed in some detail in Chap. 12.

The ternary phase stability diagram for the Li-H-O system at ambient temperature was determined [67] by using chemical thermodynamic data from Barin [68], and assuming that all relevant phases are in their standard states. An updated version is shown in Fig. 21.24.

Using the methods discussed in Chap. 11, the calculated voltages for the potentials of all compositions in the sub-triangles are shown relative to pure lithium.

If one considers an electrochemical cell with pure lithium at the negative electrode, the potential of water that is saturated with LiOH · H_2O will be 2.23 V when hydrogen is present at one atmosphere. On the other hand, water saturated with LiOH · H_2O will have a potential of 3.46 V vs. Li if one atmosphere of oxygen is present. It can be seen that under these conditions water has a stability window of 1.23 V, as is the case in the binary hydrogen–oxygen system.

These results may seem to be in conflict with the general conclusion in the literature that the potential of lithium is −3.05 V relative to that of the standard hydrogen electrode (SHE) potential in aqueous electrochemical systems. This can be reconciled by recognizing that the values calculated here are for the case that the water is in equilibrium with LiOH · H_2O, which is very basic, with a pH of 14. The potentials of both the RHE and pure oxygen, as well as all other zero-degree-of-freedom equilibria, decrease by 0.059 V per pH unit. Thus, in order to be compared to the potential of the SHE, these calculated values have to be corrected by (14 × 0.59), or 0.826 V. Then the voltage between lithium and the SHE that is calculated in this way becomes 3.056 V, corresponding to the data in electrochemical tables.

21.5.5   Examples of Phases Containing Lithium That Are Stable in Water

A number of examples can be found in the literature that are consistent with, and illustrate these considerations. Particularly appropriate are several experimental results that were published by the group of J.R. Dahn some years ago.

They performed experiments on the addition of lithium to $LiMn_2O_4$ in a LiOH-containing aqueous electrolyte using a carbon negative electrode [69] and showed that the two-phase system $LiMn_2O_4$–$Li_2Mn_2O_4$, that is known to have a potential of 2.97 V vs. Li in nonaqueous cells [70] is stable in water containing LiOH. They used a Ag/AgCl reference electrode, referred their measurements to the SHE, and then converted to the lithium scale, assuming that the potential of the lithium electrode is −3.05 V vs. the SHE. They found that lithium began reacting with the $LiMn_2O_4$ at a potential of −0.1 V vs. the SHE, which is consistent with the value of 2.97 V vs. Li mentioned above.

As lithium was added beyond the two-phase composition limit the potential fell to that for hydrogen evolution. Their data showed hydrogen evolution at a potential 2.2 V vs. pure Li, and found oxygen evolution on a carbon negative electrode at 3.4 V vs. pure Li. It can readily be seen that these experimental results are consistent with the results of the Gibbs triangle calculations shown in Fig. 21.23.

It was also found that the phase $VO_2(B)$ reacts with lithium at potentials within the stability range of water [71]. Electrochemical cell experiments were performed in which $Li_xVO_2(B)$ acted as the negative electrode, and $Li_xMn_2O_4$ as the positive electrode [72]. These aqueous electrolyte cells gave comparable results to those with the same electrodes in organic solvent electrolyte cells.

21.5.6   Materials That Have Potentials Above the Stability Window of Water

At normal pressures materials with potentials more positive than that of pure oxygen will tend to oxidize water to cause the evolution of electrically neutral molecular oxygen gas. For this to happen there must be a concurrent reduction process. One possibility is the insertion of positively charged ionic species, along with their charge-balancing electrons, into the material in question. The insertion of protons or lithium ions and electrons into high-potential oxides is one possible example of such a reduction process. When this happens, the potential of the material goes down toward that of pure oxygen.

21.5.7   Absorption of Protons from Water Vapor in the Atmosphere

A number of materials that are used as positive electrode reactants in lithium battery systems have operating potentials well above the stability range of water. Cells containing these materials and carbon negative electrodes are typically assembled in air in the uncharged state. It is generally found that the open circuit cell voltage at the start of the first charge is consistent with lithium–air equilibrium, i.e., along the Li_2O/O_2 edge of the ternary phase stability diagram in Fig. 21.1. This can be calculated to be 2.91 V vs. pure lithium. This can be explained by the reaction of these materials with water vapor in the atmosphere. Protons and electrons enter the crystal structures of these high potential materials, reducing their potentials to that value. This is accompanied by the concurrent evolution of molecular oxygen.

21.5.8   Extraction of Lithium from Aqueous Solutions

An analogous situation can occur if a material that can readily insert lithium, rather than protons, has a potential above the stability range of water. If lithium ions and electrons enter the material's structure the potential will decrease until the value in equilibrium with oxygen is reached. Such a material can thus be used to extract lithium from aqueous solutions. This was demonstrated by experiments on the use of the λ-MnO_2 spinel phase that absorbed lithium when it was immersed in aqueous chloride solutions [73].

References

1. Yao YFY, Kummer JT (1967) J Inorg Nucl Chem 29:2453
2. Weber N, Kummer JT. (1967) Proceedings Annual Power Sources Conference 21, 37
3. Coetzer J (1986) J Power Sources 18:377
4. Galloway RC (1987) J Electrochem Soc 134:256
5. Bones RJ, Coetzer J, Galloway RC, Teagle DA (1987) J Electrochem Soc 134:2379
6. Huggins RA (1999) J Power Sources 81–82:13
7. Vissers DR, Tomczuk Z, Steunenberg RK (1974) J Electrochem Soc 121:665
8. Whittingham MS (1976) Science 192:1126
9. Whittingham MS (1976) J Electrochem Soc 123:315
10. Whittingham MS (1993) Intercalation Compounds. In: Scrosati B, Magistris A, Mari CM, Mariotto G (eds) Fast Ion Transport. KluwerAcademic Publishers, Dordrecht, p 69
11. Dickens PG, French SJ, Hight AT, Pye MF (1979) Mater Res Bull 14:1295
12. Mizushima K, Jones PC, Wiseman PJ, Goodenough JB (1980) Mater Res Bull 15:783
13. Goodenough JB, Mizushima K, Takada T (1980) Jpn J Appl Phys 19(Supplement 19-3):305
14. Nagaura T, Tozawa K (1990) Progress in Batteries and Solar Cells, vol 9. JEC Press, Inc, Brunswick, OH, p 209

15. Nagaura T (1991) Progress in Batteries and Solar Cells, vol 10. JEC Press, Inc, Brunswick, OH, p 218
16. Yamada A, Hosoya M, Chung SC, Kudo Y, Liu KY. (2001) Abstract No. 205, Electrochemical Society Meeting, San Francisco
17. Nanjundaswamy KS, Padhi AK, Goodenough JB, Okada S, Ohtsuka H, Arai H, Yamaki J (1996) Solid State Ionics 92:1
18. Padhi AK, Nanjundaswamy KS, Masquelier C, Goodenough JB (1997) J Electrochem Soc 144:2581
19. Okada S, Ohtsuka H, Arai H, Ichimura M (1993) Electrochem. Soc Ext Abstracts 93–1:130
20. Okada S, Takada T, Egashira M, Yamaki J, Tabuchi M, Kageyama H, Kodama T, Kanno R. (1999) Presented at Second Hawaii Battery Conference, Jan. 1999
21. Sadadone I, Delmas C (1996) J Mater Chem 6:193
22. Bruce PG, Armstrong AR, Gitzendanner R (1999) J Mater Chem 9:193
23. Grincourt Y, Storey C, Davidson IJ (2001) J Power Sources 97–98:711
24. Paulson JM, Donaberger RA, Dahn JR (2000) Chem Mater 12:2257
25. Spahr ME, Novak P, Schneider B, Haas O, Nesper RJ (1998) J Electrochem Soc 145:1113
26. Ohzuku T, Makimura Y (2001) Chem Lett 8:744
27. Lu Z, MacNeil DD, Dahn JR (2001) Electrochem Solid-State Lett 4:A191
28. Kang K, Meng YS, Breger J, Grey CP, Ceder G (2006) Science 311:977
29. Liu Z, Yu A, Lee JY (1999) J Power Sources 81–82:416
30. Yoshio M, Noguchi H, Itoh J-I, Okada M, Mouri T (2000) J Power Sources 90:176
31. Whittingham MS (2004) Chem Rev 104:4271
32. Thackeray MM, David WIF, Bruce PG, Goodenough JB (1983) Mater Res Bull 18:461
33. Thackeray MM, Johnson PJ, de Piciotto LA, Bruce PG, Goodenough JB (1984) Mater Res Bull 19:179
34. Thackeray MM. (1999) In: Handbook of Battery Materials. JO. Besenhard (ed.), Wiley-VCH p. 293
35. Guyomard D, Tarascon JM (1994) Solid State Ionics 69:222
36. Amatucci G, Tarascon J-M (2002) J Electrochem Soc 149:K31
37. Gummow RJ, De Kock A, Thackeray MM (1994) Solid State Ionics 69:59
38. Guyomard D, Tarascon J-M. US Patent 5,192,629, (March 9, 1993)
39. Guyomard D, Tarascon J-M (1994) Solid State Ionics 69:293
40. Sigala C, Guyomard D, Verbaere A, Piffard Y, Tournoux M (1995) Solid State Ionics 81:167
41. Ein-Eli Y, Howard WF (1997) J Electrochem Soc 144:L205
42. Ein-Eli Y, Howard WF, Lu SH, Mukerjee S, McBreen J, Vaughey JT, Thackeray MM (1998) J Electrochem Soc 145:1238
43. Ein-Eli Y, Lu SH, Rzeznik MA, Mukerjee S, Yang XQ, McBreen J (1998) J Electrochem Soc 145:3383
44. Ohzuku T, Takeda S, Iwanaga M (1999) J Power Sources 81–82:90
45. Ariyoshi K, Iwakoshi Y, Nakayama N, Ohzuku T (2004) J Electrochem Soc 151:A296
46. Colbow KM, Dahn JR, Haering RR (1989) J Power Sources 26:397
47. Ohzuku T, Ueda A, Yamamoto N (1995) J Electrochem Soc 142:1431
48. Goodenough JB, Hong HY-P, Kafalas JA (1976) Mater Res Bull 11:203
49. Padhi AK, Nanjundaswamy KS, Masquelier C, Okada S, Goodenough JB (1997) J Electrochem Soc 144:1609
50. Barker J, Saidi MY. US Patent 5,871,866 (1999)
51. Saidi MY, Barker J, Huang H, Swoyer JL, Adamson G (2002) Electrochem Solid-State Lett 5: A149
52. Padhi AK, Nanjundaswamy KS, Goodenough JB (1997) J Electrochem Soc 144:1188
53. Ravet N, Goodenough JB, Besner S, Simoneau M. Hovington P, Armand M (1999) Electrochem. Soc. Meeting Abstract 99-2, 127
54. Chung S-Y, Bloking JT, Chiang Y-M (2002) Nat Mater 1:123
55. Amin R, Maier J (2008) Solid State Ionics 178:1831

56. Herle PS, Ellis B, Coombs N, Nazar LF (2004) Nat Mater 3:147
57. Ong SP, Wang L, Kang B, Ceder G. Presented at the Materials Research Society Meeting in San Francisco, March, 2007
58. Meethong N, Huang H-YS, Speakman SA, Carter WC, Chiang Y-M (2007) Adv Funct Mater 17:1115
59. Barker J, Saidi MY, Swoyer JL (2004) J Electrochem Soc 151:A1670
60. Barker J, Gover RKB, Burns P, Bryan AJ (2007) Electrochem Solid-State Lett 10:A130
61. Barker J, Gover RKB, Burns P, Bryan AJ (2006) Electrochem Solid-State Lett 9:A190
62. Barker J, Gover RKB, Burns P, Bryan AJ (2007) J Electrochem Soc 154:A882
63. Dahn JR, Fuller EW, Obrovac M, von Sacken U (1994) Solid State Ionics 69:265
64. Godshall NA, Raistrick ID, Huggins RA (1984) J Electrochem Soc 131:543
65. Stotz S, Wagner C (1966) Ber Bunsenges Physik Chem 70:781
66. Wagner C (1968) Ber Bunsenges Physik Chem 72:778
67. Huggins RA (2000) Solid State Ionics 136–137:1321
68. Barin I. Thermochemical Data of Pure Substances. 3rd Edition, VCH 1995, Published Online 24 Apr 2008, ISBN 9783527619825
69. Dahn JR, von Sacken U, Al-Janaby H, Juzkow MW (1991) J Electrochem Soc 138:2207
70. Li W, McKinnon WR, Dahn JR (1994) J Electrochem Soc 141:2310
71. Tarascon JM, Guyomard D (1993) J Electrochem Soc 138:2864
72. Li W, Dahn JR (1995) J Electrochem Soc 142:1742
73. Kanoh H, Ooi K, Miyai Y, Katoh S (1993) Sep Sci Technol 28:643

Chapter 22
Energy Storage for Medium- to Large-Scale Applications

22.1 Introduction

Most of the highly visible applications of advanced energy storage technologies are for relatively small applications, such as in portable computers or implanted medical devices, where the paramount issue is the amount of energy stored per unit weight or volume, and cost is not always of prime importance. Such energy storage components and systems have occupied much of the attention in this text, especially the later chapters related to electrochemical cells and systems.

As discussed in Chap. 1, there are several types of large-scale energy storage applications that have unique characteristics, and thus require storage technologies that are significantly different from the smaller systems that are most common at the present time. These include utility load leveling, solar and wind energy storage, and vehicle propulsion. They play critical roles in the transition away from the dependence upon fossil fuels.

More than for smaller scale applications, the important factors in large systems are the cost per unit energy storage, e.g., per kWh, efficiency of the energy storage cycle, which has a large influence upon operating costs, and the lifetime of the critical components. Investors generally expect large systems to be in operation for 25 years or more. In addition, great attention is paid to safety matters.

Several of the storage technologies that are particularly interesting and important for larger-scale applications are described in the early chapters of this book. Some others are discussed in this chapter.

© Springer International Publishing Switzerland 2016
R.A. Huggins, *Energy Storage*, DOI 10.1007/978-3-319-21239-5_22

22.2   Utility Load Leveling, Peak Shaving, and Transients

The requirements of the large-scale electrical distribution network, or grid, are discussed in Chap. 1. The major problem is to match the energy available to the needs, which typically undergo daily, weekly, and seasonal variations. In addition, there are short-term transients that can lead to instabilities and other problems with the electrical power grid. The amelioration of these problems requires not only better technology, but also an intelligent control system to couple energy generation, transmission, and storage. Major factors include cost, reliability, lifetime, efficiency, and safety.

The energy storage method that is most widely used to reduce the longer-term variations in some areas involves the use of pumped-hydro facilities discussed in Chap. 6. However, this is only possible in specific locations, where the required geological features are present. Large-scale underground compressed air storage systems also have important location requirements. Although they are often discussed, very few are in actual operation at the present time.

Other technologies can be useful in reducing the impact of short-term transients, which are now handled primarily by variation of the AC output frequency. One of these, also discussed in Chap. 6, involves the use of very large flywheels. Some of these are now available with power values up to 100 kW. The integration of a number of such units to provide total power up to 20 MW is being investigated. The Department of Energy has estimated that 100 MW of flywheel storage could eliminate 90 % of the frequency variations in the State of California.

Additional approaches that are being explored at present involve reversible high power electrochemical systems. Here, the amount of energy stored per unit cost is of prime importance. In contrast to other uses of electrochemical systems, the size and weight are generally not important for this type of application. Several of these are discussed later in this chapter.

22.3   Storage of Solar- and Wind-Generated Energy

Solar and wind energy sources are often viewed as technologies that can be both employed to satisfy transient local needs, and to supply energy into the electricity distribution grid. However, their output generally only roughly matches the time-dependent requirements of the grid. Thus energy storage mechanisms are required to assist their integration into that large-scale system. Short-term transients in their output, such as when a cloud passes over a solar collection system, or the wind drops in velocity, are generally not of great importance for that application.

For applications such as matching the time dependence of the needs and supplies of energy in the large-scale electricity grid, some relatively low cost electrochemical systems, that are not interesting for portable applications because of their size or weight, can be advantageous.

22.4 Several Recent Developments That May Be Useful for These Applications

22.4.1 Hybrid Lead-Acid Batteries for Large Scale Storage

Pb-acid batteries, due to their relative ease of manufacture, and favorable electrochemical characteristics, such as rapid kinetics and relatively good cycle life, are commonly used for automotive starting. The low cost per unit energy stored is a particularly attractive feature of this type of technology. Large groups of them are being used to support solar and wind generation systems.

As described in Chap. 17, conventional Pb-acid cells are typically kept at or near a full state of charge, but they tend to deteriorate quickly when operated at a partial state of charge. This limits their utility in a number of other applications.

This situation has changed recently, due to the development of lead-acid cells with a double negative electrode design that leads to a device that has properties that are a combination of a standard lead-acid battery and an ultracapacitor. The result of this approach is that these cells can operate for long periods of time at a partial state of charge. The label "UltraBattery" is being used by the suppliers of this technology.

The UltraBattery was invented in 2003 by Dr. Lan Lamm and his team at the Commonwealth Scientific and Industrial Research Organization (CSIRO) in Australia, where some initial production began in 2005. The firm Ecoult was formed by CSIRO in 2007 to commercialize this technology.

In 2010 Ecoult was purchased from CSIRO by East Penn Manufacturing, Inc. in the United States. The Japanese firm Furukawa Battery acquired a license for this technology in Japan and Thailand from CSIRO, but East Penn has a license for the rest of the world.

The UltraBattery is a hybrid energy storage device that combines a single positive electrode, comparable to that used in normal Pb-acid cells, with a double negative electrode—one part containing lead, and the other part carbon, in a common sulfuric acid electrolyte. This is illustrated schematically in Fig. 22.1.

This is important in applications such as their use in hybrid vehicles, where both braking (high rate charging) and acceleration (high rate discharging) can occur in rapid repetition for many thousands of cycles.

It is predicted that it is reasonable to expect that these batteries will last more than 40,000 cycles, some 10 times longer than the shallow cycles of conventional Pb-acid batteries, and more than 4 times longer than a Pb-acid battery designed specifically for idling-stop-start vehicles.

This hybrid technology provides more power and a longer lifespan than standard Pb-acid batteries and it is claimed that this technology can provide performance similar to that of metal hydride/nickel batteries, but at a significantly lower cost.

This modified version of the Pb-acid system also suffers significantly less from the development of permanent (or hard) $PbSO_4$ deposits (typically called

Fig. 22.1   General layout of an UltraBattery

"sulfation") on the negative battery electrode—a problem commonly exhibited in conventional lead acid batteries when they are used for stand-by applications.

The standard solution to this problem is the use of occasional "refresh cycles," in which they are charged, typically at a 1C rate, followed by a lengthy period of lower-rate charging at a "float" voltage so that all cells reach 100 % state of charge. This helps restore the physical state of the electrodes and allows individual battery cells to attain consistent voltages and state of charge, when otherwise they might diverge during an extended period of cycling. A refresh cycle concludes when the battery is returned to the state of charge required by the application it is serving.

During a refresh cycle, therefore, the battery is not in operation, so it is desirable to minimize this downtime. It has been shown that an UltraBattery can operate for more than ten times as many cycles between recovery charges than standard Pb-acid batteries.

Tests on hybrid electric vehicles (HEVs) showed a range of more than 100,000 miles on a single battery pack without significant degradation.

A major disadvantage of conventional Pb-acid batteries is that they are typically designed for uses in which they are kept close to fully charged. Long periods in a partial state of charge cause severe decay in their capacity.

Because they can operate at a partial state of charge, UltraBatteries are also ideally suited to provide frequency regulation services to the grid. They can respond in both directions by charging or discharging rapidly and can ramp much faster than any conventional generator, following the regulation control signal accurately.

This type of battery can be used in advanced hybrid vehicles that have an "idling-stop" function that stops the internal combustion engine when the vehicle stops. Such vehicles also typically use regenerative braking to recover some of the vehicle's energy of motion to help charge the battery. With these innovations the fuel efficiency is typically increased by approximately 10 % compared to a non-hybrid vehicle

Fig. 22.2 Longevity under high-rate partial state of charge cycling tests [1]

To accept charge from regenerative braking efficiently, the battery is kept at a partial state of charge, typically about 90 % of its full capacity, and it must be able to withstand both high discharging and charging rates. This contrasts with non-hybrid vehicles, in which the battery floats on full charge, using the alternator to provide mild charging rates. If the engine must be restarted when the driver releases the brake pedal after stopping, the number of large-current discharges can be many more than is typical in non-hybrid vehicles.

Some hybrid vehicles also use their batteries to provide electric propulsion to assist the internal combustion engine during operation in addition to idling-stop and regenerative braking. This can increase the fuel efficiency by approximately 20–25 % compared to a non-hybrid vehicle.

A further area of application of this enhanced Pb-acid battery technology involves the support of solar and wind systems to relieve the inherent variability in their energy generation.

Reserve battery capacity can be used to alleviate output shortfalls for short time scales: minutes or even seconds. This reduces the need to have large "spinning reserve" generators, which are typically used for this purpose (Fig. 22.2).

Field driving tests have demonstrated that there is no difference between the driving performance of a hybrid vehicle using an UltraBattery pack and one using a metal hydride/nickel battery pack. But the cost of an UltraBattery pack is dramatically less than the metal hydride/nickel pack.

To follow up and further quantify the road test results, laboratory cycle-life tests were conducted on 2 V cell flooded type UltraBatteries based on the power-assisting EUCAR profile [2, 3]. The tests were started at a 60 % state of charge, and no recovering charging was performed. The life of the UltraBatteries was more than 40,000 cycles, representing a cycle life more than ten times that of a

Fig. 22.3 Cycling performance of a conventional Pb-acid battery, an ISS battery, and an UltraBattery when used under a standard driving profile [3]

conventional lead-acid battery, and more than four times longer than that of a lead-acid battery specifically designed for idling-stop-start (ISS) vehicles. This is shown in Fig. 22.3.

22.5 Batteries with Open Framework Crystal Structure Electrodes

22.5.1 Introduction

As mentioned in Chap. 13, it has been known for some 25 years that charge can be stored in some important battery electrodes by the insertion of ionic species from the electrolyte. Insertion reactions play an especially important role in current versions of lithium batteries, where lithium cations are typically the inserted species in both electrodes. This is also true of a number of other lithium-containing materials. Hydrogen cations (protons) are the inserted guest species during the operation of other types of battery electrodes, including the $Ni(OH)_2/NiOOH$ electrode, the "MnO_2" electrode, and RuO_2 in aqueous systems.

Whereas most attention has been given to materials in which the guest species are cations, it is also possible to have anion insertion into some crystal structures. Materials in which the structure can accommodate *either* cations or anions are especially interesting.

There also are some cases in which *both* cations and anions can be inserted into a crystal structure. One example of this type is briefly discussed in this chapter, ternary materials with the hexagonal transition metal bronze structure.

Most attention is given here to the hexacyanometallate family of materials, however. These materials have structures which are variants of the cubic ReO_3 type of crystal structure which have rather large inter-cell windows. They can accommodate a wide variety of guest ions of both charges. Cations can be inserted into the structure at relatively low potentials, and anions can be inserted at more positive potentials. This can lead to a number of interesting features and properties.

22.5.2 Insertion of Guest Species Into Materials with Transition Metal Oxide Bronze Structures

It has long been known that a number of ions can be readily inserted into the structures of ternary oxides such as the tungsten, molybdenum and vanadium bronzes. These bronze families can exist in several different crystal structures, depending upon the identity and concentration of the lower-charge cations present. If the inserted ions are relatively small, these materials often have the cubic ReO_3 structure.

Especially interesting, however, are materials with the hexagonal tungsten bronze structure with the general formula M_xWO_3, in which there are two types of crystallographic tunnels. There is a hexagonal array of rather large linear tunnels that penetrate through the structure parallel to the c-axis. This structure is only obtained when M is a large cation, such as K^+, Rb^+, or Cs^+. K_xWO_3, where $x = 0.3$ and the K^+ ions partly occupy the positions in the large tunnels. They can be readily prepared by either solid state or electrochemical methods at elevated temperatures. This structure is shown schematically in Fig. 22.4.

This material is dark blue-black, due to the presence of both W^{5+} and W^{6+} species. If it is heated to intermediate temperatures (e.g., 400 °C) in air O^{2-} anions are introduced into the large tunnels. These balance the charge of the K^+ ions, causing all the tungsten ions to become W^{6+}. The material is thus bleached, becoming white.

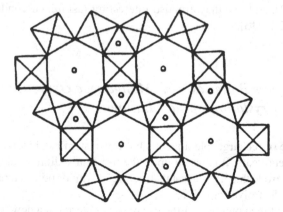

Fig. 22.4 Representation of the hexagonal tungsten bronze structure in the c-direction, small circles show the presence of both large and small tunnel sites

Li$^+$ cations can subsequently be inserted into this structure at room temperature. They go into the set of smaller tunnels that are oriented in the cross-direction, rather than into the large tunnels. The presence of the lithium ions causes the reduction of some of W^{6+} ions to W^{5+}, and the material becomes dark. Thus the low temperature insertion of Li$^+$ cations, which is both very rapid and reversible, can be employed to make this an interesting electrochromic material [4–7]. However, because of its cost and weight, it is not a practical alternative for use in batteries.

22.5.3 Materials with Cubic Structures Related to Rhenium Trioxide

Another crystal structure that can have interesting properties of this type is the ReO$_3$ (or BX$_3$) structure, which has cubic symmetry. This structure can be thought of as a simple cubic arrangement of corner-shared octahedral BX$_6$ groups. There is empty space in the center of each of these cubes that is interconnected by a three-dimensional set of tunnels through the centers of the cube faces. It is possible for this cube-center space to be either empty, partly, or fully occupied by cations, assuming that the charges of the other species present are adjusted so as to maintain overall charge neutrality

If there is a cation in the center of every cube, the nominal formula is then ABX$_3$, and this is the well-known perovskite structure, which is adopted by many oxides. In order for it to be stable, the A ions must be relatively large. The more highly charged B ions are quite small.

It has been shown that a variety of ions can reside on the A sites of this structure. In addition, there may be mixed occupation by more than one type of ion. An example of this is the family of Li-La titanates, in which lithium ions, lanthanum ions, and vacancies are distributed among the available A sites. It has been shown that these materials can have a relatively high lithium ionic conductivity at positive potentials [8–12], and that they are also interesting fast mixed-conductors at more negative potentials [13].

22.5.4 Aqueous Batteries with Manganese Oxide Electrodes with Crystallographic Channels

A different type of rechargeable aqueous battery system has received considerable attention in recent years. It is based upon the use of a sodium manganese oxide in the positive electrode that has a crystal structure in which sodium ions can enter and leave from the electrolyte rapidly.

Recognition of the potential for the use of this material in a new type of practical and inexpensive rechargeable battery by Prof. J. Whitacre at Carnegie Mellon

Fig. 22.5 Model of the crystal structure showing the locations of the sodium ions in the channels between the manganese-centered octahedra and square pyramidal units in the structure [14]

University led to the formation of the firm Aquion in 2010 to commercialize this technology.

The key element is the use of the phase whose composition can be written either as $Na_4Mn_9O_{18}$, or more simply, as $Na_{0.44}MnO_2$, as the positive electrode reactant. This material has an orthorhombic crystal structure that can be described as containing both MnO_6 octahedra and MnO_5 square pyramids. These form a special arrangement that contains interconnected channels with two different types of atomic-sized locations for the sodium ions. In one case, the sodium ions sit in sites that are interconnected. As a result, they are mobile, and can be reversibly extracted from the crystal structure and reinserted. Those in the other type of location in the structure are not mobile.

The characteristics of this structure can be seen in Fig 22.5. Examples of mobile ion positions are indicated as Na1 and Na2. Ions in Na3 type positions cannot readily move through the crystal structure [14].

There was some earlier interest in the behavior of this material as an electrode reactant in lithium ion systems [15, 16], but its properties as an electrode in sodium systems are much more attractive [17–21].

This material has an inherently low cost. Fine particle electrodes with a high surface area can be produced from it inexpensively using processes that are readily scalable. They can be made into rather thick porous electrode structures through which the high conductivity aqueous electrolyte can readily transport the transport of sodium ions.

An aqueous solution of sodium sulfate, which has a very high ionic conductivity, is used as the electrolyte. Since this electrolyte has negligible electronic leakage,

Fig 22.6 Variation of the cell voltage and individual electrode potentials with the state of charge [17]

self-discharge is quite low. Electronic short circuits between the electrodes are prevented by the presence of an inexpensive porous nonwoven cellulose separator.

The negative electrode in the Aquion batteries is produced from inexpensive finely divided activated carbon. This material stores ions reversibly deposited from the electrolyte onto its surface, and has the characteristics of a simple capacitor, with an electric potential that is essentially a linear function of the charge.

The variation of the electrode potentials, as well as the full cell voltage, with the state of charge in this system is shown in Fig. 22.6. It can be seen that there are two semi-plateau regions in the potential of the positive electrode in different potential regimes. This indicates the presence of two reactions, centered at different potentials. This conclusion is consistent with the results of cyclic voltammetric experiments shown in Fig. 22.7, which clearly shows reactions in two distinct potential regions.

Cell discharge curves at different rates are shown in Fig. 22.8. It is seen that the total capacity—down to a minimum value of 0.3 V—varies significantly with the discharge rate, as does the output voltage.

The manufacturing processes involved in this technology are inexpensive and readily scalable, and projected costs are low, less than $ 300 per kWh. Aquion is constructing a manufacturing plant to produce 500 megawatt-hours of these batteries per year near Pittsburgh, PA

Fig. 22.7 Cyclic voltammetry measurements on Na$_{0.44}$MnO$_2$ [17]

Fig 22.8 Cell potential as a function of the state of charge from experiments at different rates of discharge [17]

Their initial commercial target is to produce multi-cell systems for small niche markets, such as off-grid solar and wind installations. A cycle life over 5000 cycles at 60 % depth of discharge, with an energy efficiency over 85 % has been demonstrated, which should be satisfactory for this purpose.

22.6   Hexacyanometallate Electrode Materials

22.6.1   Introduction

There is a family of materials with crystal structures that are analogous to the BX_3 rhenium trioxide and ABX_3 perovskite materials, but in which the X positions are occupied by cyanide anions, which are appreciably larger than oxide ions. These materials are sometimes called hexacyanometallates, and the B positions are often occupied by transition metal ions. The transition metal hexacyanometallates are examples of the large family of insoluble mixed-valence compounds with interesting properties [22, 23]. An earlier, and more extensive, overview of some aspects of these materials can be found in the paper by Robin and Day [24].

The prototype material is "Prussian blue," which is also sometimes called "Berlin blue." Its nominal formula is $KFe_2(CN)_6$, or $K_{0.5}Fe(CN)_3$. It has a dark blue-black color, has been known for a very long time, and has been widely used as a dyestuff. It was evidently the first coordination compound reported in the scientific literature [25]. An account of the early work on the preparation and chemical composition of materials in the Prussian blue family can be found in ref. [26]. They have been studied extensively because of their electrochromic properties, and there has been renewed interest in them in recent years in connection with their use in "modified electrode surfaces" that are interesting for catalytic purposes.

22.6.2   The Structure of The Prussian Blues

The general formula for materials in this family is $A_x P^{3+} R^{2+}(CN)_6$, where the P^{3+} and R^{2+} species are distributed in an ordered arrangement upon the B sites of the $A_x BX_3$ structure. The value of x, which specifies the amount of A present, can often be varied from 0 to 2. When $x = 0$ the material has the ReO_3 (BX_3) structure, and when $x = 2$, the structure is analogous to the ABX_3 perovskites. In the case of Prussian blue, the A sites are half full, and x is nominally equal to 1.

The structure of Prussian blue was first discussed by Keggin and Miles in 1936 [27], on the basis of powder x-ray diffraction results. They found it to be cubic, like the ReO_3 and perovskite materials, with a simple cube edge length about 5.1 A. In normal Prussian blue K^+ ions fill half of the A positions, and Fe ions are in the B positions. In order to keep overall charge balance, half of the Fe ions have a formal charge of 3+ (and thus can be described as P^{3+} ions), and half have a formal charge of 2+ (and can be described as R^{2+} ions). The carbon ends of the CN^- ions point toward the Fe^{2+} ions, and the nitrogen ends toward the Fe^{3+} ions. Thus one can think of the P^{3+} ions being in a nitrogen-surrounded hole, and the R^{2+} being in a carbon-surrounded hole, in the structure. This structure is shown schematically in Figs. 22.9 and 22.10.

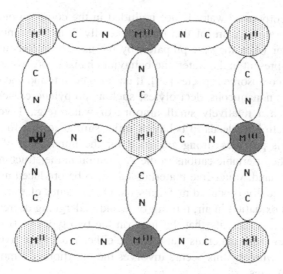

Fig. 22.9 Schematic representation of one plane in the structure of the hexacyanometallate host lattice

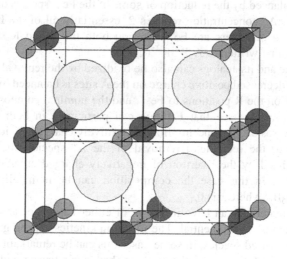

Fig. 22.10 Schematic representation of the structure of Prussian blue, in which half of the A sites are filled

It can readily be seen that these two general formulas lead to different structural interpretations. If the composition of the "insoluble" alternative, $Fe_4[Fe(CN)_6]_3$, is correct, it can be written as $Fe^{3+}_4Fe^{2+}_3(CN)_{18}$ or $(Fe^{3+})_{1/3}(Fe^{3+}Fe^{2+})(CN)_6$. This would indicate that 1/6 of the A positions are occupied by Fe^{3+} ions, instead of their being half-filled with K^+ ions in the "soluble" version. The presence of hydrated Fe^{3+} ions at the A sites has been reported by [28].

Although incorporated water is not included in the compositional formulas of materials in this family, are of them are generally found to contain substantial amounts of water when they are prepared by the use of aqueous methods. When some ethanol is present in the water, the behavior changes somewhat, indicating an influence of the co-absorbed species [29]. It has also been found that the behavior is very different in nonaqueous electrolytes, such as propylene carbonate [30]. The presence of even a relatively small amount of water (e.g., 1 vol %) in such nonaqueous electrolytes leads to behavior quite similar to that found in aqueous electrolytes. This provides strong evidence for the importance of the hydration sheath around the insertable cations in the A position in the structure.

On the other hand, water-free materials can also be produced using solid state methods [31]. The dry-produced materials with high values of x are very sensitive to moisture and oxidation in air, turning blue and undergoing decrepitation.

Because of the possibility that the ions on the two types of B sites may have variable valences these materials can often be either reduced or oxidized, or both. The mechanism whereby this occurs involves the insertion or removal of charged species on the A sites.

Reduction can occur by increasing the concentration of A^+ ions in the structure. In the case that both P and R species are Fe ions, the additional positive charge in the A sites is balanced by the reduction of some of the Fe^{3+} species on the P sites to Fe^{2+}. When the A^+ concentration reaches 2, essentially all of the Fe^{3+} ions have been reduced, so that there are Fe^{2+} ions on both the P and R sites. Thus the composition can be written as $A_2P^{2+}R^{2+}(CN)_6$.

Prussian blue and its analogs can also be oxidized by the removal of A^+ species. In this case the decreased positive charge on the A sites is balanced by the oxidation of Fe^{2+} species on the R positions to Fe^{3+}, and the nominal composition becomes $P^{3+}R^{3+}(CN)_6$. Thus we see that Fe ions can participate in both reduction and oxidation processes as the concentration of positively charged A^+ ions is varied.

In addition to the (complete) removal of the A^+ species, the structure can be further oxidized by the insertion of negatively charged anionic species (B^-) into the A sites. In this case the composition can be nominally described as $A_xP^{3+}R^{3+}(CN)_6B_y$, where x is 0.

Thus these materials can have insertion of either cations or anions into the A sites, depending upon the potential. The insertion kinetics depend greatly upon the identity of the inserted species. In some cases this can be remarkably rapid. Due to the large size of the CN^- anions, the openings between adjacent unit cells are quite large. Thus it is possible for relatively large ions to move throughout the structure by jumping through these windows. Species in the A sites are often highly mobile, and can be accompanied by an appreciable amount of water of hydration.

These hexacyanoferrates are generally stable in water at low to moderate values of pH. In acidic electrolytes containing K^+ ions they can be reduced and reoxidized many times, and show excellent reversibility. Over 10^7 cycles have been demonstrated in some cases [32–34].

22.6.3   *Electrochemical Behavior of Prussian Blue*

Prussian blue and its analogs can be readily reduced and oxidized electrochemically. Much of the experimental work in the literature on the electrochemical behavior of Prussian blue has been performed using potassium-containing aqueous electrolytes with a pH value of about 4, and potentials are generally referenced to the standard calomel electrode (SCE), which is about 478 mV positive of the reversible hydrogen evolution electrode potential at that value of pH. Although very common, this choice of reference electrode is unfortunate, as the definition of its potential relative to neutral chemical species requires knowledge of the pH, in accordance with the Gibbs Phase Rule, as discussed in Chap. 13.

Prussian blue, which has a dark blue-black color, can be both reduced and oxidized electrochemically. Reduction occurs at a potential about 195 mV positive of the SCE potential, and thus about 678 mV positive of the reversible hydrogen potential. The reduction product is white, and is generally called "Everitt's salt," although it is sometimes also designated as "Prussian white." As already mentioned, its composition can nominally be described as $K_2FeFe(CN)_6$ when K^+ ions are in the A sites.

Oxidation of Prussian blue occurs when the potential is made more positive. This occurs at about 870 mV vs. the SCE reference, and thus 1.348 mV positive of the reversible hydrogen potential. The product is only lightly colored, and is generally called "Berlin green." Its composition can be nominally described as $FeFe(CN)_6$.

A second oxidation reaction, involving the insertion of anions into the A position, is sometimes found at about 1,100 mV vs. the SCE potential, where a yellow product called "Prussian yellow" is formed. This material is unstable in water, as would be expected from its potential, some 1.8 V positive of the reversible hydrogen potential. Its nominal compositions can be written as $FeFe(CN)_6A_x$.

All of these materials have essentially the same basic cubic unit cell, with a lattice parameter of about 10.2 A. Although incorporated water is not included in these nominal compositions, these materials are generally found to contain substantial amounts of water of hydration around the A species.

Electrochemical experiments have often been made using cyclic voltammetry. A typical example is shown in Fig. 22.11. It is seen that there are reversible current peaks in two quite different potential regions, relating to the reduction and oxidation reactions described above. The second oxidation reaction at even more positive potentials is not shown in that figure because the scan rate was rather low and the potential was not increased sufficiently for it to be seen. Another voltammogram is shown in Fig. 22.12 under conditions that made it possible to see the formation of the chemically unstable Prussian yellow at more positive potentials.

The critical potentials in the Prussian blue system are shown schematically in Fig. 22.13. One can translate the semi-quantitative dynamic data obtained from cyclic voltammetry experiments into the results that would be expected if electrochemical potential spectroscopy experiments were performed. Likewise, they could be expressed as an equilibrium titration curve, as indicated schematically in

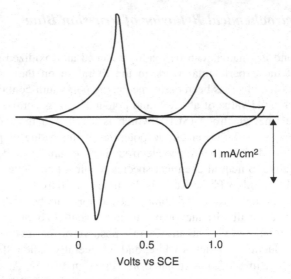

Fig. 22.11 Typical voltammogram of Prussian blue, After [34]

Fig. 22.12 Voltammogram that also shows the reaction to form Prussian yellow at more positive potentials

Fig. 22.14. In this case the formation of Prussian yellow will not be included, as it is not stable in water, as indicated earlier.

Experiments of this type provide more quantitative information about the potentials at which the reduction and oxidation reactions take place than can be obtained from the more common dynamic cyclic voltammetry experiments. However, in the case of Prussian blue, the kinetics of the relevant phenomena are so fast that there is not much difference between the information obtained from dynamic and static experiments.

Fig. 22.13 Schematic representation of the potentials at which the several reactions occur in Prussian blue

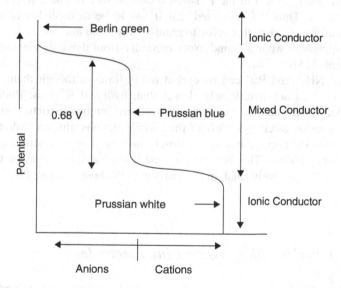

Fig. 22.14 Schematic equilibrium electrochemical titration curve for the Prussian blue system

Table 22.1 Assumed radii of hydrated cations

Cation	Radii (A)
Li^+	2.37
Na^+	1.83
K^+	1.25
NH_4^+	1.25
Rb^+	1.18
Cs^+	1.19
Ba^{2+}	2.88

22.6.4   Various Cations Can Occupy the A Sites in the Prussian Blue Structure

A number of different cations can be present in the A positions in the hexacyano-metallate structure. Monovalent examples include Li^+, Na^+, K^+, NH_4^+, Rb^+, and Cs^+ ions. The ability of these various ions to reversibly enter the structure has been interpreted in terms of their size when hydrated. These are shown in Table 22.1 below, which also includes the hydrated divalent Ba^{2+} ion.

The hard-sphere model of the Prussian blue structure has an intercell window radius of 1.6 A. Thus it is expected that it would be difficult for cations with hydrated radii greater than this value to readily move into and out of the structure. This corresponds to what is found experimentally from dynamic electrochemical measurements [25].

Hydrated NH_4^+ and Rb^+ can be cycled many times, although their insertion/extraction kinetics are significantly slower than hydrated K^+ ions. Both smaller (Cs^+) and larger (Li^+ and Na^+) hydrates showed much more restricted behavior.

In non aqueous electrolytes where the mobile cations are not hydrated these insertion reactions occur much more slowly and are not so interesting from a practical point of view. This is typical of the ionic transport behavior when the mobile species is relatively small in comparison to the host structure through which it moves.

22.6.5   Batteries with Prussian Blue Electrodes

There has been interest in the use of materials in the Prussian blue family as electrodes in batteries for some time. In 1985 V.D. Neff [35] made a cell with Everitt's salt (or "Prussian white") on the negative side, and "Prussian yellow" on the positive side, which gave an initial voltage of 0.93 V in an acid solution of 1 N K_2SO_4. The voltage across this configuration gradually decayed to 0.68 V, due to the instability of Prussian yellow in water, and its replacement by "Berlin green." Two years later Honda and Hayashi [36] showed that a rechargeable battery could be produced from Prussian blue family materials using Nafion as a solid electrolyte. In this case, a stable output voltage of 0.68 V was also observed. Upon discharge, both electrode materials change to Prussian blue, and the voltage dropped to zero in those cells.

There has been a considerable amount of work done recently on the development of practical batteries based upon materials in the Prussian blue family.

An interesting group of materials with such open framework crystal structures are the mixed-valence hexacyanoferrates, which are often called ferrocyanides. They readily intercalate a number of different hydrated ions, including Li^+, Na^+, K^+, and NH_4^+ from aqueous electrolytes [37–41].

Prussian Blue, the oldest and most studied material of this type, has the basic hexacyanometalate metal-organic framework. Materials with this structure may be described in terms of the general formula $A_xPR(CN)_6$. Nitrogen-coordinated transition metal cations (P) and hexacyanometalate complexes ($R(CN)_6$) form a face-centered cubic open framework containing large interstitial A sites, which may be partially, or fully, occupied by a number of different, generally hydrated, ions in this structure. The ionic occupancy of these A sites may vary, with corresponding valence changes in one or both of the P and R species.

As mentioned earlier, the high electrochemical reversibility of materials in this family has been known for some time. For example, the robust framework structure of Prussian Blue and its analogues has been shown to allow thin film electrochromic devices to operate for 10^5 to 10^7 cycles at high cycling rates [32, 33].

These relatively inexpensive materials possess remarkable electrochemical performance, operate in safe, inexpensive aqueous electrolytes, and may be synthesized using bulk processes at modest temperatures. Hence, they are especially attractive for use in large-scale stationary batteries to provide storage capacity for use with the electrical power grid.

Materials in the Prussian Blue family can be easily synthesized by the use of simple ambient temperature precipitation from aqueous solutions. Two examples are copper hexacyanoferrate, (CuHCF) [38] and nickel hexacyanoferrate (NiHCF) [39]. Both CuHCF and NiHCF can readily be synthesized as nanopowders in this way. Simultaneous, dropwise addition of 40 mM copper or nickel nitrate, and 20 mM potassium ferricyanide into deionized water produces controlled co-precipitation of uniform fine particles of either CuHCF or NiHCF. The synthesis of CuHCF is readily done at room temperature, while the synthesis of NiHCF is performed at $70°$ C. These solid products are then filtered, washed with water, and dried in vacuum at room temperature. The products can have a high degree of crystallinity.

Several different common alkali metal ions can be reversibly inserted into these materials from aqueous electrolytes, Li^+, Na^+, K^+, and NH_4^+, and they can be readily synthesized as nanopowders. In each case, the crystallographic lattice parameter varies only slightly, and linearly, with the concentration of the inserted ions.

However, the stiffness of the structure and the size of the interstitial sites result in only minor dimensional changes when the concentrations of the inserted ions are modified during charging and discharging. This is an important factor, for it leads to the unusually long cycle lives observed in this family of materials.

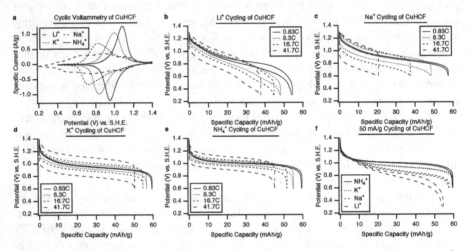

Fig. 22.15 Variation of the potential during the discharge and recharge of (CuHCF) at a 0.83 C rate [38]

Fig. 22.16 Electrochemical performance of CuHCF. (**a**) Cyclic voltammetry of CuHCF with Li^+, Na^+, K^+, and NH_4^+ ions, and (**b**) through (**e**): The potential profiles of CuHCF during galvanostatic cycling of Li^+, Na^+, K^+, and NH_4^+, respectively, at several current densities. (**f**) The potential profiles of CuHCF during galvanostatic cycling of Li^+, Na^+, K^+, and NH_4^+, at 50 mA/g (0.83C) [38]

As an example, potassium ions can be reversibly reacted with $KCuFe^{3+}(CN)_6$ (CuHCF) to produce $K_2CuFe^{2+}(CN)_6$ [38]. The discharge–recharge data at a rate of 0.83C for this case are shown in Fig. 22.15. It can be seen that this behavior is very unusual. There is very, very little hysteresis, so that the energy absorbed per cycle is very small. The shape of the curve is what is expected for a single-phase solid solution reaction.

Materials in this family have shown remarkable cycling behavior: for they can be fully charged and recharged at unusually high rates for a very large number of cycles. This is illustrated in Fig. 22.16 for the case of CuHCF, and in Fig. 22.17 for the analogous NiHCF [39].

Fig. 22.17 Electrochemical performance of NiHCF. Cyclic voltammetry of NiHCF with Li$^+$, Na$^+$, K$^+$, and NH$_4^+$ ions, and (**b**) through (**e**): The potential profiles of NiHCF during galvanostatic cycling of Li$^+$, Na$^+$, K$^+$, and NH$_4^+$, respectively, at several current densities. (**f**) The potential profiles of NiHCF during galvanostatic cycling of Li$^+$, Na$^+$, K$^+$, and NH$_4^+$, at 50 mA/g (0.83C) [39]

Fig. 22.18 Rate capability, cycle life, and effect of insertion ion size of CuHCF and NiHCF. (**a**) and (**b**) The capacity retention of CuHCF and NiHCF at various current densities. (**c**) and (**d**) The cycle life of CuHCF and NiHCF during cycling of Li$^+$, Na$^+$, K$^+$, and NH$_4^+$, and (**e**) The reaction potentials CuHCF and NiHCF as functions of the Stokes radius of the insertion ion [40]

The rate capability and cycle life, as well as the influence of the identity of the inserted ion upon the half-charge potential are illustrated in Fig. 22.18 [40].

22.6.6   Investigations of the Use of Polyvalent Prussian Blue Electrodes in Aqueous Systems

Because of the possibility that the capacity of Prussian Blue could be increased if the inserted species could carry more than one charge, experiments have been reported [42] in which the insertion of alkaline earth divalent cations : Mg^{2+}, Ca^{2+}, Sr^{2+} and Ba^{2+}, into nickel hexacyanoferrate was investigated.

There has also been some recent work on the investigation of the insertion of divalent and trivalent ions into copper and nickel hexacyanoferrate [43–46].

Water molecules, which typically are present in such materials, screen the larger charge. Such groups, rather than single ions, may move into the structure, replacing one of the Fe ions in the normal ferrocyanides.

22.6.7   Work Toward the Commercialization of Aqueous Electrolyte Batteries Containing Prussian Blue Electrodes

Work on the development of commercial batteries based upon the use of materials in the Prussian Blue family is currently being pursued by the firm Alveo Energy, which Colin Wessells formed after the completion of his PhD program at Stanford University.

22.6.8   Prussian Blue Electrodes in Organic Electrolytes

Although the discussion of the Prussian blue family of materials here has thus far involved their behavior in aqueous electrolytes, they can also be used in appropriate organic electrolytes. The incentive for this work is the fact that some organic electrolytes can be stable over significantly greater ranges of potential than aqueous electrolytes. This can lead to batteries with substantially greater voltages than is possible with aqueous electrolytes.

As described elsewhere in this book, lithium is the typical anode material in current organic electrolyte rechargeable batteries, and it readily inserts into typical current positive electrode materials. However, it cannot be reversibly inserted into the Prussian blue structure from aqueous electrolytes, due to the large size of solvated lithium ions. As a result, lithium insertion and extraction in the Prussian blue family of materials in aqueous electrolytes is not considered practical. Instead, the other alkali metal ions, which have smaller solvation shells, are more favorable.

The first experiments on the behavior of Prussian blue materials in organic electrolytes were reported by A. Eftekhari [45]. He used Prussian blue as a cathode, with a potassium anode, in an EC/EMC electrolyte containing 1 M KBF_4.

The behavior is illustrated in Fig. 22.19.

Fig. 22.19 Cyclic voltammetric behavior of a Prussian blue electrode in a nonaqueous electrolyte at a scan rate of 10 mV/s. From ref. [46]

In organic electrolytes the potential of potassium is only 0.12 V positive of that of lithium, whereas the potential of sodium is 0.32 V positive of lithium. As a result, potassium negative electrode cells have greater voltages than equivalent sodium cells. Potassium is also more attractive for kinetic reasons, for it has a significantly smaller solvation shell than sodium. This leads to quite favorable cycling behavior in nonaqueous electrolytes.

The difference between the cyclability of Prussian blue containing either potassium or lithium in a nonaqueous electrolyte is shown in Fig. 22.20, and the relatively minor change in the capacity of a potassium-containing cell after 500 cycles is illustrated in Fig. 22.21.

Investigations in the Goodenough laboratory at the University of Texas have been reported more recently on the behavior of Prussian blue electrodes with compositions $KMFe(CN)_6$ in which the M species was Mn, Fe, Co, Ni, Cu, and Zn. Although these materials initially contained potassium, they were cycled in an organic electrolyte containing equal amounts of EC and DEC and 1 M $NaClO4$ using sodium as the anode material [47]. They were able to obtain a reversible capacity of over 70 mAh/g in the range of 2.0 to 4.0 volts vs. Na at rates of C/20 in some cases (Fig. 22.22).

Subsequent work in that laboratory [47] involved the investigation of a sodium-only Mn–Fe ferrocyanide using a carbonate electrolyte. Two different compositions resulted in slightly different capacities. In Fig. 22.23 it is seen that these materials still showed appreciable capacities when discharged at very high rates in organic electrolytes.

Fig. 22.20 Difference in the cyclability of lithium and potassium in nonaqueous solutions of KBF$_4$ and LiBF$_4$. The *open circles* are data for lithium, and the *solid circles* for potassium, insertion/extraction. From ref. [45]

Fig. 22.21 Relatively small change in the charge–discharge characteristics of a Prussian blue cell with a potassium anode and a Prussian blue cathode at a C/10 rate after 500 cycles. From ref. [45]

Fig. 22.22 Charge–discharge data for Prussian Blue electrodes with several different compositions in an organic electrolyte [46]

Fig. 22.23 Behavior of Mn-Fe ferrocyanide electrodes at high rates (after [47])

22.7   A New Class of Composite Anodes

The simplest way to make a useful cell with these attractive Prussian blue cathode materials would be to use large surface area activated carbon (AC) as the anode. However, when using such a capacitive anode the cell voltage varies significantly with the state of charge. This is illustrated in Fig. 22.24.

To avoid this disadvantage, a new class of anodes that are compatible with the open framework CuHCF and NiHCF materials in aqueous electrolytes have been developed [48–50]. These anodes are based on a hybrid microstructure that operates

Fig. 22.24 Variation of the potential during galvanic cycling of a cell with CuHCF and activated carbon electrodes at a 1 C rate [48]

by a new concept: by combining an electrode material (polypyrrole, PPy) that undergoes a faradaic anion doping/dedoping reaction at a fixed potential with a capacitive electrode (activated carbon), the potential of the entire electrode can be controlled.

Fundamentally different from either traditional capacitive or active battery electrodes, this new hybrid electrode has the high rate capability of a capacitor, but with the well-defined electrochemical potential of a battery electrode. It has an attractive open circuit potential (OCP), tunable to -0.2 V versus SHE, a shallow charge–discharge profile and leads to no cell self-discharge. Furthermore, a full cell with this hybrid anode and a CuHCF cathode has shown performance that is promising for grid-scale and other stationary storage applications: high power and energy efficiency, and a lifetime of thousands of cycles.

It has been shown that the potential of this type of negative electrode can be modified by reducing the polypyrrole with $NaBH_4$. This increases the voltage of the cell, as illustrated in Fig. 22.25.

The resultant properties of the full cell at a 10 C rate are illustrated in Fig. 22.26.

The negligible variation of the capacity and Coulombic efficiency with cycling is illustrated in Fig. 22.27. It can be seen that there is no measurable capacity loss after 1,000 cycles at the 10 C rate.

Full cells with this ferrocyanide/stabilized carbon electrode combination have also been shown to have very attractive kinetic properties. The influence of the charge–discharge rate upon the cell voltage and capacity is illustrated in Fig. 22.28.

The energy efficiency and capacity retention during cycling at different rates are shown in Fig. 22.29.

Fig. 22.25 Variation of the potential during galvanic cycling of a cell with CuHCF and activated carbon electrodes containing 10 % polypyrrole at a 1 C rate [48]

Fig. 22.26 Potential profiles of copper hexacyanoferrite (CuHCF) positive electrode, 10 % polypyrrole (PPY)/activated carbon (AC) negative electrode and full cell voltage measured at rate of 10C [48]

Fig. 22.27 Variation of the specific capacity and Coulombic efficiency with cycle number [48]

Fig. 22.28 Full cell potential profiles at 1 C, 10 C, and 50 C rates [48]

22.8   An Alternative, Extension of the Stability Range of Aqueous Electrolytes

Although the stability range of aqueous electrolytes is often quoted as 1.23 V, based upon the Gibbs free energy of formation of H_2O, that value is actually only applicable to pure water at ambient temperature. If the activity of water at the

Fig. 22.29 Energy efficiency and fractional capacity retention as a function of the C rate [48]

electrochemical interface is decreased, the stability range can be appreciably extended, allowing larger cell voltages. This can be an attractive alternative to the use of organic electrolytes, for aqueous electrolytes are inherently nonflammable, and can have significantly higher ionic conductivity, and appreciably lower costs than the common organic electrolytes.

There are two general ways in which this can be accomplished. One is to have an additional electrolyte present that acts in series with the aqueous electrolyte. As mentioned in Chap. 18, the potential of a Zn electrode in aqueous batteries is 0.43 V negative of the theoretical potential for the evolution of hydrogen, due to the presence of a thin ionically-conducting, but electronically-insulating layer of ZnO on its surface. Similarly, as shown in Chap. 17, lead-acid cells typically operate at voltages between 2.0 and 2.15 V, and hydrogen and oxygen do not evolve until 2.4 V because of a dense corrosion film of electronically-insulating, but ionically-conducting $PbSO_4$. Metal hydride/nickel cells operate at 1.34 V, and oxygen evolution does not start until 1.44 V, due to the formation of an electrically-conducting, but proton-conducting layer of $Ni(OH)_2$ in contact with the electrolyte, as discussed in Chap. 19. In all of these cases the operating voltage can exceed the thermodynamic stability range of pure water.

A second situation that can also lead to the use of larger cell voltages is the reduction of the chemical potential of water by dissolving other species into it. Concentrated salt solutions can have practical stability ranges significantly greater than that of pure water. This is because a large fraction of the water molecules in solution participate in the hydration shells of the various ions, greatly reducing the mobility of the protons and hydroxyl ions present in the water. This is a kinetic, rather than thermodynamic, effect, but it reduces the effective chemical activity of the water, and extends its range of practical stability.

Fig. 22.30 (**A**) full cell voltage, (**B**) working electrode and (**C**) counter electrode potentials for Li_2SO_4 and $LiNO_3$ as a function of their concentrations at a current density of 50 µA/cm^2 [51]

The resultant low leakage current densities across concentrated aqueous salt solutions allow the practical use of larger cell voltages than is possible with pure water. At low leakage current densities, e.g.10 µA/cm^2, such electrolytes can have useful voltage windows of about 2 V. If leakage current densities up to 50 µA/cm^2 can be tolerated, the useful cell voltage can be up to 2.3 V.

If the applied voltage is above the equilibrium stability limit of an aqueous electrolyte, a leakage current will be present that will act to produce self-discharge of any aqueous electrolyte battery. The magnitude of the self-discharge rate depends upon the relationship between the electrode potentials and the stability limits of the electrolyte.

As the span of the useful electrolyte window depends upon the current density, its practical value depends upon the allowable rate of self-discharge. Given the proper choice of electrode and electrolyte materials, aqueous electrolyte batteries may successfully operate at voltages well above the nominal thermodynamic stability range of pure water, 1.23 V.

The location of the electrolytic stability range of pure water depends upon the pH, varying approximately 0.059 V per pH unit.

Concentrated $LiNO_3$ and Li_2SO_4 have been used in aqueous lithium battery experiments [51, 52]. They have neutral values of pH and have stability ranges that vary with the salt concentration. They have a stability range greater than 2.3 V at a leakage current density of 50 µA/cm^2, as shown in Fig. 22.30.

Experiments have shown that there is a linear relationship between the potential stability range and the logarithm of the current density in both of these concentrated salt solutions. In Fig 22.31 it can be seen that the total stability range of water with 5 M $LiNO_3$ is slightly greater than that with 2 M Li_2SO_4 at all current densities.

Similar experiments were performed with a number of other salts, which also showed that the practical stability range can be extended appreciably beyond the value for pure water.

Fig. 22.31 Semi-logarithmic plots of the full cell voltage, and working (W.E.) and counter electrode (C.E.) potentials of 5 M LiNO$_3$ (**a**) and 2 M Li$_2$SO$_4$ (**b**) as a function of current density [51]

All these electrolytes exhibited minor values of leakage current, which varied with the applied voltage. As mentioned above, a linear relationship was found between the logarithm of the current and the voltage. This is what is expected if there is minority electron or hole leakage through a liquid electrolyte [53, 54] and is also observed in experiments on minority electronic transport in solid electrolytes [55].

This behavior is also consistent with the empirical Tafel approximation of the general Butler-Volmer "activated complex" model of current transport across the "electron-transfer-limited region" of the electrolyte/electrode interface.

During operation, the voltages of batteries generally decrease from their maximum values as they are discharged. The result is that their rates of self-discharge also decrease. Thus this is often not a serious problem with high voltage aqueous cells.

22.9   Batteries With Liquid Electrodes

In discussions of batteries it is commonly assumed that they have solid electrodes and a liquid electrolyte. One exception to this is discussed in Chap. 12, however, the "Zebra" cell. In that case one of the electrodes is liquid sodium, and the other is solid NiCl$_2$ that is permeated by liquid NaAlCl$_4$. That arrangement could be described as a L/S/L,S configuration.

In this chapter some other battery types are discussed in which one or both of the electrode reactants are liquids.

22.10   Sodium/Sulfur Batteries

A type of battery that is beginning to be used for storing energy in large scale systems is the so-called *sodium/sulfur battery* that operates at 300–350 °C. This electrochemical system is best described as a Na/Na$_x$S cell. These batteries are different

Fig. 22.32 Schematic view of Na/Na$_x$S cell

from the common systems that most people are familiar with, for the electrodes are liquids, instead of solids, and the electrolyte is a solid sodium ion-conducting ceramic solid, NaAl$_{11}$O$_{17}$, called *sodium beta alumina*. This is thus a L/S/L configuration, which is the inverse of the conventional S/L/S, arrangement.

The sodium ion conductivity in this ceramic material, discovered by Yao and Kummer, is remarkably high at the operating temperature [53, 54, 56], about $4\,\Omega$ cm at 350 °C. The possibility that this material could be used to construct the revolutionary sodium/sulfur battery was soon pointed out by Weber and Kummer [55]. A general reference that contains a lot of information about sodium/sulfur cells is [57].

In this case the negative electrode is molten sodium, and the positive electrode is the product of the reaction of sodium with liquid sulfur. Thus the basic reaction can be written as

$$x\mathrm{Na} + \mathrm{S} = \mathrm{Na}_x\mathrm{S} \tag{22.1}$$

The general construction of such batteries is shown schematically in Fig. 22.32.

Sodium from the negative electrode passes through the surrounding solid beta alumina cylinder, and reacts with a liquid solution of sodium in sulfur, Na$_x$S. This liquid, which is not a good electronic conductor, is contained in a porous carbon "sponge" to provide electrical contact to the positive current collector.

The capacity is determined by the composition range of this sodium-sulfur liquid phase. The relevant portion of the Na-S phase diagram is shown in Fig. 22.33. It is seen that at about 300 °C only a relatively small amount of sodium can be dissolved in liquid sulfur. When this concentration is exceeded, a second liquid phase, with a composition of about 78 atomic percent Na, is nucleated. This has a composition that extends to roughly Na$_{0.4}$S. As more sodium is added the overall composition

Fig. 22.33 Part of the sodium-sulfur phase diagram

traverses the two-phase region, and the amount of this liquid phase increases relative to the amount of the sulfur-rich liquid phase. Thus a potential plateau is expected over this two-phase composition range. When the sodium concentration exceeds that corresponding to about $Na_{0.4}S$ the overall composition moves into a single-phase liquid range, and thus the potential varies with the composition. The maximum amount of sodium that can be used in this electrode corresponds roughly to $Na_{0.67}S$.

At higher sodium concentrations a solid second phase begins to form from the liquid solution. This tends to form at the interface between the solid electrolyte and the liquid electrode, and prevents the ingress of more sodium, thus blocking further reaction.

The potential of the elemental sodium in the negative electrode is constant, independent of the amount of sodium present. The potential of the positive electrode, and thus the voltage of the cell, changes as the sodium concentration varies by its transport across the cell. The variation of the potential of the positive electrode with its composition is shown in Fig. 22.34.

Early work in both the United States and Europe on this type of cell was aimed toward its potential use for vehicle propulsion. In that case, safety is especially

Fig. 22.34 Voltage versus pure sodium as a function of composition

important, and extensive testing related to the use of these cells in vehicles under crash conditions was performed in Europe in 1990s. The results of these tests were discouraging, and all of those development programs were discontinued.

On the other hand, activities in Japan were aimed at a different application, utility power storage, where they can be kept in a protective environment so that safety considerations can be minimized. A large effort was undertaken by a consortium of NGK Insulators and Tokyo Power (TEPCO) in 1983, and after extensive large scale testing, large sodium/sulfur cells became commercially available in 2000.

The large individual cells, roughly the size of a baseball bat, are enclosed in steel casings for safety reasons, and they can be arranged into parallel and series groups in order to provide the required voltage and capacities.

However, a major fire occurred in a large sodium/sulfur battery system in Tsukuba, Japan in September, 2011. It involved the leakage of hot liquids from one cell into a number of nearby ones, and then propagated into a serious problem for a large system, which, fortunately, was in an outside location. This caused widespread concern about the safety of even stationary sodium/sulfur battery systems, and NGK asked that all large batteries of this type be shutdown until the cause of the problem was identified, and a solution developed that would prevent a recurrence of this problem. This was done, and a number of design changes were made, including the installation of fuses on all individual cells.

Modified systems with power values up to 6 MW at 6.6. kV are now being produced and used in Japan. This technology is also beginning to be installed in the United States, with facilities currently up to a 1 MW size.

22.11 Flow Batteries

22.11.1 Introduction

Except for the Na/Na_xS cell, and the Zebra cell, that was discussed briefly in Chap. 12, all of the electrochemical cells that are generally considered have electrodes that are solids. In those two cases liquid electrodes can be used because the electrolyte is a solid, resulting in an L/S/L configuration.

There is another group of cells that have liquid electrode reactants, although their electrode structures contain porous solid current collectors. These are generally called "flow batteries," since the liquid reactant is stored in tanks and is pumped (flows) through the cell part of the electrochemical system. Thus such systems can also be considered to be rechargeable fuel cells.

Early work on flowing electrode systems was done by Lawrence Thaller at the NASA laboratory in Ohio during the 1970s. He used an iron–chromium electrochemical combination. The effective valence of iron changed from Fe^{2+}/Fe^{3+}, and that of chromium from Cr^{2+}/Cr^{3+} at a nominal voltage of 1.18 V. However, these early cells suffered from severe cross-contamination, which resulted in rapid capacity decay.

This problem can be mitigated somewhat by using a premixed electrolyte on both sides [58]. This general approach is currently being pursued by the firm Deeya Energy, Inc.

Following this early work a number of chemical systems have been explored, and in some cases rather fully developed. However, most of them have not been commercially successful to date. As seen below, this could well change in the near future. A total of 13 programs that involve development efforts on various types of flow batteries are currently receiving financial support from the US government agency ARPA-E.

The general physical arrangement is shown in Fig. 22.35, whereas the configuration of the cell portion of the system is shown schematically in Fig. 22.36.

It can be seen that this is also a type of L/S/L configuration. The electrolyte is a proton-conducting solid polymer, and the electrode reactants are liquids on its two sides. In the Zebra cell the reactants are both electronically-conducting, whereas in the flow cells the electrode reactants are ionic aqueous solutions that are electronic insulators. In order to get around this problem and provide electronic contact to an external electrical circuit, the liquid reactants permeate an electronically conducting graphite felt. This felt provides contact, both to the polymer electrolyte and to a graphite current collector.

The electrode reactants are typically acidic, e.g., 2 M H_2SO_4 aqueous solutions of ions that can undergo redox reactions. The function of the polymer electrolyte is to transport protons from one side to the other, thus changing the pH and charges on the dissolved redox ions.

An important difference from the Na/Na_xS and Zebra cells is that the reactant materials, the redox ion solutions, can be pumped into and out of the electrode

FLUID FLOW   FLUID FLOW

TANK   CELL   TANK

−   +

ELECTRICAL OUTPUT

Fig. 22.35 General physical arrangement of a flow battery

Fluid Flow Through Felt

Polymer Electrolyte

Carbon Felts
Containing
Redox Solution

Current Collectors

Fig. 22.36 The cell portion of the system. In some cases there are multiple bipolar cell configurations

compartments. This means that the capacity is not fixed by the cell dimensions, but is determined by the size of the liquid electrode reactant tanks. This can result in very large capacities, and is one of the potential advantages of flow battery systems. Thus flow batteries deserve consideration for relatively large stationary applications, such as remote solar or wind installations, whose outputs are dependent upon the time of day and/or the weather.

Table 22.2 Various redox systems used in flow batteries

System	Negative electrode reactant	Positive electrode reactant	Nominal voltage
V/Br	V	Bromine	1.0
Cr/Fe	Cr	Fe	1.03
V/V	V	V	1.3
Sulfide/Br	Polysulfide	Bromine	1.54
Zn/Br$_2$	Elemental Zinc	Bromine	1.75
Ce/Zn	Zn	Ce	<2

The open circuit voltage across the electrolyte is determined by the difference in the chemical potentials on its two sides. As current passes through the cell protons are transferred, changing the pH, so that the ionic compositions of the two electrode reactant fluids gradually change. Thus the cell potential varies with the state of charge. The change in the voltage with the amount of charge passed depends, of course, upon the size of the tanks.

Some of the redox systems have been explored are listed in Table 22.2.

A general discussion of these various alternatives can be found in reference [58].

There is some confusion in the terminology used to describe these systems, for the liquid reactants on the two sides are sometimes called "electrolytes," even though they do not function as electrolytes in the battery sense. Additionally, the liquid reactant on the negative side of the cell is sometimes called the "anolyte," and that on the positive side of the cell the "catholyte."

One of the most attractive flow systems involves the vanadium redox system [59–65]. In this case the negative electrode reactant solution contains a mixture of V^{2+} and V^{3+} ions, whereas the positive electrode reactant solution contains a mixture of V^{4+} and V^{5+} ions. Charge neutrality requirements mean that when protons (H^+ ions) are added or deleted from such liquids by passage through the polymer electrolyte in the cell, the ratio of the charges on the redox species is varied. This changes the state of charge of the system.

These systems are generally assembled in the uncharged state, in which the chemical compositions of the two liquid reactants are the same. In the vanadium system, this is done by adding vanadyl sulfate to 2 M H_2SO_4, which gives an equal mixture of V^{3+} and V^{4+} ions. The system is then charged by passing current, causing the transport of protons through the polymer electrolyte, so that the ion contents on the two sides become different.

22.11.2 Redox Reactions in the Vanadium/Vanadium System

One can write the reactions in the electrode solutions of the vanadium system as

$$VO_2^+ + 2H^+ + e^- = VO^{2+} + H_2O \qquad (22.2)$$

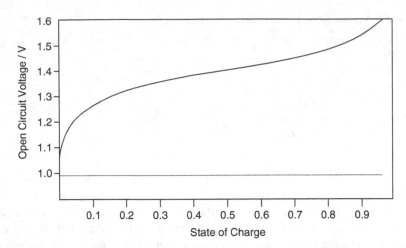

Fig. 22.37 Variation of the open circuit potential versus state of charge for the case of a V/V flow cell at 298 K

or, in terms of the vanadium ions:

$$V^{5+} + e^- = V^{4+} \tag{22.3}$$

in the positive electrode reactant solution, and

$$V^{2+} = V^{3+} + e^- \tag{22.4}$$

in the negative electrode reactant solution.

So that the overall reaction is

$$VO_2' + 2H^+ + V^{2+} = VO^{2+} + H_2O + V^{3+} \tag{22.5}$$

or

$$V^{5+} + V^{2+} = V^{4+} + V^{3+} \tag{22.6}$$

The variation of the open circuit cell potential with the state of charge in the case of the V/V system with concentrations of 2 mol/liter of each V species is shown in Fig. 22.37. Typical operation would involve cycling between 20 and 80 % of capacity, and thus at voltages between 1.3 and 1.58 V.

Since each cell produces a relatively low voltage, such batteries generally contain a number of cells arranged in series in order to produce a greater overall output voltage. Parallel configurations can be used to provide higher currents. Depending upon the application, it may be desirable to permit relatively rapid charging, which may not be necessary during discharge. Thus it may be advantageous to include a mechanism to change the number of cells and their series/parallel arrangements during different operating conditions.

If all of the cells are fed from a common liquid supply, this can result in a large voltage applied across the liquid reactants and in the passage of a considerable amount of current. This is a form of self discharge, and is sometimes called shunt or bypass current. It is different, however, from the self discharge that results from neutral species, or neutral combinations of species, traveling through or around the electrolyte in other types of electrochemical systems.

Because there are no solid-state volume changes during charging and discharging, as are typical for electrochemical cells with solid electrodes, the components of the cells, as well as the total system, can have long lives. Thus long cycle life is generally not a problem, even with repeated deep charges and discharges.

The vanadium redox system can be used over a temperature range from 10 to 35 °C, and typically operates at or near ambient temperature. The solubility of (VO) SO_4 limits the energy capacity at lower temperatures. At higher temperatures the current density increases, but the cell voltage is reduced somewhat. The overall result is that the available power is greater at somewhat elevated temperatures. But care must be taken to not let the temperature go over 40 °C to avoid the precipitation of V_2O_5.

The electrode kinetics are good, and additional catalysts are not required. The coulometric and voltage efficiencies are high, except for the self-discharge mechanism mentioned above.

The specific energy and energy density are determined primarily by the electrode reactants themselves, which are the major components in these systems. Typical values are 15 Wh/kg and 18 Wh/l, and round trip efficiencies are typically 70–75 %.

Since the electrode reactants both consist of vanadium sulfate solutions in aqueous sulfuric acid, only differing by the oxidation states of the vanadium ions, contamination by leakage across the electrochemical cell membranes only results in some capacity loss, and is fully reversible. In flow batteries in which the ions are different on the two sides this can become a significant, and irreversible, problem.

Because the cell voltage is a function of the state of charge, it is possible to determine the state of charge of such systems remotely, which may be an advantage in some system installations. The cell design also makes monitoring of the voltage across each cell possible.

Since the cells can be configured in a variety of different series/parallel arrangements, the charging and discharging cycles can operate at different voltages. As a result, such a system can be used as a DC/DC converter.

The general attractive features of redox flow batteries include their long lifetime, which can involve an unlimited number of cycles of charging and deep discharge cycles. The typical reaction time of less than 100 milliseconds means that they can be used to support solar and wind systems that sometimes suffer large sudden transients in their output. The all-vanadium systems typically suffer only 1 % energy loss per year, and have a high level of safety, as they are nonflammable and nonexplosive. They require relatively little maintenance, and have the advantage that the output and storage capacity are scalable independently of each other. Vanadium is environmentally friendly and recyclable.

The leading firm in this area is Cellstrom in Austria. As of the end of 2013 more than 50 CellCube systems had been sold to customers all over the world. This firm was founded in 2000 by DDr. Martha Schreiber, and was sold to the large German manufacturing firm Gildemeister in 2010. It is now named DMG MORI SEIKI.

22.11.3　Flow Batteries with a Modified Chemistry

A different approach to vanadium/vanadium flow batteries was introduced a few years ago by work at the Pacific Northwest National Laboratory led by Zhenguo (Gary) Yang. It involves the use of a sulfate/chloride mixed electrolyte, rather than the sulfate-only electrolyte [66, 67].

It was found that all four valence states of vanadium can be stable at concentrations up to 2.5 molar in a mixed solution containing 2.5 M SO_4^{2-} and 6 M Cl^- ions. This electrolyte produces a significantly higher energy capacity than in the standard sulfate-only electrolyte. Also, the operating temperature range can be increased from 10–40 °C to −5 to 50 °C by the use of the mixed acid electrolyte. The result is a significantly (about 70 %) increased energy density compared to the normal sulfate-only system. The voltage window of this mixed electrolyte is essentially the same as that of the standard 0.5 to 1.35 V, and there is no significant capacity fading upon cycling.

The attractive features of this new approach to flow batteries have led to the formation of the new firm UniEnergy Technologies, which intends to commercialize this technology.

22.12　All-Liquid Batteries

A relatively new concept that could be useful for large stationary storage applications is the use of an all-liquid three-component system that evolved from the PhD research of David Bradwell under the supervision of Prof. Donald Sadoway at MIT. The basic concept involves the use of two electronically-conducting liquid electrode materials that have densities different from that of an elevated temperature molten salt electrolyte, one lower, and the other greater. This leads to a three-layer configuration, with the electrolyte in the middle. The container, perhaps with a graphite insert, can act as the current collector for the lower electrode, and the upper electrode material can be contacted by an electronic conductor protruding from above. An inert gas cover, either nitrogen or argon, is needed to prevent reaction with air. This general configuration is illustrated schematically in Fig. 22.38.

Such a device can operate in the same way as conventional batteries with solid electrodes, with a charged species leaving one liquid electrode, travelling across the molten salt electrolyte, and reacting with the other liquid electrode material.

Fig. 22.38   Schematic Representation of an all-liquid cell

There are several potentially important advantages in such an all-liquid configuration, in which the components can be quite large, leading to high power. One is that there are no problems due to the dimensional changes in the electrode reactants, which can cause problems in common systems with fine solid particle battery electrodes. On the other hand, it is necessary to be aware of potential safety problems. Care must be taken to avoid contact of high temperature molten salts and highly reactive metals, such as magnesium, with air, for the results could be catastrophic.

Molten salt electrolytes can have conductivities much greater than both the sulfuric acid and KOH used in aqueous batteries, and the organic solvent electrolytes in lithium batteries. And with the possibility that all-liquid cells can be designed to have large area electrodes in contact with high conductivity electrolyte materials, it is reasonable to expect that such all-liquid cells can operate at very high power values.

An early example of this type of cell was the use of a heavy positive electrode material such as antimony, with a density of 6.5 g/cm^3, on the bottom, and magnesium, a relatively light negative electrode material with a density of 1.6 g/cm^3, on the top, with a mixed chloride electrolyte that has an intermediate density, 4.0 g/cm^3, in the middle [68]. This configuration can be written as a Mg/Sb cell. The reaction product was the solution of Mg into the liquid antimony. These cells were operated at 700 °C, and the open circuit voltage was only about 0.44 V.

Another design alternative can be used when both electrode materials are heavier than the electrolyte. An example of this type would be the use of Zn (density of 7.1 g/cm3) as negative electrode, Te (density 6.3 g/cm3) as positive electrode, and an electrolyte of ZnCl$_2$ (density 2.9 g/cm3). This is shown schematically in Fig. 22.39.

After the exploration of a number of other possible chemical systems, recent development efforts have focused on the use of molten lithium as the negative electrode, and an antimony–lead alloy, on the positive side of the cell [68, 69]. The advantage of the use of the antimony–lead alloy is that it can lead to a substantial reduction in the melting point relative to antimony itself, reducing the minimum operating temperature of such cells. This can be seen from the Pb-Sb phase diagram in Fig. 22.40.

Insulating Separator

Fig. 22.39 Schematic view of a cell in which both electrode materials have greater densities than the liquid electrolyte

Fig. 22.40 Antimony–lead phase diagram

Fig. 22.41 Charge–Discharge Behavior of Li/Pb-Sb Battery at 450 °C. From ref. [69]

Cells with the Li/Pb-Sb chemistry operating at 450–500 °C have an open circuit voltage of about 0.9 V and operate between 0.75 and 0.65 V at a 5-h discharge rate. At discharge rates of C/2 the Coulombic efficiency is over 98 %.

Although this salt electrolyte is liquid in the charged state, two phases are present on discharge, so that care must be given to the discharge rate to avoid transient local freezing. While this may not be a significant problem for relatively short pulses, it would be expected to limit performance at higher rates over an extended period of time. That does not seem to be a problem, however, if the battery is charged until the electrolyte is fully liquid after each cycle. This is due to the fortuitous fact that the transient solid product is Li_3Sb, which has been known for a long time to have an unusually rapid rate of chemical diffusion.

The charge–discharge behavior over a range of current densities at 450 °C is shown in Fig. 22.41.

The cost of such systems will depend to a large extent upon the identity of the materials used in both the electrodes and the electrolyte, which are much less expensive than those used in many other current battery systems. Thus they should be quite low, perhaps between those of pumped hydro and compressed air systems.

References

1. Hunt T, Clark N, Baca W. 18th International Seminar on Double Layer Capacitors and Hybrid Energy Storage Devices, Fort Lauderdale, FL (2008)

2. Furukawa J (2013) Development of Ultra Battery. Furukawa Review No. 43, Furukawa Electric Co
3. Furukawa J, Takada T, Monma D, Lam LT (2010) J Power Sources 195:1241
4. Joo S-K, Raistrick ID, Huggins RA (1985) Mater Res Bull 20:897
5. Joo S-K, Raistrick ID, Huggins RA (1985) Mater Res Bull 20:1265
6. Joo S-K, Raistrick ID, Huggins RA (1985) Solid State Ion 17:313
7. Joo S-K, Raistrick ID, Huggins RA (1986) Solid State Ion 18/19:592
8. Inaguma Y et al (1993) Solid State Commun 86:689
9. Kawai H, Kuwano J (1994) J Electrochem Soc 141:L78
10. Inaguma Y, Chen L, Itho M, Nakamura T (1994) Solid State Ion 70/71, 196:203
11. Inaguma Y, Yu J, Shan Y-J, Itho M, Nakamura T (1995) J Electrochem Soc 142:L8
12. Robertson AD, Garcia Martin S, Coats A, West AR (1995) J Mater Chem 5:1405
13. Birke P, Scharner S, Huggins RA, Weppner W (1997) J Electrochem Soc 144:L167
14. Sauvage F, Laffont L, Tarascon J-M, Baudrin E (2007) Inorg Chem 46:3289
15. Doeff MM, Richardson TJ, Kepley L (1996) J Electrochem Soc 143:2507
16. Doeff MM, Anapolsky A, Edman L, Richardson TJ, DeJonghe LC (2001) J Electrochem Soc 148:A230
17. Whitacre JF, Tevar A, Sharma S (2010) Electrochem Commun 12:463
18. Tevar AD, Whitacre JF (2010) J Electrochem Soc 157:A870
19. Cao Y, Xiao L, Wang W, Choi D, Nie Z, Yu J, Saraf LV, Yang Z, Liu J (2011) Adv Mater 23:3155
20. Whitacre JF, Wiley T, Shanbhag S, Wenzhuo Y, Mohamed A, Chun SE, Weber E, Blackwood D, Lynch-Bell E, Gulakowski J, Smith C, Humphreys D (2012) J Power Sources 213:255
21. Zhou X, Guduru RK, Mohanty P (2013) J Mater Chem A 1:2757
22. Huggins RA (1997) Ionics 3:379
23. Huggins RA (1998) Solid State Ion 113:533
24. Robin MB, Day P (1967) Adv Inorg Chem Radiochem 10:247
25. Brown J (1724) Philos Trans 33:17
26. Weiser HB (1938) Inorganic Colloid Chemistry, Vol 3, Colloidal Salts. John Wiley & Sons, New York, p 343
27. Keggin JF, Miles FD (1936) Nature 577
28. Wilde RE, Ghosh SN, Marshall BJ (1970) Inorg Chem 9:2512
29. Siperko LM, Kuwana T (1983) J Electrochem Soc 130:396
30. Crumblis AL, Lugg PS, Morosoff N (1984) Inorg Chem 23:4701
31. Armand MB, Whittingham MS, Huggins RA (1972) Mater Res Bull 7:101
32. Oi T (1986) In Annual Review of Materials Science R. A. Huggins (Eds), 16, p. 185
33. Itaya K, Uchida I, Neff VD (1986) Acc Chem Res 19:162
34. Itaya K, Ataka T, Toshima S (1982) J Am Chem Soc 104:4767
35. Neff VD (1985) J Electrochem Soc 132:1382
36. Honda K, Hayashi H (1987) J Electrochem Soc 134:1330
37. Wessells CD. PhD Dissertation, Stanford University (2012)
38. Wessells CD, Huggins RA, Cui Y (2011) Nat Commun 2:550
39. Wessells CD, Peddada SV, Huggins RA, Cui Y (2011) Nano Lett 11:5421
40. Wessells CD, Peddada SV, McDowell MT, Huggins RA, Cui Y (2012) J Electrochem Soc 159:A98
41. Wessells CD, McDowell MT, Peddada SV, Pasta M, Huggins RA, Cui Y (2012) ACS Nano 6:1688
42. Wang RY, Wessells CD, Huggins RA, Cui Y (2013) Nano Lett 13:5748
43. Lee H-W, Pasta M, Wang RY, Ruffo R, Cui Y (2014) Faraday Disc 176:69
44. Lee H-W, Wang RY, Pasta M, Lee SW, Liu N, Cui Y (2014) Nat Commun 5:5280
45. Eftekhari A (2004) J Power Sources 126:221
46. Lu Y, Wang L, Cheng J, Goodenough JB (2012) Chem Commun 48:6544

47. Wang L, Lu Y, Liu J, Xu M, Cheng J, Zhang D, Goodenough JB (2013) Angew Chem Int Ed 52:1964
48. Pasta M, Wessells CD, Huggins RA, Cui Y (2012) Nat Commun 3:2139
49. Huggins RA (2013) J Electrochem Soc 160:A3020
50. Pasta M, Wessells CD, Liu N, Nelson J, McDowell MT, Huggins RA, Toney MF, Cui Y (2014) Nat Commun 5:3007
51. Wessells C, Ruffo R, Huggins RA, Cui Y (2010) Electrochem Solid-State Lett 13:A59
52. Wessells C, Huggins RA, Cui Y (2011) J Power Sources 196:2884
53. Radzilowski RH, Yao YF, Kummer JT (1969) J Appl Phys 40:4716
54. Whittingham MS, Huggins RA (1971) J Chem Phys 54:414
55. Weber N, Kummer JT (1967) Proc Ann Power Sources Conf 21:37
56. Yao YFY, Kummer JT (1967) J Inorg Nucl Chem 29:2453
57. Sudworth JL, Tilley AR (1985) The Sodium Sulphur Battery. Chapman and Hall, London
58. Gahn RF, Hagedorn NH, Ling JS. DOE/NASA/12726-21 (1983)
59. Ponce de Leon C, Frias-Ferrer A, Gonzales-Garcia J, Szanto DA, Walsh FC (2006) J Power Sources 160:716
60. Sum E, Skyllas-Kazacos M (1985) J Power Sources 15:179
61. Sum E, Rychcik M, Skyllas-Kazacos M (1985) J Power Sources 16:85
62. Skyllas-Kazacos M, Rychcik M, Robins R, Fane A, Green M (1985) J Electrochem Soc 133:1057
63. Rychcik M, Skyllas-Kazacos M (1987) J Power Sources 19:45
64. Rychcik M, Skyllas-Kazacos M (1988) J Power Sources 22:59
65. Skyllas-Kazacos M, Grossmith F (1987) J Electrochem Soc 134:2950
66. Li L, Kim S, Wang W, Vijayakumar M, Nie Z, Chen B, Zhang J, Xia G, Hu J, Graff G, Liu J, Yang Z (2011) Adv Energy Mater 1:394
67. Wang W, Luo Q, Li B, Wei X, Li L, Yang Z (2013) Adv Funct Mater 23:970
68. Bradwell DJ, Kim H, Sirk AHC, Sadoway DR (2012) J Am Chem Soc 134:1895
69. Kim H, Boysen DA, Newhouse JM, Spatocco BL, Chung B, Burke PJ, Bradwell DJ, Jiang K, Tomaszowska AA, Wang K, Wei W, Ortiz LA, Barriga SA, Poizeau SM, Sadoway DR (2013) Chem Rev 113:2075
70. Wang K, Jiang K, Chung B, Ouchi T, Burke PJ, Boysen DA, Bradwell DJ, Kim H, Muecke U, Sadoway DR (2014) Nature 514:348

Chapter 23
Storage of Energy for Vehicle Propulsion

23.1 Introduction

Most vehicles are propelled by internal combustion motors that consume liquid fuels, either gasoline or diesel fuel. In those cases, the energy storage mechanism is a simple tank to hold the liquid fuel.

Over the years, the commercial introduction of electrically powered automobiles has not generally been successful, due to their high cost and limited performance, compared to what is typical of those with internal combustion motors. This is due primarily to the weight and cost of the batteries required in order to provide what is perceived to be sufficient driving performance and range.

The characteristics needed to meet the requirements for electric vehicle propulsion depend greatly upon the type of duty cycle that is assumed. Extensive measurements of actual vehicle usage were undertaken, and from them, models corresponding to typical usage patterns were established. One of these, known as the ECE-15 cycle, was developed for all-electric vehicles. It was composed of two parts, an urban part that simulated the needs during local travel, and a suburban part that required higher power levels, such as what is needed for travel at higher velocities and greater distances. These were both expressed in terms of power—time profiles, and are shown in Figs. 23.1 and 23.2.

Several auto manufacturers undertook programs to develop electrically powered vehicles in the 1990s. This was in response to a mandate of the California Air Resources Board that required auto manufacturers to develop and produce zero-emission vehicles. The requirement was that at least 2 % of the vehicle fleet sold by any manufacturer in the state of California must be a zero-emission type by 1998, and the zero-emission fraction was expected to increase in later years.

One of the most visible responses was the manufacture of almost 1000 electric vehicles, a model called the EV1, by General Motors. Other auto companies developed prototype vehicles, and some, including Honda, actually put a number on the road.

© Springer International Publishing Switzerland 2016
R.A. Huggins, *Energy Storage*, DOI 10.1007/978-3-319-21239-5_23

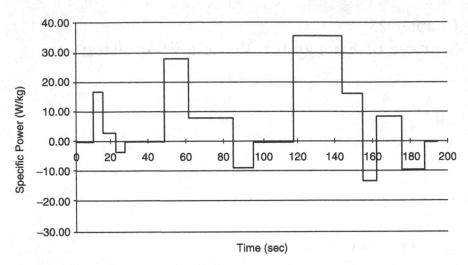

Fig. 23.1 Power demand profile for the ECE-15 reference vehicle in urban travel simulation

Fig. 23.2 Power demand profile for the ECE-15 reference vehicle in suburban travel simulation

The EV1 cars were leased, rather than sold. The early ones were powered by lead-acid batteries, and had a range of 80–100 miles under light duty conditions. Hydride/nickel batteries were installed in a later model, which gave a range of 100–140 miles.

After much political pressure from the auto industry and some changes in the composition of the Air Resources Board, the California mandate was eliminated in 2002, and EV1 production was terminated. All the existing cars were repossessed by General Motors and destroyed. The company clearly did not want to have them on the road.

It can be seen in both of the ECE-15 profiles that there are periods in which the required power is negative. This means that the propulsion system can absorb kinetic energy from the motion of the vehicle. Hybrid systems, in which energy can be temporarily stored in a capacitor or battery have been developed for this purpose. This is briefly discussed in Chap. 6.

A number of manufacturers are now successfully selling hybrid vehicles, in which the propulsion is provided by a combination of a modest internal combustion engine and a relatively small battery. The control system allows the engine to operate at relatively high efficiency, and the battery gets some charging during braking, as well as from the motor. A number of the automobile manufacturers are now producing such vehicles.

Batteries for use in these hybrid vehicles are relatively small, with energy capacities typically about 3 kWh. On the other hand, they are designed to emphasize high power performance so that they can rapidly absorb, and deliver, energy. Some provide up to 60 kW, giving a power to energy ratio of 15–20. Because of their modest size, they are not particularly expensive.

Another type of hybrid vehicle is beginning to appear, fuel cell-battery hybrids. Increasingly successful development efforts have been underway by Daimler Benz and BMW in Europe and Honda in Japan, for some time. The most advanced of this type at the present time is the Honda FCX Clarity. These autos, which are mid-size sedans, get their basic propulsion from a fuel cell, which also charges a relatively small high rate lithium battery. When more power is needed for a modest time, the battery kicks in to supplement the fuel cell output. When this is not the case, part of the fuel cell output is used to recharge the battery. A sophisticated control system is used to integrate these two systems to produce very impressive performance.

It is interesting that the Department of Energy in the USA recently decided to discontinue the support of activities aimed at the use of fuel cells in vehicles.

There is currently rapid growth of another type of hybrid alternative, plug-in hybrid vehicles. In this case, the battery part of the system can be electrically recharged, perhaps overnight at home, to provide sufficient energy to propel the vehicle a modest distance—say 20–50 miles—without the use of the internal combustion engine at all. It has been found that a large fraction of the total mileage travelled by vehicle owners in the USA is due to short distance trips, such as going to work and back. Thus a short electrical range will be sufficient to handle much of the total transportation need. This sounds attractive, as the cost of electricity per mile of travel is less than the cost of the equivalent amount of liquid fuel.

This movement toward electrically-propelled vehicles is also present in other countries. The German government, which has been heavily promoting the use of both solar energy and wind energy, has set a goal of having 1 million electric cars on its roads by 2020.

There is increasing interest in start–stop operation, in which the engine is turned off when it is not needed for propulsion, such as at stop lights, and restarted when the vehicle is put ion motion again. Start-up technology has been greatly improved in a number of vehicles in recent years, often requiring less than one revolution of the motor.

In start–stop operation there are actually two types of loads on the battery. Restarting the engine typically requires about 300–400 Amp secs, but there are also "hotel loads" on the electrical system in automobiles that are not directly connected with the operation of the motor, such as lighting, instruments, and electrical equipment used to provide comfort to the passengers. In a typical automobile, these can amount to up to 3000 Amp secs during start–stop engine off time.

The result is that during a series of start–stop sequences without time to recharge the battery has to provide much more energy than for a single start–stop. One estimate was that for a sequence of 15 start–stops the battery has to do 100 times as much work as for a single start–stop event.

There is also movement toward the use of other new technologies. The German company Bosch has been developing a start–stop system in which the engine is turned off when the vehicle is in a coasting mode, and restarted when needed.

23.2   ZEBRA Batteries

There was a brief discussion of the ZEBRA battery, which is based upon the Na–Ni–Cl ternary system, and is sometimes called the $Na/NiCl_2$ battery, in Chap. 12. This system, which evolved from earlier work on the Na/Na_xS battery, was invented in South Africa [1, 2], and has had a long and tortuous road toward commercialization [3, 4]. This involved work at BETA Research and Development Ltd. in England, and a joint effort of AEG (later Daimler) and Anglo American Corp., AEG Anglo Batteries, GmbH started pilot line production. After the merger of Daimler and Chysler, this activity was terminated, and the technology sold to MES DEA S.A. in Stabio, in southern Switzerland near the Italian border, in the late 1990s. MES DEA was sold to FZ Sonick S.A. in February, 2010. The name ZEBRA stands for Zeolite Battery Research Africa, and is a holdover from the initial idea that the ceramic solid electrolyte would be a zeolite material. General Electric is now beginning to work on this system, which they call "Durathon" in the USA.

From the start, it was intended that these cells would be used for vehicle propulsion. Modest numbers have now been produced, and used in the Twingo and the Panda autos in Switzerland and Italy, and the Think City in Norway.

The general configuration is similar to that of the sodium/sulfur cells in that the negative electrode is liquid sodium, and the electrolyte is a solid electrolyte, sodium beta alumina. However, the positive electrode contains both the solid $NiCl_2$ reactant and a second liquid electrolyte, $NaAlCl_4$. However, the positive electrode is on the inside, and the negative electrode on the outside in this case.

ZEBRA cells are produced in the discharged state, with all of the sodium present as NaCl on the positive side. They are constructed with excess sodium, so the amount of $NiCl_2$ determines the capacity. The operating temperature is kept within the range 270–350 °C, and the open circuit voltage is 2.59 V, in accordance with thermodynamic data, as discussed in Chap. 12. The theoretical specific energy of individual cells is 790 Wh/kg, which is slightly greater than that of Na/Na_xS cells, 760 Wh/kg.

Groups of batteries are encased in a temperature-controlled container, and the configuration is designed to produce a ratio of power to energy of about two, 50 kW peak power, and 25 kWh energy in one model. On a weight basis, the complete ZEBRA battery system stores about 120 Wh/kg specific energy. An attractive feature is that these cells are fully reversible, with 100 % ampere hour efficiency. It is claimed that at this stage of development the life cycle costs are less than those of lead-acid batteries, despite higher initial costs, due to their much longer lifetime.

Safety tests in Europe have indicated that these batteries are significantly safer than Na/Na_xS cells, and do not represent a significant risk under simulated crash conditions. Both details of the design and several features of the chemistry provide protection against both overcharge and overdischarge were discussed in [4].

23.3 General Comments on Hybrid System Strategies

There is a great variation in the requirements for transient power sources, and in some cases no one type of device, or any one design, will be able to optimally fulfill such diverse needs.

Hybrid systems can include components that meet two different types of needs, a primary energy source, and a supplemental source that can meet transient requirements for higher power levels than can be handled by the primary source, but has a relatively small energy capacity. This combination can be represented schematically in terms of the commonly used Ragone type of diagram, in which the specific power is plotted versus the specific energy, both on logarithmic scales, as shown in Fig. 23.3.

A possible strategy to consider in order to accomplish this is to use a high energy system that operates at a relatively high voltage when the power demand is low. The output voltage of such energy sources typically falls off as the output current is increased. If a second high-power, but lower-energy source that operates at a lower voltage is placed in parallel, it will take over under the conditions that drive the

Fig. 23.3 Typical hybrid system characteristics

Fig. 23.4 Schematic representation of possible hybrid system strategy

output voltage of the primary high-energy system down into its range of operating voltage, and meet the high-power demand for a short period. When this demand is no longer present, the voltage will again rise, and the high-power component of the system will be recharged by the higher-energy component. This combination is shown schematically in Fig. 23.4.

The way that the properties of batteries are typically described, such as by a graphical display of the discharge (voltage vs. state of charge) curves at different constant current densities, or in terms of the change of extractable capacity as a function of the number of discharge cycles, cannot be considered to provide a satisfactory description of behavior in these very different types of applications. Likewise, the value of the capacitance at a single frequency is also certainly not a satisfactory description of the behavior of a capacitor over such a wide range of potential uses.

In order to approach the development of useful devices for this type of application one should consider the several types of charge storage mechanisms that can be employed, their thermodynamic and kinetic characteristics, and the basic properties of candidate materials, as well as the relationships that determine system performance.

References

1. Coetzer J, in Proceedings of the 170th Meeting of Electrochemical Society, San Diego, CA, USA, October 1986, Extended Abstract No. 762
2. Galloway RC (1987) J Electrochem Soc 134:1
3. Sudworth JL (2001) J Power Sources 100:149
4. Dustman C-H (2004) J Power Sources 127:85

Chapter 24
A Look at the Future

24.1 Introduction

When considering what changes and new developments are possible, or even likely, in the years ahead, it is realistic to consider a quotation attributed to Thomas A. Edison:

> Making predictions can be rather precarious, especially when they have to do with the future.

Nevertheless, several things are rather obvious. One is that the need for energy storage will certainly grow substantially. This is not just due to the natural growth of all the technologies in which storage is an important component, but also to important changes in the energy production and utilization landscape. There is a greatly increased emphasis upon energy production methods based upon energy sources other than the various fossil fuels. But energy production from such sources is typically periodic, or at least intermittent, rather than continuous. Chief among these are the various solar technologies, and those based upon the use of the wind and tidal flows. Their use will surely increase in future years.

But in addition, the pressure for increased efficiency in the use of current energy sources is growing rapidly. An obvious example is the push toward the development of hybrid and plug-in hybrid, vehicles.

As an aside, it is interesting that official US Department of Energy targets typically assume that new technologies will need to meet the same requirements as those of the current technologies that they are expected to supplant. An example of this is that electrically and fuel cell-powered vehicles are expected to meet the long-range ability of current large, internal combustion vehicles. They also often assume that driving habits in the future will be about the same as those at the present time.

On the other hand, more and more of the public's attention is being given to the fact that a very large fraction of the actual vehicle use involves relatively short range daily commuter trips, for which limited range vehicles, whether battery or fuel cell-powered, would be perfectly satisfactory. Occasional longer trips would require the use of a different type of vehicle, of course. It is not unreasonable to think in terms of either two-car families, or the occasional rental of a long-distance vehicle when necessary.

The picture is not the same in all parts of the world. As can be seen in the discussion of several technologies in previous chapters, there are a number of directions in which progress has been greater in other countries than in the USA.

A large fraction of the government financial support of research and development activities related to energy technologies comes through, or is greatly influenced by, the Department of Energy, and especially, ARPA-E, in the USA. This leads to a concentration of work in a relatively small number of directions.

Recently the US Department of Energy decided to terminate efforts to develop hydrogen fuel cell – powered vehicles. In contrast, significant progress in that direction has been made in Japan and Europe, with significant numbers of demonstration vehicles on the road. This is mentioned in Chap. 22.

Some time ago it was decided to terminate work in the USA on both solid electrolyte and molten salt electrolyte elevated temperature batteries. As mentioned in Chap. 22, large sodium/sulfur solid electrolyte batteries are now being produced in Japan for use in large-scale storage facilities connected to the electrical distribution grid, and ZEBRA cells, which also have solid electrolytes, are now being produced in Switzerland for use for vehicle propulsion.

A large fraction of the long-range research in both government and university laboratories in the USA is aimed at advanced energy storage technologies that seem to be more applicable to small, high-tech portable, rather than larger-scale stationary, applications. Work on the latter is primarily concentrated on demonstration projects.

24.2 Emerging Technological Directions

Although most of the attention given to energy storage technology at the present time seems to be focused upon needs related to portable devices, such as computers and telephones, further development and increased use of larger systems is imperative.

Several of these are discussed in Chap. 22, and it can be seen that there is a regeneration of interest in elevated temperature battery systems for use in both large stationary applications and vehicles. Recent progress in both directions has been made outside of the USA, in Japan and Switzerland.

Flow batteries are now being commercially produced and sold by Gildemeister in Germany (who bought Cellstrom in Austria), Redflow in Australia, and Prudent Energy in China, as well as several other firms. The one American company producing such systems, VRB, went bankrupt several years ago, and its technology was purchased by Prudent Energy. Increased research and development activities in this area are now underway in the USA, primarily as the result of the stimulus provided by governmental funding through ARPA-E.

New alternatives are also emerging. One of these, that is actually still in the research stage, is the concept of the use of multilayer liquid battery systems that is mentioned in Chap. 22. It is too early to judge its significance.

Another approach to very large scale energy storage that has begun to get a lot of attention in the last few years, initially in Europe, but now also in the USA [1], may become very important. It involves the use of the sensible heat in relatively inexpensive molten salts as thermal storage media in conjunction with large solar systems. It can be used to periodically supply large amounts of energy to the electrical distribution grid so as to reduce the time-dependent variations in the demand placed upon the major electrical utilities. This is a type of load leveling, and could have a major effect on the cost of electrical energy, especially in areas such as the state of California, in which the demand varies by up to 50 %, depending upon the time of day.

Such a system involves the use of long parabolic reflectors to focus the sun's radiation upon tubes that carry a moving fluid. This fluid transfers the heat to a large molten salt bath, whose sensible heat acts as the storage system. This heat is then fed into a Rankine cycle steam turbine to produce electricity when needed.

The material that is initially heated by sunlight is sometimes called "solar oil," and is typically a synthetic organic material, a 50/50 mixture of the organic materials diphenyl oxide and biphenyl oxide. It has a low freezing point, 12 °C, so there is little danger than it might solidify, and it can be used up to about 400 °C. It transfers heat to a less expensive molten salt, such as the 50/50 eutectic mixture of $NaNO_3$ and KNO_3, that is sometimes called "solar salt." This salt melts at 221 °C, is stable up to about 500 °C, and has a heat capacity about half of that of water. It can be stored in large tanks, and supplies heat as needed to the steam turbine.

Typical prices are 0.5–1 dollar per kg of nitrate salts, and 3–4 dollars per kg for the low-melting organic heat transfer oils. As might be expected, efforts are being undertaken to find less expensive heat transfer media to replace the organic *solar oil*, or even a single material that can be used to handle the total thermal transfer and storage system in order to avoid the need for oil-to-salt heat exchangers. In addition, it would be desirable to be able to operate at higher temperatures, where the steam turbine is more efficient. It is important that the heat transfer material does not freeze inside the solar collector system or associated piping, of course. These nitrate salts are not corrosive, and can be readily contained in a number of metals and alloys.

Data on the compositions and minimum operating temperature of some of the nitrate molten salt materials that have been investigated are included in Table 24.1.

The important factors in the consideration of new technological approaches and systems related to large scale applications are different from those that are important in the smaller, and perhaps more high-tech applications. Both initial and lifetime costs are of great importance. In addition, as systems get larger, there will inevitably be more emphasis on safety, for larger problems can evolve into major disasters.

Table 24.1 Compositions and liquidus temperatures of several nitrate salts

Mol% Li	Mol% Na	Mol% K	Mol% Ca	Liquidus temp. (°C)
	66	34		238
	50	50		221
	21	49	30	133
30	18	52		120
31		58	11	117

24.3   Examples of Interesting New Research Directions

24.3.1   Organic Plastic Crystal Materials

The use of organic phase change materials for the storage of thermal energy is discussed in Chap. 3. The examples that were mentioned all involved the use of their heat of fusion. There are also some organic materials that undergo solid-state reactions, and exhibit *plastic crystal* behavior. They include some amines and polyalcohols that have large values of solid state phase transition enthalpy and low enthalpies of fusion [2, 3]. This topic is discussed in [4].

24.3.2   Organic Electrode Materials for Lithium Batteries

Present approaches to lithium ion batteries involve the use of metal alloys and inorganic materials as electrode reactants, as discussed in Chaps. 18 and 19. There have been several recent investigations of the potential of the use of organic materials for in this application [5–7]. A recent example is the use of polycarbonyl materials [8]. One of the advantages of these materials is that it is possible to tune the reaction potential. On the other hand, their solubility in electrolytes can be a problem. However, it is believed that this can be alleviated by increasing the molecular weight and increasing the magnitude of negative charge.

24.3.3   New Materials Preparation and Cell Fabrication Methods

As in a number of other areas of both science and technology, there is currently a lot of interest in the synthesis of nano-sized materials, and their potential use in connection with energy storage technologies.

The advantages of the use of small-dimensioned particles as electrode reactants in batteries are quite obvious in situations in which either the large surface area or the solid-state diffusion distance play an important role in controlling the kinetic behavior of electrodes.

But small nanowires can have an additional advantage in the case of some electrode materials that can have very large capacities. A particularly interesting example is the lithium-silicon alloy system. Under equilibrium conditions at elevated temperatures up to 4.2 lithium atoms can react per silicon atom [9], resulting in a theoretical electrode capacity of 4200 mAh g^{-1}. But the volume changes by about 400 % upon insertion and extraction of lithium of so much lithium, and this results in pulverization and capacity fading [10].

However, synthesizing silicon in the form of nanowires that are spaced apart makes it possible to accommodate such large volume changes without mechanical damage. This has been done by [11] and [12], who used the vapor–liquid–solid (VLS) method.

Using this method, it is possible to grow silicon nanowires on metallic substrates, such as stainless steel, so that each wire is attached to the current collector, avoiding the problem of the loss of electronic contact often found with particulate reactants.

The VLS method was first used in connection with the growth of whiskers for entirely different purposes [13]. It has subsequently been used for the growth of a number of other materials [14–19].

Another method that can be used for the formation of large numbers of nanowires employing a special chemical etching procedure has also been recently reported [20]. This is done by electrochemically etching silicon to form macropores, followed by uniform chemical etching to increase their diameter to the point that adjacent pores touch. The result is the formation of a large number of parallel fine nanowires, or pillars, of silicon. Galvanic deposition of copper onto the substrate results in a structure in which the wires are encased in copper at the bottom.

Innovative methods are also being pursued for the synthesis of positive electrode materials and their incorporation in novel electrode structures in high-energy batteries in a number of laboratories. These often involve variants of wet chemistry. One particular interesting method involves the formation of very fine particle oxides by use of a polymer precursor decomposition method [21–23], and a modification involving the use of citric acid [24].

In addition to the synthesis of fine reactant materials, there is a significant interest in methods to coat them with protective, yet electrochemically transparent layers. Another variant is the use of nanofibers to support thin layers of reactant material. One example is the deposition of amorphous silicon coatings onto carbon nanofibers [25].

24.3.4 Batteries with Physically Moving Electrode Structures

A very different approach to reversible energy storage has emerged in the last few years that involves a type of lithium ion battery containing a mechanically flowing semisolid material containing fine reactive particles [26–28]. It can be considered to be a type of flow battery in which the reactant materials are actually fine particles suspended in a viscous liquid. This novel approach is being developed by students and associates of Prof. Yet-Ming Chiang at MIT, and commercialization is being pursued by the new company 24M Technologies.

The label "sludge" has been used to describe the slurry-type of electrode structure that is made electronically conductive by co-suspending nanoscale conductive carbon black particles with the reactant particles.

One example is an aqueous lithium—ion system based upon the $LiTi_2(PO_4)_3$—$LiFePO_4$ couple in a 1 M $LiNO_3$ aqueous electrolyte with a pH of 11–12. In this case the maximum voltage is limited by the range of stability of water (roughly 1 V).

In this type of configuration there is a rather complex relationship between the mass flow behavior and the electrochemical efficiency, and there are inevitable mechanical pumping losses when using high viscosity semisolid electrodes.

Flow suspensions with high viscosity and non-Newtonian rheology benefit from use of an "intermittent flow" mode, in which the electrochemically active region of the cell is replenished in discrete steps, followed by electrochemical cycling. This results in lower mechanical energy dissipation than the use of a continuous flow mode operation.

A principal advantage of this novel approach, which is illustrated schematically in Fig. 24.1, is the ability to provide volumetric capacities an order of magnitude or more greater than that of conventional aqueous chemistries.

This is thus a type of flow battery that combines the high energy density of rechargeable batteries using solid storage electrodes with the architectural advantages of redox flow batteries.

Fig. 24.1 Schematic diagram of a semisolid flow cell, and a model of a laboratory configuration showing the anode and cathode flow channels, separators, and reference electrode [28]

24.3.5 Alternate Electrolytes

There is a growing interest in the use of aqueous electrolytes in lithium systems, primarily for application in moderate-to-large systems in which low cost, high rate, and safety are of particular interest [29–34].

New electrolytes are also being investigated. One group of these that is drawing a lot of attention includes materials called *ionic liquids*. These are molten salts that have low melting temperatures [35–39]. This is accomplished by having one or both of the ions have complicated high entropy structures that are hard to crystallize. These typically contain large organic groups with rather low symmetry. It appears that some of these materials are stable in the presence of lithium battery electrode components.

24.3.6 Interesting New High Power, Long Cycle Life Battery

Early work on a new type of battery chemistry suddenly became visible in April, 2015 [40]. It involved an aluminum metal foil anode and either a pyrolytic graphite foil cathode or a 3-dimensional graphitic foam cathode. The electrolyte used was a nonflammable ionic liquid: $AlCl_3$/1-ethyl-3-methylimidazolium chloride, [(EMIm) Cl], that transports aluminum chloride anions. It was vacuum-dried so that it contained less than 500 ppm of residual water.

This system operates by the electrochemical deposition and dissolution of aluminum at the anode, and the reversible insertion of chloroaluminate anions into the graphite at the cathode. This mechanism is illustrated schematically in Fig. 24.2.

The reaction equations are:

$$4\, Al_2Cl_7^- + 3\, e^- = Al + 7\, AlCl_4^- \tag{24.1}$$

and

$$C_n + AlCl_4^- = C_n(AlCl_4) + e^- \tag{24.2}$$

Where n is the molar ratio of carbon atoms to intercalated anions in the graphite.

On the cathode side, $AlCl_4^-$ is reversibly intercalated into the graphite structure up to a capacity of 60–66 mAh per gram of graphite mass. On the anodic side of the cell metallic aluminum and $AlCl_4^-$ are transformed into $Al_2Cl_7^-$ during discharging, and the reverse reaction takes place on charging.

This cell has been shown to have relatively flat charge and discharge curves at about 2 V, a high coulometric efficiency, and a capacity of 60–70 mAh per gram, as shown in Fig. 24.3.

$$Al + 7AlCl_4^- \rightarrow 4Al_2Cl_7^- + 3e^- \quad C_n[AlCl_4] + e^- \rightarrow C_n + AlCl_4^-$$

Fig. 24.2 Schematic representation of the operation of the aluminum/graphite battery [40]

Fig. 24.3 Galvanostatic charge–discharge curves of an Al/pyrolytic graphite cell at a current density of 66 mA/g [40]

The consistency of the charging and discharging behavior of the aluminum/pyrolytic graphite cell up to 200 cycles at a current of 66 mA per gram is demonstrated in Fig. 24.4.

However, such cells showed reduced capacities at rates higher than 1C, and this was thought to be due to slow transport of the relatively large chloroaluminate anions through the graphite layer structure.

Fig. 24.4 Cycling behavior at 66 mA/g [40]

Fig. 24.5 Porous structure of the graphitic foam electrode [40]

Subsequent experiments were performed using very fine-structured porous, and flexible, graphitic foam that was produced by chemical vapor deposition upon a nickel foam template. The microscopic structure of this material is shown in Fig. 24.5. This allows the reaction to take place over a very large (internal) surface area, and makes very high rates of charge and discharge possible.

The very high surface area cathode makes it possible to operate at surprisingly high currents, with total charging times as short as 1 min at a current density of 4 A/g, which is equivalent to a specific power of about 3 kW per kg. This behavior is shown in Fig. 24.6. A cycle life of more than 7500 cycles without appreciable decay is shown in Fig. 24.7.

Fig. 24.6 Galvanic charge and discharge curves at a current density of 4 A/g [40]

Fig. 24.7 Long-term stability of an Al/graphitic foam cell [40]

The aluminum dissolution and deposition efficiencies were reported to be very high, 98.6–99.8 %. No dendrite formation was found upon charging the aluminum electrode, even at these very high rates, up to a charging voltage cutoff of 2.45 V.

However, it was reported that reduced efficiency was observed when charging to higher voltages. The presence of larger amounts of water in the electrolyte also caused a reduced Coulombic efficiency. These observations are consistent with earlier reports on aluminum electrode behavior in a similar electrolyte [40].

The specific energy of this new type of battery is not especially high, about 40 Wh/kg, which is comparable to current lead-acid and Ni-MH batteries. But when using the highly porous graphitic foam in the cathode, the power density can be unusually high, about 3 kW/kg, which is in the range of supercapacitors.

24.4 Final Comments

Energy storage is becoming increasingly important. There are two general reasons for this. One is the recognition of the inevitable depletion of nonrenewable fossil fuels such as oil, and the need to shift, at least partially, away from today's dependence upon them as the primary energy source, and toward the use of alternate energy sources.

In addition, there is growing concern about the pollution resulting from the use of the major current sources. This may be relieved, at least in part, by the use of some of the alternative sources.

On the smaller scale, there are an increasing number of relatively small portable electrically powered devices that have to carry their energy sources with them. This results in the need for improvement in electrochemical battery or portable fuel cell technology.

The author hopes that this book will be helpful in providing an understanding of the different methods by which energy can be stored.

He also wishes to applaud, and cheer on, all those who have contributed to the current state of knowledge of energy storage science and technology.

References

1. R.W. Bradshaw and N.P. Siegel, ES2008-54174, in Proc. of Conf. on Energy Sustainability 2008, ASME (2008)
2. Murrill E, Breed L (1970) Thermochemica Acta 1:239
3. Benson DK, Burrows W, Webb JD (1986) Solar Energy Mater 13:133
4. Chandra D, Chien W-M, Gandikotta V, Lindle DW (2002) Phys Chem 216:1433
5. Umemoto T (2004) US Patent 6737193 B3
6. Chen H, Armand M, Demailly G, Dolhem F, Poizot P, Tarascon J-M (2008) Chem Sus Chem 4:348
7. Chen H et al (2009) J Am Chem Soc 131:8984
8. Armand M, Grugeon S, Vezin H, Laruelle S, Ribiere P, Poizot P, Tarascon JM (2009) Nat Mater 8:120
9. Wen J, Huggins RA (1981) J Solid State Chem 37:271
10. Boukamp BA, Lesh GC, Huggins RA (1981) J Electrochem Soc 128:725
11. Chan CK, Peng H, Liu G, McIlwrath K, Zhang XF, Huggins RA, Cui Y (2008) Nat Nanotechnol 3:31
12. Laïk B, Eude L, Pereira-Ramos J-P, Cojocaru CS, Pribat D, Rouviere E (2008) Electrochem Acta 53:5528
13. Wagner RS (1970) In: A.L. Svitt (ed.), Whisker Technology. Wiley Interscience, New York, p. 47
14. Morales AM, Lieber CM (1998) Science 279:208
15. Huang MH et al (2001) Adv Mater 13:113
16. Dick KA et al (2005) Adv Funct Mater 15:1603
17. Pan ZW, Dai ZR, Wang ZL (2001) Science 291:1947
18. Wang Y, Schmidt V, Senz S, Gosele U (2006) Nature Nanotech 1:186
19. Hannon JB, Kodambaka S, Ross FM, Tromp RM (2006) Nature 440:69

20. Föll H, Hartz H, Ossei-Wusu E, Carstensen J, Riemenschneider O (2009) Phys Status Solidi Rapid Res Lett 9999:4
21. Hamling BH (1968) U.S. Patent 3,385,915
22. Hamling BH (1973) U.S. Patent 3,736,160
23. Deshazer HD, LaMantia F, Wessells C, Huggins RA, Cui Y (2011) J Electrochem Soc 158: A1079
24. Shen P, Jia D, Huang Y, Liu L, Guo Z (2006) J Power Sources 158:608
25. Cui L-F, Yang Y, Hsu C-M, Cui Y (2009) Nano Lett 9:3370
26. Duduta M, Ho B, Wood VC, Limthongkul P, Brunini VE, Carter WC, Chiang Y-M (2011) Adv Energy Mater 1:458
27. Li Z, Smith KC, Dong Y, Baram N, Fan FY, Xie J, Limthonkul P, Carter WC, Chiang Y-M (2013) Phys Chem Chem Phys 15:15833
28. Brunini VE, Chiang Y-M, Carter WC (2012) Electrochimica Acta 69:301
29. Li W, Dahn JR, Wainwright DS (1994) Science 264:1115
30. Li W, McKinnon WR, Dahn JR (1994) J Electrochem Soc 141:2310
31. Li W, Dahn JR (1995) J Electrochem Soc 142:1742
32. Zhang M, Dahn JR (1996) J Electrochem Soc 143:2730
33. Wessells C, Huggins RA, Cui Y (2011) J Power Sources 196:2884
34. Wessells C, La Mantia F, Deshazer H, Huggins RA, Cui Y (2011) J Electrochem Soc 158:A352
35. Auborn JJ, Barberio YL (1985) J Electrochem Soc 132:598–601
36. Wilkes JS, Levisky JA, Wilson RA, Hussey CL (1982) Inorg Chem 21:1263–1264
37. Lai PK, Skyllas-Kazacos M (1988) J Electroanal Chem Interfacial Electrochem 248:431–440
38. Jiang T, CholierBrym MJ, Dube G, Lasia A, Brisard GM (2006) Surf Coat Tech 201:1–9
39. Borg RJ, Dienes GJ (1988) An Introduction to Solid State Diffusion. Academic, New York
40. Lin M-C, Gong M, Lu B, Wu Y, Wang D-Y, Guan M, Angell M, Chen C, Yang J, Hwang B-J, Dai H (2015) Nature 520:324

Index